THE MATERIAL BASIS OF EVOLUTION

RICHARD GOLDSCHMIDT

WITH AN INTRODUCTION BY
STEPHEN JAY GOULD

SILLIMAN MILESTONES IN SCIENCE

THE MATERIAL BASIS OF EVOLUTION

RICHARD GOLDSCHMIDT

with an Introduction by
STEPHEN JAY GOULD

NEW HAVEN AND LONDON:
YALE UNIVERSITY PRESS

YALE UNIVERSITY
MRS. HEPSA ELY SILLIMAN
MEMORIAL LECTURES

Printed in the United States of America by
Halliday Lithograph Corporation,
Hanover, Massachusetts.

Library of Congress Cataloging in Publication Data

Goldschmidt, Richard Benedict, 1878–1958.
The material basis of evolution.

(Silliman milestones in science)
Based on 8 Silliman lectures delivered by
the author at Yale University in Dec., 1939.
Reprint. Originally published: New Haven : Yale
University Press, 1940. (Yale University, Mrs. Hepsa
Ely Silliman memorial lectures ; 28)
Bibliography: p.
Includes index.
1. Evolution. 2. Genetics. I. Title.
II. Series. III. Series: Mrs. Hepsa Ely Silliman
memorial lectures ; 28.
QH366.2.G64 1982 575 81–15969
ISBN 0–300–02822–9 AACR2
ISBN 0–300–02823–7 (pbk.)

10 9 8 7 6 5 4 3 2 1

THE SILLIMAN FOUNDATION LECTURES

On the foundation established in memory of Mrs. Hepsa Ely Silliman, the President and Fellows of Yale University present an annual course of lectures designed to illustrate the presence and providence of God as manifested in the natural and moral world. It was the belief of the testator that any orderly presentation of the facts of nature or history contributed to this end more effectively than dogmatic or polemical theology, which should therefore be excluded from the scope of the lectures. The subjects are selected rather from the domains of natural science and history, giving special prominence to astronomy, chemistry, geology, and anatomy. The present work constitutes the twenty-eighth volume published on this foundation.

PREFACE

THIS book presents an elaboration of the material which was used, in less technical and less detailed form, for presentation in eight Silliman Lectures delivered in December, 1939. The manuscript was completed in September, 1939, and the author has tried to include important literature up to October 1. He is greatly indebted to Mr. Richard Blanc, Mrs. Laura G. Rauch, and Dr. D. F. Poulson for assistance in editing the text; further, to a number of publishers for permission to reproduce some of the illustrations. Services rendered by the personnel of Works Progress Administration Official Project No. 465–03–3–192 are hereby acknowledged.

R. G.

Berkeley, California,
October, 1939

CONTENTS

THE USES OF HERESY:
AN INTRODUCTION TO RICHARD GOLDSCHMIDT'S *THE MATERIAL BASIS OF EVOLUTION*

STEPHEN JAY GOULD

I. BACKGROUND

In his autobiography, published posthumously in 1960, Richard Goldschmidt wrote of the work here reprinted, "I am confident that in 20 years my book, which is now ignored, will be given an honorable place in the history of evolutionary thought" (1960, p. 324). We are within the limits of statistical error on Goldschmidt's prediction and would have met it exactly had I honored an initial deadline. This reprint—since presses, even university presses, are not, at least by choice, eleemosynary institutions—commemorates the strong reawakening of interest in Goldschmidt's views among evolutionary biologists. If this attention be the harbinger of "an honorable place," then so be it—though I suspect that a self-assured, professional iconoclast like Goldschmidt would relish the renewed argument more than any potentially favorable outcome.

Richard Goldschmidt, one of the world's great geneticists, delivered the prestigious Silliman lectures at Yale in 1939 and published his remarks in 1940 as *The Material Basis of Evolution*. He wrote just as the Darwinian paradigm was coalescing into a confident general theory of evolution. The foundations of population genetics had been established and Dobzhansky (1937) had initiated the "modern synthesis" of traditional disciplines around this Darwinian core, with its central emphasis upon continuity in process and cause for all evolutionary events from the spread of alleles in local populations to major trends in the history of life. The observable and operational realm of microevolutionary change would become a model for all levels of evolution; speciation, as Darwin had argued, is a smooth extension of adaptive change in local populations. "Races, species, genera, and families are nothing more than different degrees of phylogenetic divergence" (Dobzhansky, 1951, p. 266). ''Adaptation to local conditions and evolutionary change are two aspects of the same genetic phenomenon, the continuous adjustment of an integrated gene complex to a changing environment'' (Mayr, 1963, p. 332).

Amidst this incipient chorus of consensus, so welcome after decades of strife and fruitless disagreement within evolutionary theory, Goldschmidt's views injected an especially disharmonious note. He seemed bent on undoing this unity and invoking the shade of de Vries to assert once again that macroevolution was a thing apart, abrupt in its occurrence, and unilluminated in principle by processes of microevolution that could be studied directly. He even invented an unforgettable and disturbing (if whimsical) name for his independent macroevolutionary agent—the ''hopeful monster.''

Of his reception by neo-Darwinians, Goldschmidt claimed (1960, p. 324) that he ''certainly had struck a hornet's nest. The Neo-Darwinians reacted savagely. This time I was not only crazy but almost a criminal.'' Ernst Mayr recalls (1980, p. 420), ''Even though personally I got along very well with Goldschmidt, I was thoroughly furious at his book, and much of the first draft of *Systematics and the Origin of Species* was written in angry reaction to Goldschmidt's total neglect of such overwhelming and convincing evidence.''

The counterattack was successful, if not triumphant—though Goldschmidt (as I shall argue later) had largely himself to blame for burying his gems so deeply in unacceptable and overextended claims. Indeed, he suffered the worst fate of all: to be ridiculed *and* unread. Textbooks of the past forty years have generally included a polite and perfunctory paragraph of dismissal. The following is typical:

> Some biologists have suggested that the origin of major groups of animals and plants, such as phyla, classes, or orders, may at times have come about through single mutations involving large and complex changes that happen to be successful. Such creatures, called by Richard Goldschmidt ''hopeful monsters,'' seem with the advance of our knowledge, to be less and less necessary to explain the beginning of new adaptive advances. A monster is far more likely to be hopeless than hopeful. The gaps formerly present in our knowledge of many groups are being filled today by evidence of the usual, gradual transformation of characters under natural selection. [Eaton, 1970, p. 45]

I have witnessed widespread dogma only three times in my career as an evolutionist, and nothing in science has disturbed me more than ignorant ridicule based only upon a desire or perceived necessity to follow fashion: the hooting dismissal of Wynne-Edwards and group selection in any form during the late 1960s and most of the 1970s, the belligerence of many cladists today, and the almost ritualistic ridicule of Goldschmidt by students (and teachers) who had not read him. I do not know how often it happened, and my memories have been pooh-poohed both by my older colleagues who knew and revered Goldschmidt and by younger colleagues trained in our present era of reawakening interest, but I know that I experienced it in classroom after classroom as a graduate student in the mid-1960s. My experiences are neither unique nor misremembered, for Frazzetta (1975, p. 85) reports from the same time and different place, "No one stopped to consider whether in all of Goldschmidt's assailable propositions, there existed anything worth thinking about. There was no time for such consideration as long as there was so much merry mayhem to be carried out. In my university classes, the name 'Goldschmidt' was always introduced as a kind of biological 'in' joke, and all we students laughed and snickered dutifully to prove that we were not guilty of either ignorance or heresy." An eminent senior colleague and former professor told me that he went to his library to consult *The Material Basis of Evolution* after reading an article that I had written in Goldschmidt's defense. He could not find it and was frustrated until he remembered—and then he was merely angry—that he had thrown it out several years ago as containing nothing of value. I do not think I exaggerated (Gould, 1977a) in comparing Goldschmidt with Goldstein, the object of daily "two minute hates" in Orwell's *1984*.

Who then was this Goldschmidt whom so many reviled in ignorance? First of all—and this must be a general principle for objects of intense ridicule—he could not have been a minor or second-rate thinker, for such scientists are not worth the emotional energy devoted to Goldschmidt's intellectual persecution. No one likes to waste time on a nonentity. Goldschmidt was one of the premier geneticists of our century, a shoo-in on anybody's list of the top ten. He was born in Frankfurt in 1878, the son of a wealthy merchant and scion of an old, prosperous, and intellectual Jewish family. He studied under the greatest teachers from the golden age of late nineteenth-century German biology: with Otto Bütschli and Karl Gegenbaur in Heidelberg and with Richard Hertwig in Munich. Along with T. H. Morgan and Th. Dobzhansky, he thus obtained the greatest gift and preadaptation that any future geneticist could receive for evolutionary studies: a rigorous training in classical morphology and embryology. He became Hertwig's assistant in 1903 (after defending a thesis on maturation, fertilization, and embryonic development in the trematode *Polystomum integerrimum*) and remained in Munich until 1913. When Theodor Boveri and Carl Correns organized the Kaiser Wilhelm Institute for Biology in Berlin, they invited Goldschmidt to become director of genetics, a post he held until 1935.

Goldschmidt was a professor in the stereotypical Germanic mode. He was arrogant and haughty, but invariably kind and even courtly. He wrote of a Nazi poster citing the genealogy of his family as a testimony to Jewish danger, "It could well be used as a chart demonstrating the effect of long selection of favorable hereditary traits upon the improvement of human families" (1960, p. 5). He enjoyed the prerogatives of office and expected the "deference due to a man of pedigree" (as Gilbert said of Pooh-Bah). Viktor Hamburger (personal conversation) recalls that his fellow students called Goldschmidt "the Pope," perhaps not so peculiar a title for an offspring of established German-speaking Jews who often surpassed many Prussians in their loyalty and assimilation (I can still hear the acid words of my Yiddish-speaking peasant grandmother, recalling the snubs of well-bred Viennese girls). He was erudite and enormously cultured, an expert on oriental art, and a man who never really accommodated to the incipient "California life-style" of prehippie Berkeley.

The two world wars affected Goldschmidt in the reverse order of Marx's famous statement that all events in history occur twice, the first time as tragedy, the second as farce. When the Great War broke out (so called until a greater occurred twenty-five years later), Goldschmidt was in Hawaii on a travelling fellowship. He went, virtually by necessity, to the American mainland and spent the war with us, first as a teacher, but finally, in the orgy of jingoism that engulfed us as we entered the war in 1917, in an internment camp for enemy aliens. Repatriated to Germany after the war, he resumed his old post in Berlin, only to relinquish it by force in 1935 as Hitler destroyed German science by purging Jews and other racial and ideological "undesirables." Goldschmidt again travelled to America, this time as a refugee, and became a professor at Berkeley in 1936, a post he held until he entered a continually active retirement. He died in 1958, his powers—and his faith in them—undiminished. The last words of his autobiography do not proffer cosmic advice, but merely state, "It is my greatest intellectual happiness that I can still work in my laboratory and even make interesting discoveries in the field of chemically induced phenocopies" (1960, p. 326).

Goldschmidt's scientific writings spanned a range often included in several university departments. (Goldschmidt's autobiography, 1960, contains a complete list of his publications. A short summary of his life and works, and a list of primary and secondary sources can be found in the entry by A. Tétry in the *Dictionary of Scientific Biography,* vol. 5, pp. 453–55.) He spent most of the first decade of his career working in classical morphology. He helped to establish and explicate the fascinating phenomenon of cell constancy in nematodes and made many important contributions to the histology and embryology of protochordates. From this phase of his life, though written late in 1921, comes Goldschmidt's most charming work, a classic that spread his name to thousands who would never encounter the rancor of his professional opposition: *Ascaris,* a popular book ostensibly about a "lowly" nematode, but extending outward (as its parasitic subject must) into all areas of zoology. Still read and admired in Europe, its English translation (1938) never attained the popularity that Goldschmidt had anticipated.

When Goldschmidt moved into genetics, he chose the gypsy moth, *Lymantria dispar,* as his primary subject. (I write this essay in May 1981, as caterpillars of the gypsy moth swarm, locustlike, over New England.) This work led him in two major directions. First, he studied several problems in microevolution from a Darwinian standpoint. He travelled throughout the world and performed one of the most inclusive and elegant of early studies on geographic variation in phenotypes with known genetic bases. He also elucidated the genetics of industrial melanism and proposed scenarios that sought the adaptive value of dark color not directly in crypsis against visual predators, but indirectly as the by-product of a metabolic change that permitted the caterpillars to feed upon plants loaded with industrial chemicals.

Secondly, Goldschmidt studied the genetics of sex determination and recognized that normal sexuality is a quantitative phenomenon produced by a balance of male and female sex determiners. He developed a series of graded intersexes by altering these balances experimentally. During this decade between 1910 and 1920, most geneticists devoted their attention to the principles of transmission. Goldschmidt, on the other hand, maintained a primary interest in gene function, and his work on intersexes in *Lymantria* led to his most important insight and to the profession that he called physiological genetics. He recognized that genes work by controlling the rates of chemical processes. Normal development requires a proper balance and definite timing of substances; evolutionary change occurs when the timing of development alters. If genes affect timing, then experimental manipulations of temperature and chemical surroundings might induce changes identical to those found in mutants, thus confirming the rate hypothesis. Goldschmidt produced these mutant phenotypes without mutations and christened them phenocopies. We see, in this work, the germs of Goldschmidt's later apostasy on macroevolution, for he recognized that a small genetic change, operating early enough in ontogeny, might engender a cascade of effects producing a large phenotypic ''jump'' in a single genetic step. In Berkeley, Goldschmidt turned his attention to the physiological genetics of *Drosophila*.

In the midst of this experimental effort, Goldschmidt developed his radical views about macroevolution by a gradual personal ontogeny. He began, he often stated with pride and irony, as a convinced neo-Darwinian before the neo-Darwinian school arose. Why else devote so much time to the study of geographic variation in *Lymantria?*

As a convinced Darwinian I believed geographic races to be incipient species. I hoped to prove by such an analysis the correctness of this idea. I was completely acquainted with what twenty years later was rediscovered as ''the new systematics,'' and my convictions, as expressed in 1920 and 1923, were practically the same as those of present-day Neo-Darwinians. [1960, p. 318]

In Goldschmidt's first step to heterodoxy, as mentioned above, he employed ''rate genes'' from his physiological genetics to explain the rapid origin of discontinuities that could not be bridged by smooth geographic variation:

Such conclusions forced me to think of what addition to Darwinism was needed in order to account for the macroevolutionary processes. The solution was the existence of macromutations, which, in rare cases, could affect early embryonic processes so that through the features of embryonic regulation and integration at once a major step in evolution could be accomplished and fixed under certain conditions. I spoke half jokingly of the hopeful monster in my first publication on the subject, a lecture read by invitation in 1933 at the World's Fair in Chicago. [1960, p. 318]

Had Goldschmidt stopped here, his heresy would have been mild indeed. Darwinians, with their traditional preferences for gradualism and continuity, might not shout hosannas for large phenotypic shifts induced rapidly by small genetic changes that affect early development; but nothing in Darwinian theory precludes such events, for the underlying continuity of small genetic changes remains. But Goldschmidt accelerated, shifted gears, and finally proposed that these abrupt macroevolutionary changes are products of a genetic event different in kind from the micromutations underlying geographic variation—the systemic mutation, or chromosomal repatterning:

I derived the hypothesis that in addition to small or large mutations of genic loci, there exists a completely different type of mutation. I called this ''systemic mutation,'' meaning that a reshuffling or scrambling of the intimate chromosomal architecture, which might occur rarely by chance, will act as a macromutational agent. This means that it will produce, suddenly, a huge effect upon a series of developmental processes leading at once to a new and stable form, widely diverging from the former. [1952, p. 96]

Finally, Goldschmidt came to deny the existence of the ''corpuscular gene'' altogether (1955) and to view all genetic changes as alterations in pattern. (In this system, ''micromutations'' became small and localized changes in pattern.)

Goldschmidt had made a clean and complete break between micro- and macroevolution, thus challenging the most important premises of the modern synthesis—continuity and extrapolation—and justifiably earning the enmity of a growing orthodoxy. *The Material Basis of Evolution* is the major work of his full-fledged heresy.

II. A SELECTED PRÉCIS OF *THE MATERIAL BASIS OF EVOLUTION*

MANY of the great twentieth-century books in evolutionary theory have been exhaustive general summaries of entire fields (see Mayr, 1963 for the best example). Goldschmidt's book has a similar outward appearance; it ranges widely among topics and examples. But *The Material Basis of Evolution* is a work of different form: it is a long, partisan argument—sometimes repetitive, sometimes inconsistent—for a definite view of evolutionary processes.

Goldschmidt did not invent the words micro- and macroevolution, but he did popularize them. By microevolution, he referred to changes within local populations and geographic variation—in short, to all evolutionary events occurring within species. Macroevolution designates the origin of species and higher taxa. (Goldschmidt recognized, of course, that higher taxa must begin as new species, but he believed that the morphological jumps accompanying some events of speciation are so profound that descendant species must be designated as new higher taxa from their inception.)

For most evolutionists, this contrast between micro- and macroevolution can only be intergrading and indistinct because geographic variation, by intensification, leads to the origin of new species. But not for Goldschmidt. Viewing the two phenomena as products of distinct genetic mechanisms, he envisioned an absolute break between geographic variation and speciation. If continuity from micro- to macroevolution, with unity of genetic mechanisms throughout, is the primary belief of neo-Darwinism (as I believe it is; see quotes of Mayr and Dobzhansky on p. xiv), then no claim could be more unorthodox.

Goldschmidt had rekindled an issue that extended back to the earliest days of evolutionary theory. After all, Lamarck had contrasted local adaptation, induced by "l'influence des circonstances," with progress up life's ladder, caused by "the force that tends, incessantly, to complicate organization." And Chambers, author of the anonymous *Vestiges,* had separated diversification within type from transition between types as products of different mechanisms of change. In an important sense, Darwin's greatest achievement was not merely to support evolution (as these worthy gentlemen had done before him), but to propose continuity between local changes that could be observed and made the object of controlled experiment and large-scale evolutionary changes that could not be seen directly. And now Goldschmidt, albeit in different guise, was resuscitating the old dichotomy just when modern Darwinians thought they had finally buried it for good.

The "long argument" of *The Material Basis of Evolution* can be briefly summarized:

1. Microevolution is all that Darwinians say it is: pervasive, adaptive, and integral to the success and spread of species.

2. Microevolution does not lead, by extension, to the origin of species. True species are separated by "bridgeless gaps." Microevolutionary change leads local populations into "diversified blind alleys."

3. New species arise by macromutation, not by the "accumulation of micromutations." The genetics of macromutation are different in kind from the point mutations underlying microevolutionary change. These "systemic mutations" involve fundamental repatterning of chromosomes. (Goldschmidt also attributes many abrupt shifts of phenotype to small genetic changes affecting developmental rates early in ontogeny; I shall discuss the potential inconsistency between systemic mutation and these alterations of rate below.)

4. The nature of developmental programs, with their alternate (and often discontinuous) channels of phenotypic expression and their regulative properties that direct major alterations into viable pathways, permits macromutation to be effective in the saltational genesis of higher taxa. (These arguments, the bulk of the book, have been generally ignored. I believe that they embody the part of Goldschmidt's argument with abiding value. They also counteract the caricatured dismissals of Goldschmidt's views and render them interesting and coherent, even if unacceptable today. In particular, they explain why monsters may be

hopeful and biologically well-functioning, and why macromutants may spread within local populations, even while breeding with conspecifics of normal phenotype.)

Goldschmidt's chosen title, which strikes so many people as peculiar, accentuates the differences between his view and the tendencies of Darwinian argument. Just what does he mean by the *material* basis of evolution; could it be nonmaterial? Goldschmidt wished to focus on the constraints and opportunities provided by inherited genetics and development. ''A change in the hereditary type can only occur within the possibilities and limitations set by the normal process of control of development'' (p. 1). This theme, though unavoidably part of any evolutionary theory, has always been underplayed by strict Darwinians because a purely continuationist and adaptationist perspective deemphasizes the ''internal'' contribution of organisms to their own future change. If evolution is gradual and continuous, moving in any direction dictated by selective pressures of a changing environment, then developmental constraints play little role except as hindrances to be overcome by selection or as generators of nonadaptive and unimportant by-products of primary events. But, in Goldschmidt's view, changes are abrupt and not directly superintended by environment (though environment may reject them by conventional selection). The direction and extent of these changes are set, limited, and facilitated by inherited developmental pathways—the material basis of evolution:

> What I propose to do is to inquire into the type of hereditary differences which might possibly be used in evolution to produce the great differences between groups, and the title of this book, accordingly, ought to be something like: The genetical and developmental potentialities of the organism which nature may use as materials with which to accomplish evolution. [p. 3]

Goldschmidt divided his book into two equal parts, labelled microevolution and macroevolution. The theme of the first half is ironic: although microevolution is all that Darwinians say, it has nothing whatever to do with the origin of new species. A study of this section immediately refutes one of the most common caricatures invented to refute Goldschmidt: that he was an unflinching typologist, a great ''lab man'' perhaps, but lacking the naturalist's feel for continuous variation.

Goldschmidt was always a naturalist first, a man who did as much work as anyone on geographic variation in the field. As Darwin used pigeons, so Goldschmidt invoked his own work in collecting natural populations of *Lymantria* throughout the world, recording the continuity and adaptive value of their variations, and bringing them to the laboratory to raise different races in controlled conditions for assessing the genetic basis of adaptation. We may not accept Goldschmidt's views about the restricted role of geographic variation in evolution, but they did not arise from ignorance of—but rather from immersion in—natural examples.

Goldschmidt begins by discussing single micromutants and their evolutionary effect, focusing upon industrial melanism in his beloved moths. He then moves to "local polymorphism"—the random and small-scale variation (as in breeds of domestic dogs, for example) that constitutes little more than noise at the larger and more important scale of adaptive geographic variation in rassenkreise.

The bulk of the first part, focusing upon his own work with *Lymantria,* discusses large-scale geographic variation in widespread species and the general concept of rassenkreis, or "circle of races." (This concept has been widely misunderstood by English-speaking evolutionists who have translated the word literally and then used it for a different concept to which the literal meaning seems to apply. In German, a "Kreis" need not refer to a literal, geometric circle, but to any collection of related elements, as in our English "circle of friends." Thus, rassenkreis is merely another name for a polytypic species, and that is all. It does not refer to a linear, spatial chain of races, whose adjacent members can interbreed and whose endpoints cannot, though this mistaken attribution is still common in the literature.)

Providing a foretaste of later arguments, Goldschmidt begins by stating Darwin's view that the subspecies of rassenkreise are incipient species and by announcing his later appraisal:

The taxonomists who worked upon these problems . . . have come to the conclusion that the geographic races are incipient species, that the formation of subspecies within a species over its geographic range (the rassenkreis) is the first and typical, even obligatory, step in the evolution of new species and higher categories. There is no doubt that such a view is very attractive at first sight. . . . We shall begin our discussion with the facts pertaining to the rassenkreis as such and their analysis, and shall only proceed afterwards to the decisive point; i.e., whether this type of microevolution can lead beyond the confines of the species. [p. 30]

Goldschmidt's discussion of rassenkreise focuses upon the twin Darwinian themes of continuity and adaptive value. He regards clines that correlate with graded environments as the best demonstration of adaptation and cites several examples from *Lymantria,* primarily involving the timing of larval life (shorter in regions with cold winters or hot and dry climates). "It is just this set of facts by which it could be demonstrated that the seriation parallels climatic series in nature and by which it could be proven that the genetic differences are actually adaptations of the life cycle of the animal to the seasonal cycle in nature" (p. 60). Goldschmidt then extends the range of adaptation by emphasizing how many features of morphology (larval size, for example) are correlated with the physiological bases of environmental advantage (length of larval life, for example). "These examples demonstrate that what apparently is non-adaptational may turn out to be strictly adaptational if only the proper environmental factor and the proper physiological process can be located. Therefore I am inclined to consider all subspecific characters which vary in a cline parallel to a geographical, climatological, or other environmental cline . . . as, at least indirectly, adaptational" (p. 78). He refutes nonselectionist explanations of clines and rejects Lamarckian, while strongly supporting Darwinian, explanations for the "parallelism of subspecific clines" encountered in the ecogeographic rules of Bergmann and Allen.

He then discusses the genetic basis of geographic variation and defends the accumulation of small micromutations as its Darwinian source:

The decisive differences, which must have arisen by mutation, are based on groups of extremely small but additive deviations, as revealed by multiple-factor or multiple-allelic differences. These differences accumulate, beginning with differences between colonies of such a minor order that they can hardly be described, though each investigator knows them, and aggregating into the easily distinguishable quantitative differences separating actual subspecies. The genetic picture, *within* the species, then agrees with Darwin's ideas. [p. 101, Goldschmidt's italics]

I have emphasized the sphere of Goldschmidt's Darwinism and quoted him at length because I wanted to illustrate that his apostasy at higher levels did not arise from an ignorance of Darwinian themes, but from a careful consideration of them and their defense at a level that seemed to him appropriate. But now the argument shifts.

In the next section, on "limiting features of subspecific variation," Goldschmidt asks whether extreme members of rassenkreise approach the status of separate species. In an argument that I regard as often forced or inconsistent (see next section), he argues that neither facts of fertility nor isolation lead to the belief that extreme subspecies can become separate species. He admits that extreme subspecies of a rassenkreis may exhibit lowered fertility, but argues that the causes of their partial incompatibility are not the stuff of which true species are made. He argues that many extreme subspecies (or very closely related sympatric populations) can be induced to produce fertile offspring in the laboratory even if they do not do so in nature (thus brushing by the fact that populations who don't, even if they can in artificial circumstances, are still reproductively isolated). He allows that isolation may induce less orderly and adaptive variation than that displayed by continuous clines, and he admits that isolation may accentuate the amount of variation; but "there is no reason, at least as far as the factual material goes, to suppose that isolation makes subspecies develop into species. . . . Isolation or no isolation, the subspecies are diversifications within the species, but there is no reason to regard them as incipient species" (p. 136).

Goldschmidt ends this section by reemphasizing that Darwinism "works perfectly within the limits of the species" (p. 139) and by questioning the value of population genetics not on any mathematical basis, but on the falsity of its Darwinian premise:

The most brilliant mathematical treatment is in vain if the biological rating of the material is not correct. I am of the opinion that this criticism applies also to the mathematical study of evolution. This study takes it for granted that evolution proceeds by slow accumulation of micromutations through selection, and that the rate of mutation of evolutionary importance is comparable to that of laboratory mutations, which latter are certainly a motley mixture of different processes of dubious evolutionary significance. If, however, evolution does not proceed according to the neo-Darwinian scheme, its mathematical study turns out to be based on wrong premises. [pp. 137–38]

In his final section on "the species," Goldschmidt argues for "bridgeless gaps" between true species. Microevolution works down by "diversifying the primary form either by adapting the species genetically to diverse conditions of the environment . . . or by a diversification which is more haphazard and nonadaptational. . . . In all cases the diversification could be subdivided almost without limit down to differences between individual colonies" (p. 139). But the primary forms themselves—the species—must arise all at once, for they are separated by bridgeless gaps:

If subspecies are considered to be incipient species which only need isolation to become species, such a merging of one rassenkreis into another must be observable. . . . If, however, subspecies are nothing but an intraspecific diversification which adapts the species, at least in the majority of cases, to definite conditions within its area of distribution, the limit between two species or rassenkreise ought to be in the nature of a hiatus, an unbridged cleft. . . . Most of these species are, as every earnest inquirer will find, in their natural areas of distribution rather circumscribed products, which do not live in any extensive connubium with congeners of other species. The bridgeless gaps . . . remain to be explained. [pp. 142–43]

Species, Goldschmidt argues, are separated by genetic differences distinct in kind from the accumulated micromutations that produce rassenkreise; they "turn out to involve a chromosomal reorganization" (p. 181). Moreover, "species differences are differences of the whole developmental pattern" (p. 180). New species are fundamentally different animals, not extremes of an intergrading series. Thus, Goldschmidt ends the first part of his book with a paragraph in italics:

Microevolution by accumulation of micromutations—we may also say neo-Darwinian evolution—is a process which leads to diversification strictly within the species, usually, if not exclusively for the sake of adaptation. . . . Subspecies are actually, therefore, neither incipient species nor models for the origin of species. They are more or less diversified blind alleys within the species. The decisive step in evolution, the first step toward macroevolution, the step from one species to another, requires another evolutionary method than that of sheer accumulation of micromutations. [p. 183]

In the second part, entitled "macroevolution," Goldschmidt searches for this "other evolutionary method" behind the origin of species and higher taxa. This part contains two rather different discussions; the first (pp. 184–250), shorter and more controversial, on systemic mutation and the nonexistence of "corpuscular genes"; the second (pp. 250–396), more extended and full of holistic insight about the nature of integrated organisms, on the constraints and opportunities of developmental systems and the potential macroevolutionary result of mutations affecting early development.

Goldschmidt centered the first discussion upon a pet theme—the nonexistence of corpuscular genes—that occupied more and more of his time and rendered him more and more controversial as the years wore on; (his last book, *Theoretical Genetics* (1955), contains his most developed and extreme statements). The progression was, perhaps, inevitable, but it led Goldschmidt so far beyond the pale for many colleagues that his points of value, and the real problems that elicited his novel solutions, often sank from sight.

He began with the search for a macroevolutionary mechanism other than accumulation of micromutations. Since inversions, translocations, and other chromosomal changes can exert a marked effect upon phenotypes in the absence of any alteration within genes, point mutations cannot be the only source of evolution. "In these cases there is no indication that anything has changed but the serial order of the constituent parts of the chromosome" (p. 188). If new species represent fundamentally new developmental programs and if individual point mutations exert local and minor effects (except in cases of massive pleiotropy or mutations affecting early development), then perhaps the "different" genetics behind speciation involves a fundamental reordering of chromosomal pattern. In such a mechanism evolution would find a device for the rapid and wholesale transformation that accompanies speciation. Goldschmidt named these hypothetical changes of general chromosomal patterns "systemic mutations."

For a long time I have been convinced that macroevolution must proceed by a different genetic method. . . . A pattern change in the chromosomes, completely independent of gene mutations, nay, even of the concept of the gene, will furnish this new method of macroevolution. . . . So-called gene mutation and recombination within an interbreeding population may lead to a kaleidoscopic diversification within the species, which may find expression in the production of subspecific categories. . . . But all this happens within an identical general genetical pattern which may also be called a single reaction system. The change from species to species is not a change involving more and more additional atomistic changes, but a complete change of the primary pattern or reaction system into a new one, which afterwards may again produce intraspecific variation by micromutation. One might call this different type of genetic change a *systemic mutation,* though this does not have to occur in one step. . . . Whatever genes or gene mutations might be, they do not enter this picture at all. Only the arrangement of the serial chemical constituents of the chromosome into a new, spatially different order; i.e., a new chromosomal pattern, is involved. [pp. 205–06]

With systemic mutation, Goldschmidt felt that he had escaped "the dead end reached by neo-Darwinian theory" (p. 203). "The systemic pattern mutation—as opposed to gene mutation—appears to be the major genetic process leading to macroevolution; i.e., evolution beyond the blind alleys of microevolution" (p. 245).

But Goldschmidt went further. We have definite physical evidence for the causes of repatterning, but what do we know of so-called gene mutation beyond the fact of localization on chromosomes? This localization does not imply that the gene is a physical "corpuscular entity." Why have two parallel systems—chromosomal repatterning and gene mutation—one that we understand in physical terms and one that merely exists as a name for ignorance? Citing no less an authority than William of Occam, Goldschmidt argues that, in the absence of direct evidence for genes as "things," we should restrict ourselves to the objects we know and interpret "point mutation" as localized and small-scale pattern change: Why not accept "the viewpoint, which in my opinion is daily becoming more probable, that actually no particulate genes exist, but that all mutations are based on very small pattern changes" (p. 203)? Goldschmidt cites Occam's motto as the last word in his book: *Frustra fit per plura quod fieri potest per pauciora* (In vain we do with many things what can be accomplished with fewer). Goldschmidt began by separating change in pattern from mutations in genes and ended by denying genes themselves. Of "the classical atomistic theory of the gene" (p. 209), Goldschmidt wrote: "It is this theory which blocks progress in evolutionary thought. . . . We have already foreshadowed the twilight of the gene" (p. 210).

The second discussion, entitled "evolution and the potentialities of development," is in one sense complementary, but in another curiously contradictory, to the first. It includes two topics: norms of reaction (or "reactivity" in Goldschmidt's phrase) and mutations affecting early development. After writing with such joyous abandon about the twilight of the gene, Goldschmidt brings point mutations right back to develop a different, and more acceptable, theory of macroevolution by small genetic changes that produce marked phenotypic effects by acting upon developmental rates in early ontogeny. (I wonder if Goldschmidt wrote the two parts of his second half at different times, or if he put the entire book together in haste.) Others noted this inconsistency, and Goldschmidt himself commented later (1955, quoted in Frazzetta, 1975, p. 116), "I have been reproached for not having made clear in my book *The Material Basis of Evolution* whether I was speaking of systemic mutation (scrambling of the chromosomal pattern) or of ordinary mutations of a macroevolutionary type, and of being confused myself on what I meant."

The constraints and opportunities of developmental systems is the common theme of this second discussion—and it must be so because Goldschmidt has now dropped wholesale alteration of chromosomal pattern (systemic mutation) for macroevolution by small mutations of large impact. How can ordinary mutations be amplified in their effect to produce major, discontinuous changes in phenotype? Goldschmidt answers that they can only do so by "playing off" the inherited developmental program. If that program is so constructed that small alterations in timing can shift development into radically different, but still viable, channels, then small inputs can generate cascades leading to large effects. In this theme, Goldschmidt's lifetime of work on physiological genetics, rate genes, and reaction velocities achieved its integration with evolutionary theory.

The argument based on norms of reaction invokes the idea that there exists both an enabling and a constraining force and that the interplay of these forces allows "mutations affecting early development" to produce viable and fundamentally altered phenotypes. The multiplicity of developmental pathways permitted by a single genotype defines the enabling force. Often, Goldschmidt demonstrates, discontinuous phenotypic changes of macroevolutionary magnitude can be generated by environmental manipulation of a constant genotype. Phenocopies that mimic or duplicate the effects of gene mutations can be made by perturbing environments of development, often in minor ways. Since these perturbations induce changes in rates, genes themselves must act in similar, quantitative ways. Small doses of hormones may enhance or inhibit metamorphosis and produce the macroevolutionary effect of neoteny in an unaltered genotype. In this case, the potentialities of development are not alternate pathways, but sequential and profound transformations in ontogeny, whose suppression or induction produces a radically altered adult. "In other words," Goldschmidt writes (p. 260), "within a constant genotype the potentialities of individual development may include a range of variation of the same phenotypic order of magnitude which otherwise characterizes large evolutionary steps based upon changes in the genotype. The norm of reaction thus shows what paths are available for changes in the genotype (mutations in the broadest sense) without upsetting normal developmental processes."

The creative constraining force is embodied in the last phrase—"without upsetting normal developmental processes." If shifts to alternate pathways simply discombobulate the system, then all monsters will be hopeless. Many intricately complex systems simply fall apart or change in injurious ways under the impact of perturbations. But organic systems are regulated to accommodate impacts by minimal change and to integrate changes into canalized and viable pathways. Goldschmidt cites a range of phenomena illustrating organic "buffering," from regeneration of amphibian limbs to reaggregation of sponges after dismemberment into cells. Monsters may indeed be hopeful because strongly altered phenotypes will, when regulation is effective, be integrated within still viable developmental systems. (Goldschmidt's critics often conflated systemic mutation with "hopeful monsters." But note that Goldschmidt discusses hopeful monsters (pp. 300–93) only under the heading of norms of reaction and mutations affecting early development. The phrase is not pure whimsy or nonsense. Goldschmidt granted hope to his monsters *because* regulation can integrate certain large alterations of phenotype into viable systems of development; moreover, the large alterations themselves are extensions or modifications of inherited developmental channels produced by changes in rates within established systems.)

If such "play" exists within constant genotypes, imagine what mutational change might accomplish. And so Goldschmidt reaches his final topic, "mutations affecting early development." Goldschmidt, who codified the concept of "rate genes," had long emphasized the quantitative nature of gene action. "The genetic material controls the velocities of production, and the time of action, of the determining stuffs which control differentiation. The proper timing of these processes is the decisive feature in the general control of development" (p. 263).

If mutations affect the timing of crucial early stages in development, they may initiate a cascade of consequences inherent within the norm of reaction and channeled into viable pathways by organic regulation. "A single mutational step affecting the right process at the right moment can accomplish everything provided that it is able to set in motion the ever-present potentialities of embryonic regulation" (p. 297).

Goldschmidt's general themes include D'Arcy Thompson's vision of reducing complex forms to few generating parameters (whose small quantitative alterations, beginning early in growth, can lead to major changes of adult phenotype) and the phenomena of homeosis and rudimentation discussed as small changes that induce large effects by switching an organism into developmental pathways already contained within its genome's norm of reaction. His specific examples are all of the same form (and not really adequate in missing a crucial step of the usual argument for saltation, as I shall discuss on p. xxxv): phenotypes that arise as teratologies or as products of environmental perturbation within one species have become the normal forms of related species; since these phenotypes can arise (or be experimentally induced) abruptly in species that do not normally produce them, they probably arose by macromutation in related species defined by their fixation (see pp. 304, 306, 331, 353, 356, 360, and 376). In the moth *Orgyia,* for example, the normal antennae of males are identical with those of aberrant intersexual males in the closely related *Lymantria* (p. 304). The fly *Termitoxenia* "has minute rudimentary wings of a very peculiar type"(p. 333), but they are identical with reduced wings intermediate between halteres and normal wings in two homeotic mutants of *Drosophila*.

Thus, the primary agents of macroevolution may be small mutations acting early and shunted through developmental systems adapted for responding to perturbations by channeling changes into viable alternative pathways:

> The physiological balanced system of development is such that in many cases a single upset leads automatically to a whole series of consecutive changes of development in which the ability for embryonic regulation, as well as purely mechanical and topographical moments,* come into play; there is in addition the shift in proper timing of integrating processes. If the result is not, as it frequently is, a monstrosity incapable of completing development or surviving, a completely new anatomical construction may emerge in one step from such a change. [p. 386]

Constraints of development have their creative side; small mutations may be potent forces for abrupt and extensive change if they can commandeer this creativity by acting early in development and not forcing developmental systems beyond their breaking point. The hopeful monster is a creature bearing such mutations. He, and his cousin (or twin) the systemic mutant, are agents of macroevolution and exterminating angels of the Darwinian hope that all evolution might be rendered as the promoted and accumulated product of small, adaptive changes that we observe in the field or produce by selection in the laboratory.

III. CRITIQUE AND APPRECIATION

WHATEVER private doubts Goldschmidt may have harbored, his almost overbearing self-confidence permitted no public airing. Of his heterodox

*D'Arcy Thompson's theme.

views, he wrote, "I am certain that in the end I shall turn out to have been right" (1960, p. 307). Although we are now only infinitesimally closer to that end, and although Goldschmidt's star may again be on the rise among evolutionists, I doubt that his major book will ever achieve total approbation. It contains too many inconsistencies and includes at least two poor arguments at crucial points. I do, however, believe that its general vision is uncannily correct (or at least highly fruitful at the moment) in several important areas in which conventional Darwinian theory has become both hidebound and unproductive.

The Material Basis of Evolution will always be shrouded in ambiguity, as, I suppose, every irreverent masterpiece must be. I believe that it contains a number of weaknesses gathered under two general headings. I also believe that other aspects of the same headings define its greatest strengths.

The two dubious themes both represent areas in which Goldschmidt had an important insight, but pushed and extended it well beyond the bounds of acceptability. In carrying his campaign against the "corpuscular gene" into the pages of his book, Goldschmidt introduced a red herring (made more offensive by the conscious red flag of his colorful pronouncements) that led people to angry dismissal and to a disregard for his cogent arguments about the integration of developmental and gene function. (This attempted reconstruction of theoretical genetics was, of course, central to Goldschmidt's world view; in this book, however, it played a diversionary role and alienated scores of geneticists who otherwise had no rooted antipathy to his macroevolutionary ideas.)

Goldschmidt's views were not quite as heretical as he made them sound. He did not deny the heuristic value of treating localized modifications of the genome as if they represented the alteration of particles on a string, even if (in his reality) they represented small examples of repatterning—but he didn't bother to place the disclaimer in this book. Later, in 1946, he wrote:

> There has been much misunderstanding of our conclusions. There is, of course, no doubt that the chromosome has a serial structure and that localized changes of this structure, the mutant loci, can be located by the cross-over method. There is no doubt either that these localized conditions of change can be handled

descriptively as separate units, the mutant locus or gene, and that for all descriptive purposes the extrapolation can be made that at the normal locus a normal gene exists. Further, there can be no doubt that almost all genetical facts can be described in terms of corpuscular genes, and that a geneticist who is not interested in the question of what a gene is may work successfully all his life without questioning the theory of the corpuscular gene. [Quoted in Dodson, 1960, p. 225]

Not only did he push a good insight too far, but he also brought other important points down with it. By continually conflating the unacceptable systemic mutant (though aspects of this concept have merit too) with the small mutation that has macroevolutionary impact by affecting early development, he led people to dismiss his important material about the potential for macroevolution inherent in constraints and opportunities of developmental systems. The colorful terminology of hopeful monsters also led to easy and unconsidered dismissal, as people falsely linked these putative macroevolutionary agents with rejected systemic mutants, whereas Goldschmidt specifically labelled them "hopeful" because the regulative properties of development might channel small impacts upon early embryology into viable pathways of major phenotypic effect.

As his second dubious theme, Goldschmidt carried a good insight too far by advocating an absolutely clean break in principle and kind between micro- and macroevolution. In numerous haughty phrases about "blind alleys," "dead ends," and "twilights," he argued that microevolution has nothing whatever to do with the origin of species and that a completely different style of genetic change (chromosomal repatterning versus micromutational accumulation) underlies speciation. Could any statement be more calculated to arouse Darwinian ire?

The current debate about macroevolution (Eldredge and Gould, 1972; Stanley, 1979; Lewin, 1980) indicates strong support for Goldschmidt's gut feeling that extrapolation of small-scale adaptive change within local populations will not encompass all of evolution. But I doubt that any current critic of the modern synthesis would hold that no species can ever arise by an intensification of geographic variation, that saltation is the major (or only) mode of origin for new species, and (especially) that microevolution provides no insights, but only diversions and delusions, for our understanding of evolutionary trends and the origin of new *Baupläne*.

Moreover, Goldschmidt supported his concept of an absolute break between levels by two poor arguments at key points. The first occurs in part 1, when Goldschmidt must face the intermediary cases that seem to exist in fair abundance between geographic variation and speciation—imperfectly separated local populations with impaired interfertility. By arguing that these extreme members of rassenkreise do not approach species, Goldschmidt engages in an almost frantic special pleading. When he must acknowledge impaired fertility, he argues that it is not of the sort that produces true reproductive isolation in speciation; it is "a more or less freakish type of microevolution of the nature of a blind alley within the confines of a species" (p. 128), he states in one place. When he encounters the intermediate situation of two entities that interbreed in some parts of their range but not in others (see pp. 155–68 on "the border cases"), he proclaims them mere members of a single rassenkreis because the fact that they do in some places reflects their compatibility, while the fact that they don't in others merely records a current state, not a potential. Here Goldschmidt wins his own argument by definition. He has precluded the very possibility of acknowledging intermediacy by admitting populations to full membership in an ordinary rassenkreis if they interbreed anywhere, and setting them up as separate species if they interbreed nowhere.

The second poor argument pervades part 2, for Goldschmidt fails to invoke the central claim—and still an indispensable one, I think—of the classical argument for saltation. All Goldschmidt's illustrations of potential saltation involve fixed characters of species that are present as mutants, teratologies, or environmentally induced phenocopies of related species. In other words, he shows that pathways of development *could* permit the expression of these phenotypes in single steps. But *could* isn't *must,* and the simple fixation in some species of phenotypes that represent deviant pathways of development in related taxa does not establish their saltatory origin. For as Lande (1980) and others have emphasized recently, the accumulation of modifiers can lead to the crossing of phenotypic thresholds under gradual selection.

The classical argument for saltation, on the other hand, requires a claim for the *inviability of conceivable intermediate states*. The fact that a phenotype arises discontinuously as a teratologous mutant in one species does not prove that it cannot be built gradually in other circumstances. Interesting claims for phenotypic saltation have always invoked the inconceivability of intermediary stages in an evolutionary sequence—as in the torsion of snails, Frazzetta's snakes with a split maxillary (1970), and Long's rodents with inverted cheek pouches (1976). Mivart's old argument (1871) about the inviability of "incipient stages of useful structures" seems as sound as ever, and Goldschmidt fails to use it.

The strengths of Goldschmidt's argument, and the power of his vision, also invoke the same two themes, but divested this time of Goldschmidt's extended claims for them. It must be said, first of all, that many of Goldschmidt's points are not nearly so extreme or unthinking as the usual caricatures depict them. He did not argue that all chromosomal repatterning between species must arise in a single step. Rather—and he says so repeatedly, not merely as a disclaimer in one hidden footnote—he imagines that much repatterning occurs gradually and sequentially with no (or minor) outward effect upon phenotypes. The systemic mutant is the step, usually a large one to be sure, that provokes the crossing of a phenotypic threshold. "This new genetic system, which may evolve by successive steps of repatterning until a threshold for changed action is reached, produces a change in development which is termed a systemic mutation" (p. 396). Moreover, he did not ignore, as many critics have charged, the problem that hopeful monsters must spread through populations by interbreeding with individuals of normal phenotype. In arguments still invoked by modern models of chromosomal speciation (White, 1978; Bush, 1975), Goldschmidt (p. 207) specified the conditions of inbreeding and lack of strong selection against heterozygotes that would permit the phenotypes of hopeful monsters to spread and reach fixation. Finally, since the hopeful monster is not a phenotypic absurdity reflecting a fundamentally altered genotype, but the product of small genetic changes regulated by inherited developmental pathways into viable phenotypic channels, it is not the deviant "basket case" that any "right thinking" normal form would reject, or that would invariably produce even more bizarre heterozygote offspring.

If I may epitomize the first theme as "the material basis of macroevolution" and the second as "hierarchical levels of evolution," we may begin to assess the strengths of Goldschmidt's vision. For the first, Goldschmidt's denial of the corpuscular gene may be unacceptable today,* but the problems in conventional thinking that led him to propose it remain acute. As a man trained in classical morphology and embryology, Goldschmidt developed a professional sense that animals must be viewed as integrated wholes and that our organic world of divergent *Baupläne* cannot be rendered by models of strict, gradual continuity in transformation. As a mechanist at heart (see p. 398), he groped and struggled for a holistic but material concept of the genetic basis for such integration (see p. 218).

The systemic mutant may represent a misguided attempt, but the insight that inspired it has fundamental validity. Under strict Darwinism, with its emphasis on adaptive, gradual change guided by natural selection, organisms move where selective pressures push them. In Galton's metaphor (see Gould, 1980, pp. 128–29), organisms are spheres pushed by the pool cue of natural selection along preferred paths constructed by environments (for the table is not smooth). Accumulated micromutation becomes a viable mechanism for all evolutionary change. But if organisms are polyhedrons, then they "push back" and resist change, can only alter in certain directions, and flip from one stable system to another when they do change. Natural selection may still be the only pool cue, but the "internal" factors of organic integration constrain and direct the possible paths of pushing. The polyhedron may slide on its current facet in adapting to local environments by micromutation, but the flip from facet to facet, or macroevolution, may require other styles of genetic change.

Constraints of developmental programs define the facets of Galton's polyhedron and suggest that flips and slides—or macro- and mi-

*The Watson-Crick model in its original form demolished Goldschmidt, but part of his claim may hold in our modern world of split and overlapping genes. Still, history of science must reject the anachronistic method of evaluating people according to the number of times they are vindicated by good and unanticipated later fortune.

croevolution—may not represent a pure continuum. Most of Goldschmidt's book discusses these developmental constraints and opportunities. His vision of a difference between macro- and microevolution is rooted in this theme that got lost in the strict Darwinian shuffle.

We do not now accept Goldschmidt's notion of a strict separation between the genetics of slides and flips. Galton's polyhedron implies the theme of developmental constraints, but does not require a different genetics for flips—since ordinary mutations can induce flips by exciting developmental switches, inducing a movement across thresholds, or acting early in development with cascading consequences. (Goldschmidt recognized all this, but submerged it in his later enthusiasm for systemic mutation.) Still, even the idea of different genetic styles has caught on again, though not in Goldschmidt's extreme form: witness Carson's open and closed systems (1975), the current emphasis on structural vs. regulatory genes (King and Wilson, 1975; Wilson et al., 1975), renewed emphasis upon rapid chromosomal speciation vs. micromutational change within species (White, 1978), and awakening interest in evolutionary aspects of development and such classic heterochronic phenomena as neoteny (Gould, 1977b; Alberch et al., 1979, the Dahlem conference on development and evolution held in Berlin, May, 1981).

We will not return to Goldschmidt's extreme view that all speciation is different in genetic kind from microevolution; classical Darwinian continuity works in many cases, and many species are separated by relatively minor genic differences. But the origin of new *Baupläne* requires the reorganization of developmental systems, the flips between facets that Darwinians deemphasized and Goldschmidt knew were important. Goldschmidt's vision was sound, his solution too extreme.

On the second theme of hierarchical levels, Goldschmidt was again too extreme in asserting a "bridgeless gap" (if I may borrow his phrase for another context) between micro- and macroevolution, or geographic variation and speciation. For him, microevolution led only to blind alleys and operated by genetic mechanisms playing no role in speciation; therefore it illuminated nothing of interest about macroevolution. This is a depressing claim indeed, for it renders useless for most purposes the most extensive, and the only directly observable, data that we have about evolution—small-scale change within local populations. Hopes for a unified and general theory of evolution fade.

But the only more depressing claim is unity bought at the high price of ignoring a vital aspect of evolution. I believe that the Darwinian modern synthesis achieved this spurious unity by submerging the concept of levels and opting for an extrapolationist vision that reduces all macroevolution to microevolution extended. Goldschmidt's levels were too distinct and noninteracting; but better this perhaps than no levels at all.

I believe that we may achieve a unified and more general evolutionary theory by combining parts of both visions: the acknowledgment of levels that Goldschmidt demanded, with the Darwinian belief in a unity of processes across these levels. The notion of hierarchy does not demand separate causes, for the same set of causes may produce different results in acting upon the disparate phenomena of distinct levels (Gould, 1980; Eldredge and Cracraft, 1980). Moreover, the levels are not separated by impenetrable barriers, but by interacting boundaries that permit extensive leakage and feedback. Speciation doesn't need a distinct genetics to be meaningfully different from microevolution. If microevolution is fundamentally a process of adaptation, and if reproductive isolation (speciation) is the mere by-product of divergent selection upon two isolated populations, then we have smooth continuity and the modern synthesis is vindicated. But if reproductive isolation often arises first, then speciation merely provides an opportunity for subsequent divergent adaptation—and speciation is not microevolution extended. Speciation becomes a recognizable level of evolution—distinct from adaptive diversification within populations—without requiring a distinct set of genetic causes. Eldredge and Cracraft (1980) write:

> Speciation is not, fundamentally, a process of adaptation. Therefore, a theory emphasizing adaptation at its core cannot properly be extrapolated smoothly from the level of microevolution to the level of macroevolution. The conflict with the syntheticist form of macroevolutionary theory, which is a direct, wholesale extrapolation of within-species, microevolutionary theory, arises from the necessary obliteration of species as the basic evolutionary units in the syntheticist view. [p. 326]

Likewise, if most species do not change appreciably following their geologically rapid origin, then large-scale evolutionary trends are not microevolution extended, but must represent a higher-order sorting out of speciation events themselves (Gould and Eldredge, 1977; Stanley, 1979). But again, we have no separate genetics for evolutionary trends, only a distinct operation of ordinary processes.

In conclusion, I believe that Goldschmidt made a profound (deep, not large) and interesting error in supporting his cogent vision that evolution must be viewed as a hierarchy of levels. As a mechanist, he apparently felt that a claim for difference must be supported by something concrete and material. He understood that micro- and macroevolution were different, and he therefore sought a distinct genetic basis for the two processes. He proposed one without enduring success.

But I believe that legitimate differences may arise for reasons more abstract than definite mechanistic causes; they may, as in this case, be the product of similar causes working within a set of phenomena arranged as distinct hierarchical levels. If I may make an analogy: the eighteenth-century preformationists postulated a homunculus within the egg because they correctly understood that a formless egg could not unerringly generate the same complex phenotype again and again. In their world, form meant definite concrete structure—and they had no alternative but to postulate actual parts or organs of the next generation within the ovum. Their vision was right; the egg cannot be formless. But we now know that its inherent structure, though still material of course, is a thing more abstract than actual parts—namely, coded instructions for building the parts.

The systemic mutant, the saltatory origin of nearly all new species, may be rejected, but Goldschmidt's vision was sound—and it supplied (or rather resupplied) an essential ingredient that strict Darwinism had expunged from evolutionary theory: the idea that evolution works through a hierarchy of distinct levels with important independent properties (however strong the ties of feedback that Goldschmidt denied). Thus, when I place Goldschmidt in the balance during the year of his own predicted triumph, I find him not victorious, but weighted equally with his self-proclaimed Darwinian opponents. He occupied his pan virtually alone; the other pan contained a cast of thousands. If he has truly balanced them, then he is a weighty man indeed—and his book, which I now invite you to read, will stand as an enduring document, however flawed.

BIBLIOGRAPHY

Alberch, P., S. J. Gould, G. F. Oster, and D. B. Wake, 1979. Size and shape in ontogeny and phylogeny. Paleobiology 5: 296–317.

Bush, G. L., 1975. Modes of animal speciation. Annu. Rev. Ecol. Syst. 6: 339–64.

Carson, H. L., 1975. The genetics of speciation at the diploid level. Am. Nat. 109: 83–92.

Dobzhansky, Th., 1937. Genetics and the origin of species. New York: Columbia Univ. Press.

Dobzhansky, Th., 1951. Genetics and the origin of species (3rd ed.). New York: Columbia Univ. Press.

Dodson, E. O., 1960. Evolution: process and product. New York: Reinhold.

Eaton, T. H., Jr., 1970. Evolution. New York: W. W. Norton.

Eldredge, N., and J. Cracraft, 1980. Phylogenetic patterns and the evolutionary process. New York: Columbia Univ. Press.

Eldredge, N., and S. J. Gould, 1972. Punctuated equilibria: an alternative to phyletic gradualism. In: T. J. M. Schopf, ed., Models in paleobiology. San Francisco: Freeman, Cooper and Co., pp. 82–115.

Frazzetta, T. H., 1970. From hopeful monsters to bolyerine snakes. Am. Nat. 104: 55–72.

Frazzetta, T. H., 1975. Complex adaptations in evolving populations. Sunderland, MA: A. Sinauer.

Goldschmidt, R., 1938. Ascaris: the biologist's story of life. London: English Universities Press.

Goldschmidt, R., 1952. Evolution as viewed by one geneticist. Am. Sci. 40: 84–98.

Goldschmidt, R., 1955. Theoretical genetics. Berkeley: Univ. of Calif. Press.

Goldschmidt, R., 1960. In and out of the ivory tower: the autobiography of Richard B. Goldschmidt. Seattle: Univ. of Wash. Press.

Gould, S. J., 1977a. Return of hopeful monsters. Nat. Hist., June–July, pp. 22–30.

Gould, S. J., 1977b. Ontogeny and phylogeny. Cambridge, MA: Harvard Univ. Press.

Gould, S. J., 1980. Is a new and general theory of evolution emerging? Paleobiology 6: 119–30.

Gould, S. J., and N. Eldredge, 1977. Punctuated equilibrium: the tempo and mode of evolution reconsidered. Paleobiology 3: 115–51.

King, M. C., and A. C. Wilson, 1975. Evolution at two levels in humans and chimpanzees. Science 188: 107–16.

Lande, R., 1980. Microevolution in relation to macroevolution. Paleobiology 6: 233–38.

Lewin, R., 1980. Evolutionary theory under fire. Science 210: 883–87.

Long, C. A., Jr., 1976. Evolution of mammalian cheek pouches and a possibly discontinuous origin of a higher taxon (Geomyoidea). Am. Nat. 110: 1093–97.

Mayr, E., 1963. Animal species and evolution. Cambridge, MA: Harvard Univ. Press, Belknap Press.

Mayr, E., and W. Provine, 1980. The evolutionary synthesis. Cambridge, MA: Harvard Univ. Press.

Mivart, St. G., 1871. On the genesis of species. London: Macmillan.

Stanley, S. M., 1979. Macroevolution. San Francisco: W. H. Freeman.

Wilson, A. C., G. L. Bush, S. M. Case, and M. C. King, 1975. Social structuring of mammalian populations and rate of chromosomal evolution. Proc. Nat. Acad. Sci. 72: 5061–65.

White, M. J. D., 1978. Modes of speciation. San Francisco: W. H. Freeman.

THE MATERIAL BASIS OF EVOLUTION

I. INTRODUCTION

THE major part of the genetical work which I performed during the past thirty years proceeded along three apparently very different avenues: sex determination, physiological genetics, and evolution. But even in early stages of these lines of research it was recognized that such apparently different topics were linked with each other by some generalizations. These were brought out in a group of essays which I wrote in the winter of 1917–18, while a guest at Yale University (Goldschmidt, 1920). In these essays I showed that the work on sex determination led to conclusions regarding the action of the hereditary material in development. The same work was simultaneously concerned with a problem in evolution, the problem of geographic variation, and this led to a consideration of both these problems, genic action and evolution, from the same point of view. It was recognized that a change in the hereditary type can occur only within the possibilities and limitations set by the normal process of control of development through the action of the germ plasm. These possibilities and limitations were a direct consequence of the solution of the problem of genic action which we had found in terms of reaction velocities. Actually, the discussion of all these problems was tentative, and decisive points were mentioned only briefly, even relegated to footnotes. The idea was to wait for a full discussion until my own experimental work on some evolutionary problems as well as genetic ones would have been finished. It happened that in the following decades the first two parts of this work, sex determination and the action of the gene, came to the foreground, and that I could repeatedly work out in detail my ideas, based upon my own work as well as on the ever-increasing amount of work by others. Thus this part of the old essays finally grew into four elaborate and completely documented books; two of them, with about ten years' interval between them, on sex determination (Goldschmidt, 1920a, 1931), and two others, again separated by ten years of more analytical work, on physiological genetics (Goldschmidt, 1927, 1938). I had always wished to accomplish the same for the evolutionary part of the essays and I intended to do so after my own chief experimental contribution to evolution, the analysis of geographic variation, was finished. When this finally came to pass (in 1932), other work had come to the fore and I contented myself with embodying some of the generalizations in a short paper and in occasional lectures, delivered at different meetings and in different European and American universities. Only a few of these (Goldschmidt, 1932, 1933, 1935) were published. The appointment as Silliman Lecturer—an honor for which I am deeply grateful—has finally furnished the necessary stimulus to carry out the plan, for which the material has been collected for a long time.

II. THE PROBLEM

According to the deed of this lectureship, "its general tendency . . . may be such as will illustrate the presence and wisdom of God as manifested in the Natural and Moral World." To the naturalist this means the demonstration of law and order in his chosen field. As evolution is our topic, this might mean that a full discussion of the facts, laws, and theories of evolution is to be expected. This, however, cannot be accomplished. No individual can claim such a mastery of all facts pertaining to evolution to enable him to present such a discussion. Moreover, it is not my intention to present an objective review of the present status of the problem of evolution. Though attacking the problem as a geneticist, I do not even intend to discuss evolution from the geneticist's point of view alone. What I propose to do is to inquire into the type of hereditary differences which might possibly be used in evolution to produce the great differences between groups, and the title of this book, accordingly, ought to be something like: The genetical and developmental potentialities of the organism which nature may use as materials with which to accomplish evolution. In the analysis of this problem I shall try to use whatever viewpoint seems to lead to progress. Many of the conclusions which we shall reach will be in disagreement with the views held generally by geneticists or, on a different basis, by taxonomists. I trust that negative and sterile criticism will not be found in our discussion, and that whatever doubt is cast upon established ideas will be based upon ample facts and will be the type of doubt which is the sire of progress. There are many important facts relating to evolution, genetic and otherwise, which will not be mentioned. This does not mean that I underrate their importance, but only that they are considered to be outside the sphere of those problems in evolution which are selected here for discussion. If I may compare the individual facts concerning evolution to individual glass mosaic cubes, it is not my intention to present a huge bagful of them to be used on a future day for assembly into a figure. I intend to build a smaller but finished picture, using only a selected part of the cubes in the bag. Under such circumstances I shall not try to bring together and to review all literature relevant to the subject. This would be a Herculean task and it would, in addition, tend to drown the general picture in a mass of detail. I shall, therefore, have to select my examples and to use those which best illustrate the argument. It is my wish to make this selection in as fair and open-minded a spirit as possible, and I shall try to include at least all really important facts. This book, then, is no treatise on evolution and does not intend to compete with comprehensive treatises like the brilliant texts by Haldane (1932) and Dobzhansky (1937), and the many other collections of fact presented from different angles, viz.; Berg (1926), Cuénot (1911, 1936), Guyénot (1930), Hertwig (1927), Robson (1928), Robson and Richards (1936), and others.

The problem of evolution as a whole consists of a number of subproblems, with some of which we are not concerned here at all. There is, first, evolution as a historical fact. With all biologists we assume that evolution as such is a fact. There is the problem of selection or survival of the fittest. It may be considered as established, both biologically and mathematically, that given hereditary variations, definite systems of heredity like Mendelian heredity, and differences in regard to survival value, selection may wipe out one type or isolate a new type. This means that there is no difficulty in the understanding of evolution, provided the *necessary* hereditary variations are given. There are the different aspects of adaptation, only some of which will be discussed. It is mainly the problem of the hereditary differences as the material of evolution which we shall discuss.

The information on this topic is derived from different fields of study. The basic knowledge is furnished by the taxonomist who registers the actually existing forms down to the smallest recognizable units, and states their natural affinities, their ecology, and their habitat. A different kind of information is available to the geneticist. He follows the origin of hereditary differences and locates their actual basis in the germ plasm. But it is evident, though sometimes forgotten, that the methods of evolution cannot be derived, say, from the genetics of coat colors of rabbits, without taking into account the existence of what may be called macrotaxonomy. The laws which are supposed to explain the diversification of species must also account for families, orders, and phyla: differences rat-mouse, cow-whale, horse-lizard, butterfly-snail, must all be explained. This means that the geneticist who comes to definite evolutionary conclusions with his limited material must test them within the larger field of macrotaxonomy, the origin of the higher systematic categories, and admit failure if this test fails.

The same applies, of course, to the taxonomist. He used to derive his opinions upon species formation from studies of closely related species. Nowadays he adds to this the study of the subspecies found in nature and their geographic relations. We might call this microtaxonomy. Conclusions derived from microtaxonomical studies upon the methods of evolution are valuable as generalizations only if they can explain also the facts of macrotaxonomy. It is in microtaxonomy that the geneticist and the taxonomist come together. Macrotaxonomy is practically inaccessible to genetic experimentation, but the range of the subspecies up to, or nearly up to, the limit of the species is accessible both to the geneticist and the taxonomist. The results of both, therefore, may be mutually checked, and definite conclusions seem possible.

The field of macrotaxonomy, however, is not directly accessible to the geneticist, or only to a very limited degree. Here the paleontologist, the comparative anatomist, and the embryologist are supreme. The geneticist must try to apply his findings in microtaxonomy to the materials of macrotaxonomic order which he finds in those fields, provided this can be done. This is where the geneticist faces his most difficult task.

There is, finally, another field which has been neglected

almost completely in evolutionary discussions; namely, experimental embryology. The material of evolution consists of hereditary changes of the organism. Any such change, however, means a definite change in the development of the organism. The possibility and the order of magnitude of genetic changes are therefore a function of the range of possible shifts in the processes of development, shifts which may take place without upsetting the integration of embryonic processes. From this it follows that the potentialities of individual development are among the decisive factors for hereditary change and therefore for evolution.

This statement of the problem already indicates that I cannot agree with the viewpoint of the textbooks that the problem of evolution has been solved as far as the genetic basis is concerned. This viewpoint considers it as granted that the process of mutation of the units of heredity, the genes, is the starting point for evolution, and that the accumulation of gene mutations, the isolation and selection of the new variants which afterwards continue to repeat the same process over again, account for all evolutionary diversifications. This viewpoint, to which we shall allude henceforth as the neo-Darwinian thesis, must take it for granted that somehow new genes are formed, as it is hardly to be assumed that man and amoeba may be connected by mutations of the same genes, though the chromosomes of some Protozoa look uncomfortably like those of the highest animals. It must further be taken for granted that all possible differences, including the most complicated adaptations, have been slowly built up by the accumulation of such mutations. We shall try to show that this viewpoint does not suffice to explain the facts, and we shall look for explanations which might evade these and other difficulties and simultaneously account for such facts as have to be pushed to the background to make the popular assumptions plausible. At this point in our discussion I may challenge the adherents of the strictly Darwinian view, which we are discussing here, to try to explain the evolution of the following features by accumulation and selection of small mutants: hair in mammals, feathers in birds, segmentation of arthropods and vertebrates, the transformation of the gill arches in phylogeny including the aortic arches, muscles, nerves, etc.; further, teeth, shells of mollusks, ectoskeletons, compound eyes, blood circulation, alternation of generations, statocysts, ambulacral system of echinoderms, pedicellaria of the same, cnidocysts, poison apparatus of snakes, whalebone, and, finally, primary chemical differences like hemoglobin vs. hemocyanin, etc.[1] Corresponding examples from plants could be given.

1. The important problem of the chemical differences has been emphasized in the reviews by Schepotieff (1913), Pantin (1932), Redfield (1936).

III. MICROEVOLUTION

This term has been used by Dobzhansky (1937) for evolutionary processes observable within the span of a human lifetime as opposed to macroevolution, on a geological scale. It will be one of the major contentions of this book to show that the facts of microevolution do not suffice for an understanding of macroevolution. The latter term will be used here for the evolution of the good species and all the higher taxonomic categories.

1. THE MICROMUTATIONS

When Darwin wrote his first drafts of the *Origin of Species* (essays of 1842 and 1844) he believed that sports, nowadays called mutations, played a major part in evolution. Later he changed his mind and was inclined to assume that it is the body of small variations which forms the material for selection. With De Vries' theory of mutation again the large steps came to the foreground, and though his original material, Oenothera, turned out to be of importance in quite a different direction, the awakening Mendelism took over the theory of mutants as the basic material of selection and evolution. All the earlier Mendelian studies were done with mutant types which differed rather considerably from the original form, most of them recessive and the majority hardly viable under natural conditions, if not actually monstrous. Certainly the optimism created by the discovery of the ubiquity and rather considerable frequency of the mutants ran wild. But soon a reaction set in. Some geneticists realized that the taxonomists, who looked with scorn at these mutations as of possible evolutionary significance, were right, and began to ask themselves whether no better materials were available. Thus Johannsen (1923) expressed his doubts in the following words: "Is the whole of Mendelism perhaps nothing but an establishment of very many chromosomal irregularities, disturbances or diseases of enormous practical and theoretical importance but without deeper value for an understanding of the 'normal' constitution of natural biotypes? The problem of species, evolution, does not seem to be approached seriously through Mendelism nor through the related modern experiences in mutation." Johannsen, however, did not point to any positive possibilities. In the same year Goldschmidt (1923), who had already insisted on this point in the aforementioned essays, wrote: "The extraordinary material of analyzed mutants from Drosophila work demonstrates that the type of gene mutation observed there can hardly play any role in species formation. . . . This does not mean that such mutants could not appear and hold their own in nature. . . . [example follows]. But never thus a new . . . [example] species would be formed. Recently Sturtevant has carried out a very useful comparison between natural species of Drosophila and the experimental mutants. His result, with which every expert in any group of animals will agree is: Species differ in innumerable minor characters, mutants in a few extreme differences. Experimental mutants, however, show these large differences because only these are checked. But it is to be assumed that also the very small mutational steps, which change the organism very inconsiderably and probably do not disturb its balance, are just as frequent but escape notice. If two basic differences were isolated and such micromutation would recur, finally different species could be produced, different in numerous genes. This does not remove all difficulties, as, e.g., the sterility of species hybrids shows. But one thing becomes clear, rather surprisingly to some people, that the facts have made us return again to Darwin, though with the improvement of an exact analysis of variation." We shall see that the facts which have come to light since forced me to revise my standpoint.

The replacement of the typical Mendelizing mutations by the less tangible micromutations whenever questions of evolution are involved has since reappeared many times. Two more examples from experienced authors may suffice. Baur (1925) wrote: "By exercising one's senses in the course of years one realizes that the conspicuous mutants are only extreme cases. At least equally frequently, and probably more so, minute mutants appear, which are not pathological but quite viable types. . . . These micromutations . . . are of very different types, small differences in the color of leaves or flowers, in the relative length of anthers, in the type of hairiness, size of seed, etc. . . ." It ought to be added, however, that the work on these assumed micromutations of *Antirrhinum* never went beyond such general statements.

One more recent statement of the same views may be quoted. East (1936) writes: "The situation is so peculiar that taxonomists have little interest in the characters with which geneticists deal, maintaining that they are wholly unnatural material for evolutionary processes. Professor C. T. Brues has examined the published descriptions of mutational effects in Drosophila at my request, and finds that only a limited few characteristics of similar type have ever survived in nature and these often in distant genera. . . . [There follows a very appropriate discussion of the fact that nevertheless the rate of *Drosophila* mutation is made the basis of theoretical conclusions.] I suggest that constructive mutations *are numerous* but have ordinarily remained unnoticed simply because destructive mutations are more easily described, catalogued and scored, and therefore have been more convenient in genetic research. There is evidence of a varied nature, nevertheless, in support of the idea that constructive mutations occur with remarkably high frequency." The evidence which East derives from his experience with Nicotiana is actually of the same type as Baur's from *Antirrhinum.*

These statements, then, may serve as a starting point. As far as genetics is concerned, the heritable variations which Darwinism needs as materials of evolution are available in constantly appearing mutants, the more conspicuous of which are deleterious or even monstrous, whereas the small deviations which may be even more frequent are less easy to detect and to isolate. Our task now is to find out how far these mutants and their accumulation by interbreeding and selection will explain the successive steps in evolution. For this part of our discussion it will not make any difference which theory we accept regarding the nature of mutants. We therefore use for the first part of our discussion the terminology of the classic theory of the gene, assuming the chromosome to be a string of units, the genes, each located at a definite locus of the chromosome and each playing a definite role in controlling normal development. A mutation is a localized change in one of the individual genes and therefore Mendelizes with the original form. The term "point mutation" or "mutant locus" will appear in the same sense as "gene mutation." Doubts in regard to the theory of the gene will appear only later and only where further analysis requires definite ideas about mutation.

2. THE SINGLE MUTATIONS

THE first step to be taken in order to get acquainted with the materials of evolution in nature is to start with the lowest taxonomic units and to relate natural conditions to genetic analysis. There is no doubt that the type of laboratory and field mutants with which the geneticist works occurs within wild species in nature where for one reason or another such a mutant might even become a frequent occurrence. It is known that occasional albinos are found in many groups of animals. In the year 1910 in one small area in upper Bavaria numerous albinos appeared among the field mice (*Microtus*), so numerous that I saw a dozen within an acre of land, but they were never observed again. A systematic study of *Drosophila* populations in Russia made by Dubinin and collaborators (1934) revealed that quite a number of flies were heterozygous for well-known recessive mutants, which then had a chance to appear as rare visible variations. (A long time ago I had found the same for *Drosophila* trapped in the surroundings of Berlin [Grunewald]. But I did not trust my observations because of the presence of genetics laboratories within not many miles. This source of error is excluded in the Russian work.) Many others have since reported similar findings. It is also known in the cases of other animals and plants that rare but typical mutants may be confined to definite localities where the collector may happen to find them. The taxonomists have never thought of these "varieties or aberrations" as being of importance for evolutionary problems, for they are very frequently of a type which occurs in the same way not only in different species but in different families and orders. Obviously, in these cases one rather generalized process of development is liable to be affected only in a few simple ways (see below). Thus, albinism is found in innumerable mammals and birds, and also in mollusks and insects. Melanism, or partial melanism, or progressive, graduated melanism is a very frequent type of mutant in the same groups. Wherever red pigment occurs, yellow mutants are found, and white ones arise from yellow. I have analyzed genetically many such cases in moths and butterflies showing all the different types of one-factor Mendelian inheritance. The yellow aberration of the Arctiid moth *Callimorpha dominula,* with largely red hind-wings, an example of a mutant found only in rare localities, is a simple Mendelian recessive. The dark aberration of the fritillary *Argynnis paphia,* the form *valesina,* is a recessive with sex-controlled inheritance; the aberration of the gypsy-moth caterpillar with a black dorsal band is a simple dominant mutant. These examples from my own experience could be indefinitely multiplied by adding further facts taken from the large body of published experiments which have used mutants of wild species in both animals and plants.

We have already expressed our opinion that we agree with the taxonomists that these aberrations cannot play any major role in evolution. But this does not mean that they may not contribute to microevolution, to diversification within the species. It is quite conceivable that under definite circumstances such a commonplace mutational type might establish itself either by supplanting the original form or by occupying an independent area. We shall soon meet with such cases of a little more complicated order. I do not know of any simple case in which all the relevant facts are known.

We should certainly suspect such a situation whenever we find two forms of a species occupying different areas near each other and distinguished only by a single Mendelian factor difference. Such cases, however, seem to be very rare (see below, Harrison). In addition, one has to be cautious in assuming such a difference on any other basis than an actual crossing experiment. It may happen that a racial difference looks to all purposes like a simple Mendelian difference without this being so. The following is an example from my own experience. As already mentioned, in the Arctiid moth *Callimorpha dominula* with largely red hind-wings a yellow mutant occurs in the environs of Berlin, Germany, and this is a simple recessive. In Italy a geographic race of the same species is found which closely resembles (in some of its forms, see Goldschmidt, 1924) the yellow mutant, and one might expect that here the yellow mutant has replaced the red form. But a cross between the German red and the Italian yellow form gives an intermediate orange F_1, and in F_2 the whole range from red to yellow is found, indicating a multiple-factor inheritance. This proves, therefore, a completely different origin for the two cases.

Many similar instances could probably be quoted. I wish to emphasize again their importance, as corresponding unanalyzed cases have frequently been used in erroneous generalizations. Thus Kinsey (1936), whose work on gall wasps will be quoted later on, has found forms with small wings which occur as seasonal variations within a race, as well as in the form of definite races or hereditary units at different points of a racial circle. He lists these short-winged forms simply as mutants, and whereas this wing character has formerly been considered as of generic value, he makes this a point of considerable evolutionary significance. As no genetic information is available in this case, his interpretation may be right and it may be wrong. We shall discuss in a later chapter the problem of wing rudimentation in insects, and shall see which possibilities are available. Here we want only to caution against the use of unanalyzed material for sweeping conclusions.

We turn now to another group of cases of this same type which are of significance for microevolution. There is, first, a group of interesting data, for which, however, the genetic facts are more or less unknown; namely, the "mutations" in birds studied by Stresemann (1923–26). These "mutants" are known from nature; in a few cases hybrids are known, and a few crosses made in zoological gardens are available. These mutants are different from the type in color; e.g., melanisms, albinisms, rutilisms, the presence of differently colored spots or bands. They appear in more or less large numbers within the typical population, and in some cases they tend to supplant the original form. The hawk *Accipiter n. novaehollandiae* has a white mutant in Australia; in Tasmania nowadays only the mutant is found. The white mutant of the gray snow goose *Anser caerulescens* has almost supplanted the original form. Only the black mutant of the red-breasted weaving bird *Colius asser ardens* is now found in large areas of West Africa. The egret *Demigretta sacra* is found in New Zealand in the original gray form; in other parts of its area a white mutant is also met with which some day may replace the gray form, just as has happened in other species. In a few cases very rapid suppression of the original form is known. At one time a melanic mutant of the flycatcher *Rhipidura flabellifera* was known to occur only in the southern island of New Zealand. Today it has conquered the northern island, where it first appeared in 1864. Other similar cases are reported for the West Indian bird *Coereba saccharina*, of which the black "mutant" has completely replaced the original form.

But there are also genetically known cases in which a kind of microevolution is based upon a relatively simple type of ordinary mutations. The latter belong in part to the group of the typical mutants and in part may be classed with the hardly discernible micromutations which we discussed above. One such case which is rather well known and which actually was the first evolutionary problem I personally attacked with genetical methods thirty years ago[1] is the case of the

1. Owing to the war, the results were published only seven years after the work was finished (Goldschmidt, 1920b).

melanistic nun moth, *Lymantria monacha* L. This common pest has typically white wings with black zigzag bands in a definite arrangement. Up to about eighty years ago this form was rather constant, though occasional melanic individuals were found and much cherished by collectors as rare aberrations. That this type is hereditary was proven in pre-Mendelian days by Standfuss (1896). It has apparently always existed, as it is mentioned even in eighteenth-century literature. During the second half of the last century the number of these melanic aberrations increased and, starting from certain centers within the general area of distribution, spread from there over the whole range. When I started my work in 1909 and obtained my material from large areas infested with this pest, the majority of individuals already showed the different degrees of melanism. It is known that the same phenomenon was simultaneously observed in a number of other moths, and that in all these cases the centers of distribution were found in the areas of high industrialization, both in England and in Germany. Hence the term, "industrial melanism."[2]

The genetic analysis in this case showed that this melanism was the result of at least three mutational steps of unequal value, but all of them dominant. Two autosomal gene mutations increased the breadth of the zigzag bands and produced some pigment between them. The effect of these mutations individually is so small that they might be classed as micromutations. The third gene is sex-linked and produces the deep black of the higher grades. All these genes are additive in action and their different combinations produce the complete series of gradations from white with black bands to completely black (fig. 1).

There can be no doubt that these mutational changes of a type which is found in numerous animals—melanism—led to a definite step in microevolution in this case. This means

2. For references to melanism in 1785 see P. Schulze, *Berl. Entomol. Wochenschr.*, 57. 1912. Schroeder (*Ztschr. Wiss. Insektenbiol.*, 4, 1908) could hardly find a melanic specimen, as a young man, in the same localities where they are prevalent now. Details regarding the spreading of the melanic form may be found in German textbooks on forest entomology.

that here a mutant, or a set of mutational advances in the same direction, led to a new condition in the relations of the species to its environment which permitted the new form to replace the old one, a process which of course is aided by the dominance of the mutants. Obviously, the melanistic forms had a selective value, as the relation to industrializa-

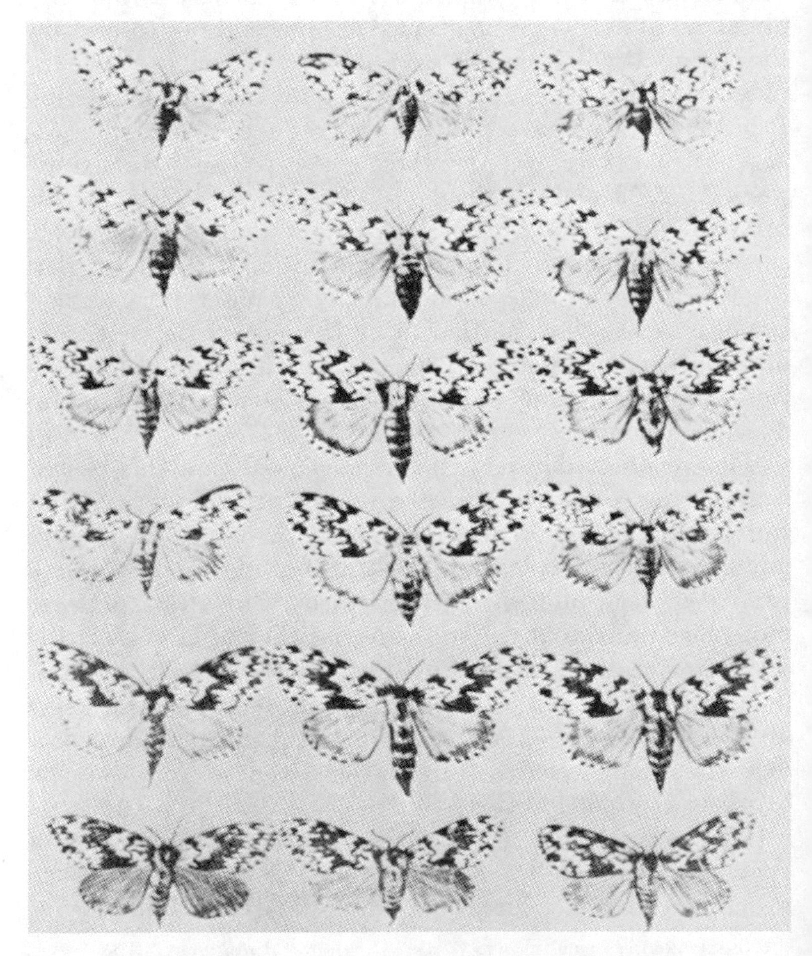

Fig. 1a. Melanic variation of *Lymantria monacha* ♀. Each row represents another combination of the two autosomal and one sex-linked dominant mutations. (From Goldschmidt.)

tion indicates, that was of a definite type. (A calculation—see Goldschmidt, 1920b—showed that the quick replacement of the original form by the melanic one required only a small selective advantage.) We tried to see whether the dark indi-

FIG. 1b. Melanic variation of *Lymantria monacha* ♀. Each row represents another combination of the two autosomal and one sex-linked dominant mutations. (From Goldschmidt.)

viduals were larger or stronger,[3] but with negative results, though these individuals were more viable in breeding, an experience also reported by Harrison (1920) for British melanic moths. Thus we concluded that the difference must be a physiological one. The relation to industrialization seemed to offer a clue. The nun-moth caterpillar feeds on fir trees, which undoubtedly deposit in their tissues various chemicals, especially metal salts, from their surroundings. The idea arose, therefore, that the melanic forms were in some way changed in regard to their metabolism, and that this enabled them to feed on the poisoned food of the industrial districts. For a considerable time we performed experiments in this direction, but the susceptibility of the animals to infectious diseases prevented a solution. I am convinced, however, that the actual explanation cannot be far from this one. (As a matter of fact, Harrison [1920] conceived the same idea in regard to cases of melanism found in England, which, however, he interpreted in a Lamarckian sense.) I am encouraged in this belief by the knowledge that these animals are rather sensitive to the chemical composition of their food. One of the close relatives of the nun moth is the gypsy moth, which is much hardier and omnivorous. For years we tried to make the breeding of a winter generation possible by using artificial food. Once we were very successful with shredded pine needles; but the following winter the same method was a complete failure. It turned out that we had taken our material from a different tree which had a higher content of rosin in its needles. The general correctness of this interpretation, aside from the special features, is demonstrated by some recent work along similar lines. A number of authors (Kühn and collaborators, 1934, for Ephestia; Timofeeff-Ressovsky, 1934, for *Drosophila*) have shown that genetic strains of different viability exist, among them also some with increased viability. More specific is the finding by Kühn and Von Engelhardt (1937) that a melanic mutant of the geometrid moth *Ptychopida seriata* is better adapted to lower temperature and higher humidity.

3. A positive claim made in Goldschmidt, 1917, was later corrected.

An interesting account of the formation of local forms (races) involving melanism, as in the former examples, as well as other traits of pattern and color, has been given by J. W. H. Harrison (1920) for local races of the geometrid moth *Oporabia autumnata*. In the short time between 1885 and 1919 a certain region in England changed its character completely in regard to the food trees of the species and corresponding microclimatic conditions. Two very different ecological habitats were formed, separated by a half mile of heather, one a coniferous wood, the other a birch wood with some alder. The two habitats now actually contain separate races distinguished by size, choice of food, color and markings, time of emergence. Breeding experiments showed the differences to be hereditary. For one of the characters, hatching time, an adaptive value to microclimatical conditions of the habitat could be demonstrated. In the same species there are also melanic forms, the evolution of which, as already mentioned, is interpreted by Harrison in a Lamarckian sense. A more probable explanation has already been presented. (A good review of industrial melanism, especially the English material, is to be found in Ford [1937].)

To return to our main topic, microevolution by mutation. A mutant or a combination of mutants with similar effect had actually changed the species in such a way as to give to the new type a physiological advantage under proper environmental conditions. When, in the case of the nun moth, these were furnished by the chemical effects of industrial smoke upon food plants, the mutant began to replace the original form. The result is that a white nun moth has become a more or less black one. *Here this type of evolution ends*. The same conclusion, *mutatis mutandis*, probably applies to the ornithological examples already mentioned.

A good example of a situation such as is discussed here can be derived from a comparison of *Lymantria dispar* and *monacha*. We have already reported on the melanic varieties of *monacha*, their adaptational value, and the positive results of natural selection which has slowly replaced the light race by the dark one within man's memory. In *dispar*, the nearest relative of *monacha* but with different ecology, melanic variations also occur. They nearly duplicate the *monacha* series (fig. 1), but for details of pattern. In addition, the genetic basis in both cases seems to be very similar though not identical, as far as information goes. But this

Fig. 2. *Spilosoma lubricipeda,* type form, melanic form (*zatima*) and hybrid in between. (From Goldschmidt.)

melanism is extremely rare and has been found only once, by Klatt (1928). It never occurred in the hundreds of thousands of individuals of all races which I have bred. Obviously, a very rare set of mutations is involved here, and it is, in addition, without positive selective value thus far.

The next question is whether this simple diversification by Mendelizing mutants could not carry an evolutionary process one step farther. We have already seen that in many instances a definite mutant is not found everywhere over the area of distribution of a species, but sometimes only in definite localities, as is well known to collectors. From a genetical point of view this might be a purely chance situation. If a mutant occurs in a well-isolated population and if there is no counterselection, it may hold its own and be present in the population. This, however, will hardly be the explanation if the mutant is present in different localities but only in those of a similar type. There is, for example, the Arctiid moth *Spilosoma lubricipeda*, a small moth with yellow wings and a few black dots, which is found all over central Europe. This has a dominant mutant called *zatima* (F_1 is actually intermediate) with almost black wings. This mutant frequently appears among the type form on the coasts of Holland, the Friesian Isles, and especially Heligoland[4] (fig. 2). There, undoubtedly, the regular presence of a definite mutant in the population has a relation to a definite geographic situation with rather definite climatic features, and the simple problem of mutations of hardly any evolutionary significance begins to widen its scope, though strictly within the sphere of microevolution.

3. LOCAL POLYMORPHISM

The next step is best introduced by a phenomenon which we might call local polymorphism. Just as a definite locality might typically harbor a definite mutant, such a locality might also be inhabited by a group of interbreeding mutants if the situation is as described before; i.e., considerable isolation and no selective value of the mutants. A group of mutants appearing under such conditions and interbreeding freely among themselves and with the original species will lead to a stable population (under average environmental

4. The genetics of the case has been worked out by Federley (1920) and Goldschmidt (1924). There is a difference of one major gene and, in addition, a series of modifiers for the degree of melanism.

Fig. 3. Polymorphic series of *Callimorpha dominula* from the Abruzzi. (From Goldschmidt.)

conditions) in which the different types and their recombinations are represented. The population thus is polymorphic. The same type of polymorphism would result if a series of mutant stocks of flies in a laboratory were mixed up. An example of this type—if I may be excused for preferring examples with which I have experimented myself—is the local polymorphism of the Arctiid moth *Callimorpha dominula*. I have already mentioned the yellow Italian form of this species. In a small area of the Abruzzi mountains at the base of the Gran Sasso d'Italia this species occurs in a series of forms ranging from the typical ones through all intermediate grades to almost black ones (fig. 3). For many years a collector who specialized in this form found during the proper season all the forms on the same spot and provided me with living material. The genetic analysis (Goldschmidt, 1924) showed a multiple-factor inheritance, different combinations of genes giving the different types.

There is only a small step from this localized polymorphism based upon recombination of Mendelizing mutants to a similar phenomenon on a larger geographic scale. A long time ago the French conchologist Coutagne (1896) noticed that each individual colony of *Cepaea (=Helix) hortensis* which he studied had its special character. In all colonies he found a considerable variation with regard to color and the types of banding. The same material has been studied since by many authors and in a general way the same results were obtained. (A fine review of the facts and a discussion of their significance have been recently presented by Diver [1939].) Today we know from Lang's genetic studies that all these diversities are based on Mendelizing mutations and that therefore a population containing many such types presents a typical case of polymorphism by interbreeding and segregation of mutants. But in Coutagne's case the individual colonies were typically different with regard to the types found as well as to their relative numbers. There can be no doubt that in view of the considerable inbreeding in these mollusks the composition of a colony may be mainly due to the mutations which by chance happen to be present and are preserved in similar proportions for lack of positive or negative selection. But it is also conceivable that occasionally one or the other mutant is better or worse adapted to physiological conditions of soil, water, etc., which would result in a selection with regard to the composition of the colony. I do not know of any positive evidence in this respect, but there are other cases of the same kind which might lead to the belief that the distribution of the mutants in the population is not purely a chance distribution.

Unfortunately, there are not many cases in which the genetic basis is known. One is the following. The different color patterns of the ladybird beetle *Harmonia axyridis* are mainly based on three pairs of genes, the different combinations of which characterize the forms which have been called *frigida*, 19-*signata*, etc. (Tan and Li, 1934). Dobzhansky (1933) studied the distribution of these forms in different populations (see fig. 4). Table 1 is an excerpt from his material.

TABLE 1

DISTRIBUTION OF COLOR TYPES IN POPULATIONS OF A LADY BEETLE

From Dobzhansky.

	succinea	*Mutant forms, per cent*			
	frigida				
Region	*19-signata*	*aulica*	*axyridis*	*spectabilis*	*conspicua*
Altai	.05		99.95		
Irkutsk	15.1		84.9		
W. Transbaikal	50.8		49.2		
Amur	100.0				
Vladivostok	85.6	.8	.8	6.0	6.8
Manchuria	79.7	.5		11.3	8.6
Japan	27.2		11.0	14.3	47.4

It is not possible to say whether the absence of some mutants in definite localities is a matter of pure chance or whether they are not fit to exist there. But it seems probable that such types of distribution which in a similar way seem to occur also in other Chrysomelid beetles (work of Timofeeff-Ressovsky, 1932, and Tower, 1918) are indicative of an underlying rule.

Such a rule, however, has hardly anything to do with problems of evolution, although it might seem to be the case if large geographic areas are checked as a whole, as was done in the foregoing example. In Coutagne's *Cepaea* colonies the different colonies may have been within easy reach of each other. Rensch also has recently (1933) made a similar survey for the garden snail *Cepaea hortensis.* Table 2 gives a part of his results.

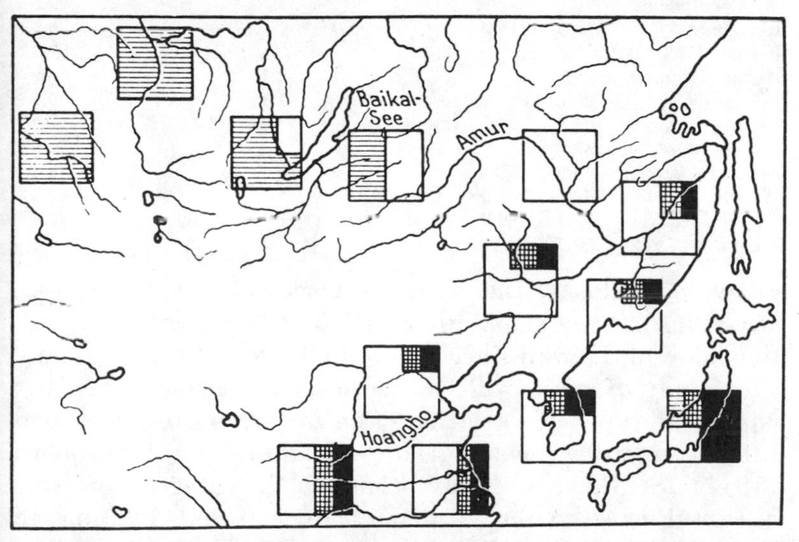

Fig. 4. Geographical distribution of four color types of the ladybird *Harmonia axyridis* in Asiatic populations. White = *signata,* shaded = *axyridis,* crosshatched = *spectabilis,* black = *conspicua.* (Diagram after Dobzhansky from Timofeeff-Ressovsky.)

This table shows no order or geographic relation of any kind. Obviously, the situation in each locality is controlled by chance presence of mutations and perhaps also by chance selection in favor of or against individual mutants or combinations. In a general way such a situation has as many or as few possibilities for microevolution as in the former examples. There is reason to believe that this type of polymorphism is a typical feature of life in more or less isolated colonies. Wherever such conditions of life were studied, espe-

TABLE 2

COLOR AND BANDING OF CEPAEA SHELLS IN ONE LOCALITY

1 2 3 4 5—0 presence or absence of bands 1–5. These are Mendelizing traits.

	Per cent individuals							
Locality	*yellow* *00000*	*red* *00000*	*yellow* *00305*	*yellow* *10305*	*yellow* *10345*	*yellow* *12045*	*yellow* *10045*	*12345*
East Prussia	32.3							67.7
Ruegen	32.9			26.5	1.2			40.3
Ruegen	36.2		.4	23.5	1.7	.4		37.8
Warnemünde	77.7							22.3
Ratzeburg	70.3	14.9	2.7	5.4				6.7
Berlin—Buch	10.9					1.4		87.7
Berlin—Spandau					1.1			98.9
Berlin—N 1	72.5				1.1			26.4
Berlin—N 2	65.3				.7			34.0
Berlin—3	45.3				1.9			52.8
Berlin—4	45.6				19.3		7.0	28.1
Weimar	52.0	5.5						42.5
Czechoslovakia	.1	22.2				.5		77.2
Vienna	93.7				.4	.6		5.3

cially in mollusks, the same phenomenon was found (see Diver, *loc. cit.*). I shall only mention at this point the Achatinellidae of Hawaii (according to Welch, 1938), because this family of snails will be used later to demonstrate other important types of variation. The facts reported by Lloyd (1912) upon rat colonies in India, and by Hagedoorn (1917) for Java, probably belong to the category of facts discussed here. A huge amount of material relating to Chrysomelid beetles (*Leptinotarsa*) is to be found in Tower's monograph (1918), though it is rather difficult to understand some of the details. But the examples already discussed illustrate the point sufficiently. In plants a parallel case can be found in Gregor's work on *Plantago* (1938).

We have already indicated that we cannot see that the different types of hereditary variation discussed thus far, the mutants of the ordinary type, have any meaning for an understanding of evolution. It is true that here we have a means for diversification within a species, which under conditions of selection might also be termed a kind of microevolution. But it is always the same little change of tune which is produced in these cases, melanisms, albinisms, rutilisms, and their like. I cannot see that they could be conceived better as the beginning of an evolution or as a model of this. They are a rather unimportant type of variation (as the taxonomists correctly interpreted) which becomes important only as material for genetic experimentation (see the quotation from Johannsen, p. 9) because it is a material taken from nature of the same type and significance as the mutants of domestic breeds or laboratory animals. If we were to use such cases as the last ones mentioned as a starting point for evolutionary deliberations it would be as though we visited numerous dog shows in different countries and took the relative representation of different breeds in these as a starting point for an analysis of species formation.

4. SUBSPECIES AND GEOGRAPHIC VARIATION

A. The Taxonomic Facts

Up to the end of the last century the species concept in taxonomy was handled in two different ways. Either any recognizable form was made a new species, thus enlarging their number immensely, or only a relatively small, well-established group of species was recognized, and aberrant forms assigned to them as far as possible. I remember distinctly the shock which it created in my own taxonomic surroundings (I was an ardent coleopterologist at that time) when Matschie claimed that the giraffes and other African mammals had many different subspecific forms characteristic for different regions which he could recognize with certainty; when Kobelt claimed that the mussel *Anodonta fluviatilis* was different in each river or brook; when Hofer stated that each Alpine lake contained a different race of the fish *Coregonus;* or when Heincke claimed the same for different schools of herring. The ornithologists, Kleinschmidt, Rothschild, and Hartert, soon followed by the lepidopterologists (Jordan), were the first to base upon these and related facts a new principle of classification and to work up their own fields accordingly. Rensch (1934), who is one of the present-day protagonists of these principles, has described their coming into being in the following words, which I quote verbatim (my translation), as I could not improve upon his statement:

"In the bird genus *Sitta* (nuthatch) 19 different species and one subspecies had been described from the palaearctic region between 1758 and 1900. . . . All these species differed in size, color, proportions of beak, tail, wings, construction of the nest, etc. There were no transitions existing from one to another form. The areas of origin were different in most cases but some species also lived in the same region. The taxonomy of the Palearctic *Sitta*, then, was clear and unequivocal. With increasing research more and more forms of *Sitta* became known which proved to be more difficult to arrange within existing species. These forms were marked off more or less clearly but they were so nearly related to named species that they were considered to be subspecies or varieties of these. Thus Reichenow (1901) described a form *caucasica* which he considered to be a subspecies of the central European *caesia*, while Witherby (1903) considered his new form *persica* to be a subspecies of the North European *europaea*. But there was no reason why one should not have made *caucasica* a subspecies of *europaea* instead of *caesia*, the more so as a very similar form *S. europaea britannica* has been described by Hartert (1900). The species concept thus turned out to be vague; the specialists were unable to differentiate *Sitta europaea* with its subspecies from *S. caesia* with its subspecies. Similar difficulties appeared with other species and it seemed as if the increase in material led to a chaos in nomenclature. In addition, different authors differed more and more in their definition of species.

"Chaos would certainly have prevailed if one had retained the old ideas concerning species. It was hardly chance that in this critical epoch a new school of thought appeared in ornithology which soon began to become dominant under the leadership of O. Kleinschmidt and E. Hartert. They put the study of the geographic distribution of the individual forms into the foreground and they looked with a systematic effort for forms which were both morphologically and geographically transitional between two 'species.' In the case of the forms of *Sitta* it was possible to unite thus not only *persica*, *britannica*, and other races, but also the Central European *caesia* with the species *europaea*. Thus the difference of viewpoint, mentioned above, was removed; *caucasica* was a geographic substitute both for *caesia* and *europaea*. In this way many 'species' were combined into groups of forms, mutually replacing each other geographically. This was called geographic rassenkreise (racial circles), or large species, or formenkreise (Kleinschmidt). In such a rassenkreis neighboring races are very similar, but geographically distant ones frequently were very dissimilar. The names of the races were added as third names to the name of the rassenkreis, i.e., the name of the form which happened to be described first. When E. Hartert began (1904) publication of *Birds of the Palaearctic Fauna* he could simplify the chaos of twenty-four known *Sitta* species into four rassenkreise, leaving three isolated species. In the interim the number of geographic races has still increased and today the rassenkreis *Sitta europaea* alone contains twenty-six geographic races (without the contested 'subtilrassen') [i.e., still lower categories].

"Taxonomy thus has become definite again: the new 'large species,' i.e., the geographic rassenkreise, are clearly defined, are natural units. There are no transitions from one rassenkreis to the other: the rassenkreis *Sitta europaea* stretches over the whole of the palaearctic region. All its races (which usually show a continuous gradation among themselves) are distinguished from the races of the rassenkreis *neumayeri* occupying a stretch from the Balkans to Belutchistan by chestnut-colored sides in the male, or from the *S. canadensis* races by the lack of a black head, etc."

The principle of rassenkreise as the natural taxonomic unit has conquered taxonomy and largely replaced the other species concepts.[5] Most of the best-known groups of animals are described in these terms. Whether all points are clear

5. Its history is more completely treated in Stresemann (1936). Criticism is found in Reinig (1938).

and beyond debate is not to be decided here. But it is certain that here a major principle of the diversification within the lowest systematic categories has been found which therefore is highly significant for the problem of evolution. Actually, the taxonomists who worked upon these problems and quite a number of others have come to the conclusion that the geographic races are incipient species (see p. 139 for quotation from Darwin), that the formation of subspecies within a species over its geographic range (the rassenkreis) is the first and typical, even obligatory, step in the evolution of new species and higher categories. There is no doubt that such a view is very attractive at first sight. The problem is to find out whether it can stand a closer scrutiny.

We shall begin our discussion with the facts pertaining to the rassenkreis as such and their analysis, and shall only proceed afterwards to the decisive point; i.e., whether this type of microevolution can lead beyond the confines of the species. In previous discussions it was, of course, tacitly assumed that the differences between the members of a rassenkreis were hereditary. Occasionally some nonhereditary modifications may have passed for geographic races, but as a rule one may assume that actually hereditary differences are involved. Wherever they could be tested they turned out to be hereditary. The most important feature of the geographic race is that it occupies its own area and that no other member of the same rassenkreis lives within the same territory. Where two such races meet, frequently a mixed population, or a slow intergradation from one to the other, is found. As far as is known, members of a rassenkreis are completely fertile *inter se* (see, however, below), and will normally breed together if there is a chance (see below). There are rassenkreise the individual members of which are so different that there is no difficulty in assigning an individual to its race without knowing the place from which the specimen came. But more frequently the different geographic races show all intermediate and transitional stages. In this case the description of a definite race is possible only from large series and in some cases only with the aid of statistical methods.

I might mention one such case in order to show that a conception very similar to the rassenkreis concept had been arrived at in a very different way prior to that taxonomic reform. The herring in the North Sea forms large schools which are found in definite localities and travel to definite spawning grounds. These localities are different over the whole area inhabited by the species, and each area has a different constant race which, however, cannot be distinguished by ordinary taxonomic methods. Only a biometric study of a series of variable characters like number of vertebrae, number of keeled scales, and about sixty others, and their evaluation by biometric methods, permitted Heincke (1897–98) to find the constant racial differences. Since that time similar work with identical results has been performed by many ichthyologists (Schmidt, Hubbs, Schnakenbeck, 1931; literature in Schnakenbeck's paper).

A very typical feature of many rassenkreise is that the distinguishing characters of the individual subspecies form a graded series along the geographic range, beginning with an extreme minus type and ending with a plus type of the character in question. When this is the case, the races at the two ends of the range are also the extremes in regard to the character in question. This chainlike arrangement of races of course requires a corresponding linear seriation of geographic areas, which is not always found (see below).

At this point of our discussion I should like to draw attention to the fact that the simplicity of the situation as represented thus far is only an apparent one. There are, no doubt, cases which agree with the description of rassenkreise as given thus far. But nature can hardly be expected to be as diagrammatic as this. In the actual study of manifold materials many a case has been found which it is difficult to fit into the general scheme. Such cases are of special importance when the problem of the limit of the species comes up. We shall report on them in that connection.

Darwin spoke of the "origin of species." He took it for granted that an explanation for the origin of species automatically also explains the origin of the higher systematic categories by the same process found to be involved in the origin of the lower categories. If, then, the geographic races within a rassenkreis are incipient species or, if we are more modest, are models of what incipient species would look like, the genetic exploration of a rassenkreis should furnish the decisive information about the materials of evolution. As I happened to be in constant contact with a progressive group of taxonomists in different fields[6] and therefore was early aware of the new developments in taxonomy, I embarked in 1909 upon such a genetic study with a few crosses from which later an extensive program for an all-round study of a rassenkreis was developed. A short preliminary report upon part of the material was published in 1917, and another in 1920. The working hypothesis which I started to prove was that geographic variation is actually a model of species formation.[7] As I happened to be the first geneticist who realized this problem and embarked upon its experimental study, and as there is no other work available in which the same investigator has done all the taxonomic, ecological, morphological, physiological, cytological, genetical, and most of the field work with all available geographical races, I may be pardoned for presenting first and in more detail my own results, which turned out to comprise practically all the aspects of the problem, which have also been emphasized by others.

B. Genetic and Biological Analysis Combined with Taxonomy

a. Analysis of the rassenkreis of Lymantria dispar *L.*[8]

Lymantria dispar, the gypsy moth, spreads over the whole Palearctic area, of course barring recent introductions by

6. In the R. Bavarian Museum, Munich.

7. The term, "species formation," is nowadays frequently supplanted by the term, "speciation." Using this term once in a discussion with one of the leading British taxonomists, I was violently rebuked for adopting this linguistic atrocity. As a scholar, raised in the traditions of the classical languages, I have to agree with my critic and shall therefore not use the contested term.

8. For details, pictures, curves, tables, and many points not mentioned here see Goldschmidt, 1917, 1920, 1924a, 1929, 1932, 1932a, 1932b, 1932c, 1933a, 1933b, 1935.

Here, then, we have at least ten types confined to definite geographic areas and clearly distinguished by a genetic character which in addition shows a continuous seriation if we go from Korea south and then northeast over the Japanese Isles. The character partly coincides with the visible racial distinctions—e.g., Hokkaido; it partly subdivides groups which can hardly be told apart morphologically (Korea, Kyushiu, northern Honshiu). In addition, certain conditions restricted to localized areas are known. The physiological meaning of this racial character is unknown, though there are indications that there is a connection between these sex valencies and certain time phases of the life cycle. Genetically the difference between the sex races is conditioned by two major features: first, a condition of the cytoplasm, inherited maternally, which is responsible for the changes in female direction; and, secondly, a series of multiple allelomorphs of the male-determining sex gene (or sex chromosome?—see below). The strangeness of this physiological character as distinguishing geographic races is somewhat diminished by the fact that at least one and probably two comparable cases are known in animals. In frogs different sex races exist with regard to definiteness of bisexual differentiation (Pflüger, R. Hertwig). Also, here a definite geographic rule, namely, relation to temperature, seems to obtain (Witschi, 1923). Another case is found in the so-called species of killifish in the West Indies. They will be discussed later. There is also a case in plants, where Oehlkers (1938) has described crosses between different species of *Streptocarpus*. (In plants subspecies are frequently given specific rank.) These crosses result in a type of intersexuality which closely parallels the *Lymantria* case, if my interpretation of the facts (Goldschmidt, 1938a) is correct.

2. Length of time of larval development

This seems at first sight not to be a very important character. But if we realize that the time from hatching in spring to pupation in summer is the most important part of the moth's life cycle (the imago does not feed), and that this span has to coincide with the proper conditions for feeding in nature, we see the great physiological importance of this character, which is actually hereditarily different in different races. But the values are frequently so near each other through transgressive variability that significant differences can be observed only for the larger groups of races, the same groups which were also distinguished in the case of the sex races. Regarding the Eurasian races we find the same two large groups as were described above for the character, abdominal wool. All the northern forms (dark wool) have a relatively short larval period; all the southern forms (Mediterranean, Turkestan) (light wool) have a slower development. (They were of course bred under identical conditions.) Again, the series of eastern Asiatic races shows a definite order, which, however, does not coincide with the order of the sex races. Starting at the northernmost Japanese island of Hokkaido, we find a short larval period of about the same length as the South European (see table 3.) Going over to the northern part of the main island we find a much longer larval time, and to the southwest this increases up to the borderline separating northeastern from southwestern

TABLE 3

AVERAGE LENGTH OF LARVAL DEVELOPMENT OF *L. DISPAR* IN EXACTLY CONTROLLED EXPERIMENTS (IN DAYS)

Racial group	♀	♂
Hokkaido	47.5	44.9
Northern Japan	53.5	52.0
Mountain region	50.8	46.7
Border near Gifu	57.3	52.1
Border near Lake Biwa	56.7	52.4
Border near Japan Sea	54.7	53.1
Western Japan	52.0	50.0
Korea	51.3	48.1
Southern Europe	47.1	45.2

Japan, passing roughly from the Ise Bay via Lake Biwa to Tsuruga Bay. In this region the maximum larval time is reached; and going southwest and finally into Korea it decreases again.

Within this general arrangement some special groups of importance are found, two of which are contained in the table. The forms living by the Japan Sea have a shorter larval time than their neighbors inland. The forms living in the mountainous parts of central Japan have a generally shorter larval time than other central or north Japanese forms. This different speed of differentiation, which in the respective races becomes visible from the beginning of development, if larval instars are plotted against time, is quite obviously a hereditary trait, which adapts each race or racial group to the seasonal cycle of the inhabited region. In a northern climate the short summer requires quicker performance of the animal's life cycle than is needed in a warmer climate. (Note that the difference in days in the experiment carried out at 25° C., with optimal conditions of food, moisture, etc., would correspond in nature to differences of weeks at least.) Actually the data on larval life parallel most closely the meteorological data on the average temperature during the period of vegetation in the respective areas. Such special cases as the shortening of larval life in the mountain region and near the Japan Sea provide a good check; these are regions with cold winters and lower average temperature. But why does the larval time decrease again from the Lake Biwa region west into Kyushiu, with temperatures actually increasing, going southwest? As a matter of fact, temperatures become extremely high in late summer in these parts. We know that *L. dispar* never spread into really warm zones, to which it is obviously not preadapted. In the hot southwest region of Japan, then, the limiting factor is not the length of the vegetation period but the onset of too high a temperature. Hence, shortening of larval time will again be adaptational.

Length of larval time is inherited in crosses. There is a considerable cytoplasmic element involved and in addition a Mendelian segregation which is difficult to analyze on account of transgressive variability. Either a few gene differences of a multiple-factor type, or a series of multiple alleles, are involved.

I do not doubt that this type of subspecific difference is frequently found in nature, though it cannot be discovered without special experimentation. In plants the average temperature seems to be the major limiting feature of distribution, and as we know that in cultivated plants hereditary differences with regard to time of development exist, it would be surprising if such a genetic difference were not one of the distinguishing traits of geographic races (see below: ecotypes).

3. The number of larval instars

The number of larval instars is a quite distinctive racial trait, to which a certain general interest is attached because the ending of larval life by pupation after a definite number of instars is controlled by a hormone (Wigglesworth, *et al.*). The time of release of a hormone is then a racial character. There may be four molts in both sexes; or five molts in both sexes; or four molts in all ♂♂, five in all ♀♀; or four in all ♂♂, four or five in the ♀♀; or five in all ♀♀, four or five in the males. These types are found in different races, and they are caused genetically by a number of multiple alleles: T_1 causes four molts in both sexes, T_2 causes four molts in ♂ and five in ♀, and T_3, five molts in both sexes. The other types found are the result of heterozygosity. The actual method of action of these genes is that T_1 causes a very fast initial growth, whereas T_3 causes a small initial growth. Obviously, the pupation hormone is shed only when a definite amount of growth has been completed, which means fewer instars with fast initial growth, and vice versa. Geographically speaking, there are two main groups of races: one containing T_1 but not T_3, which means the existence of 4-molter females; another with T_3, which means also 5-molter males. In addition, there is a third group in which T_1 T_2 T_3 have been found. Within one group the individuals may be homozygous or heterozygous for the respective alleles. All Eurasian forms contain only T_1 T_2. (In Spain only T_1 was found.) In Japan, starting north T_1 T_2

is also characteristic for the Hokkaido race. But the otherwise completely different forms from the northernmost part of the main island have the same constitution. Northern Japanese races had only T_2, and going west T_3 appears in middle Japan and remains present farther southwest. But in the southwestern island of Kyushiu T_1 reappears in addition to T_2 and T_3, and in Korea, adjacent to Kyushiu, T_3 is again missing and T_1 T_2 are present, as in Europe and Hokkaido. It is further remarkable that the Turkestan forms resemble in this respect the central Japanese, having T_2 T_3. We shall see that this distribution coincides completely with that for body size, though a different genetic basis is involved. No adaptational value of this trait is discernible.

4. The length of the diapause

The length of the diapause is one of the most characteristic traits of subspecific significance. In the egg of the gypsy moth, laid in summer, the caterpillar is immediately developed. But it hibernates within the eggshell—the diapause—and hatches only in the following spring. One might expect that this hatching is caused by the action of a definite temperature, or at least by the sum total of temperature which has acted upon the eggs during diapause. This, however, is not the case. A number of factors come into play (for details see the author's papers), and one of them is the race, the hereditary constitution. This can, of course, be ascertained only by experiments with all other conditions—temperature, moisture, etc—kept constant. In such conditions the different races, if given the same opportunity for hatching, act differently. If, for example, a number of egg batches from different races are kept near the freezing point for a definite time and are then transferred into a definite temperature, the time interval until hatching depends for all of them upon the temperature, the time of previous freezing, and similar variables. But, in addition, there is a racial difference, a slower or faster reaction, which is constant for the individual races in the different types of experiment. As these differences can be tested in a very exact way and with large numbers, this is one of the most reliable physiological characters for the study of differentiation of geographic races.

In detail the physiological elements involved in the hatching reaction which ends the diapause are rather complicated, and more than one element is involved which may be racially different. But we do not need to go into such details here, as the major part of the difference between races is due to one main factor; namely, the total temperature sum which is needed to start the hatching reaction. We consider here, therefore, only the incubation time as the decisive racial difference to be determined. Incubation time means the time elapsing between transfer of the eggs from a low temperature (which practically stops all life processes) into an appropriate hatching temperature, and the moment of eclosion. This incubation time showed actually constant relations among the different races, whether it was measured under nearly natural conditions or under diversified conditions of experiment.

The result is the following. A short incubation period characterizes the races of the northern Eurasian continent; also, the forms of Korea and Hokkaido, so different in other respects, belong here. In Japan proper a strange seriation occurs: northeastern Japan down to the border zone near Lake Biwa contains races with a long incubation period. This is shortened in the transition zone near Lake Biwa, and still more so west of this line. But in western Japan and Kyushiu the time increases again. An incubation period considerably longer than that of any of these races is found in the Mediterranean races, and the very longest known characterizes all forms from Manchuria. The seriation, then, is, starting with long incubation: 1, Manchuria; 2, Mediterranean; 3, northern Japan; 4, southwestern Japan; 5, Lake Biwa region; 6, adjacent western Japan; 7, Eurasia, Korea, Hokkaido. Within these regions of major differences, furthermore, a number of clearly different subregions have been distinguished (and more might be found). For example, the mountain region in central Japan, or the region of the Japan Sea, are different from their surroundings. Most characteristic of these subregions is a small strip at the western shore of Lake Biwa, as will be seen at once.

These differences, then, are hereditary, and genetic analysis points, as expected, to a multiple-factor inheritance, probably with one of the factors paramount in action. These hereditary differences of length of diapause, especially of the temperature sum needed to produce hatching, finally can be shown to be adaptive traits; namely, adaptation of the length of the diapause to the seasonal cycle of the respective environment. The relation is at once clear for the forms with short incubation time which inhabit the moderate regions of Europe, Korea, Hokkaido. A long winter, a rather slow onset of spring, and a short period of vegetation require that the comparatively low temperature sum, offered in winter and spring, suffice to produce hatching in time for the completion of the life cycle in summer. A climate with a mild winter, however, which offers within the same time a high temperature sum to the hibernating egg, requires a different reaction. A short incubation time, as in the former case, would bring the caterpillars out the first warm day, only to perish. A higher temperature sum, then, is required for adaptation to such a climate, and therefore a longer incubation period is to be expected in this case. Actually, the Mediterranean and southwestern Japanese forms behave thus. But in the remaining part of Japan the incubation time increases from southwest to northeast, though the average yearly temperature decreases. The meteorological data, as well as personal experience, show that in northern Japan the spring rise in temperature occurs much earlier than in moderate climates, but the trees on which the caterpillar feeds are in leaf rather late. Proper adaptation, therefore, requires a longer incubation period. It becomes more difficult to prove the point when smaller subregions are involved. In some cases like the cold mountains and the snow-laden shores of the Japan Sea, the relation can be easily established; in other cases the necessary climatological (not meteorological) knowledge is lacking. But that a relation could be established if all data were known is proven by the following case. We mentioned that all the races from the northwestern shore of Lake Biwa behaved differently from all others around the lake (shorter period). We may add

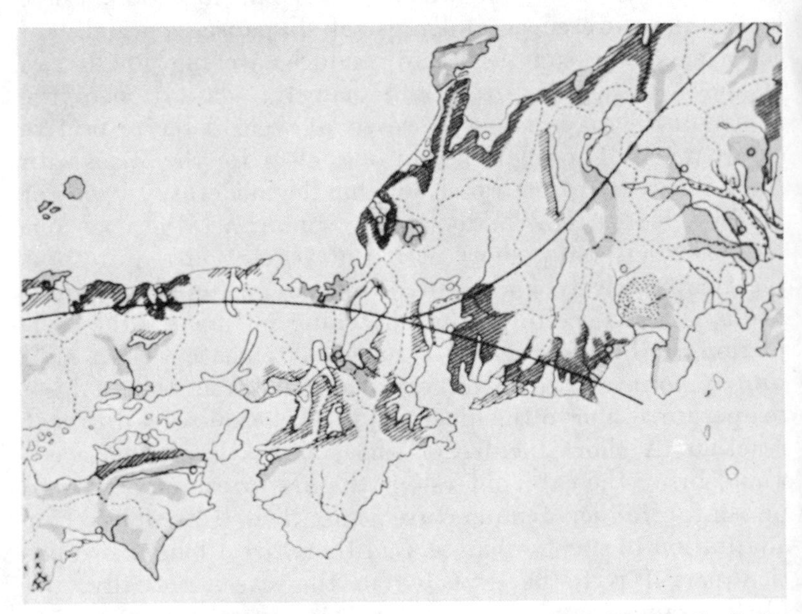

Fig. 7. Map of the borderline region in central Japan indicating the limits of three different soils. (Aftar Seki from Scheidl.)

that in this and other characters (sex genes) the forms of the Gifu region were different from those of the adjacent Nagoya region within the border region which separates northern and western races; the Gifu type belongs more to the northeastern, the Nagoya race to the southwestern group. Recent climatological work in Japan (see Goldschmidt, 1938b) has shown that the races are obviously very sensitive indicators of climatic differences. It was shown that different conditions of climatic factors, altogether making up the seasonal cycle, produce definite types of soils which characterize geographic regions. Three regions which were thus delineated by the climatologists agree very closely with the regions inhabited by the *Lymantria* races (see map, fig. 7). One line of demarcation between different soils and climates is found in the borderline region near Lake Biwa, which has been mentioned here so frequently, and this line runs exactly through Lake Biwa, separating its western shore from the rest, and it also cuts through the Pacific side of this region, separating the Gifu from the Nagoya region. Thus it is shown that the smallest details of racial traits, if adaptational, coincide with differences of environment. This point will be found to be of importance later in our discussion.

5. Growth and body size

Body size, or size of parts of the body, is one of the characters most frequently used by taxonomists. We have already seen that this is also one of the distinguishing traits of the geographic races which we are studying. Actually, the size of the imago is one of the few traits in our case which is accessible to the taxonomist and would permit him to describe four major subspecies. The adult size differences are foreshadowed in the larvae, and actually races of different final size have a divergent growth rate from the very beginning. This character is, of course, most liable to modification. In figure 39 are represented three caterpillars of a middle-sized race, sisters, bred side by side, one under normal conditions, one under optimal (temperature, food) conditions, and one under starvation conditions (all during the fourth molt). The imagines would be of corresponding sizes. A collection from one locality with bad conditions during the year in question might put a whole series of dwarfs or giants into the hands of a taxonomist, who has to be rather careful, therefore, with such modifiable characters. Breeding under identical conditions, however, permits the real genetical differences to be seen. But even these are detectable only to a certain degree, owing to transgressing variability and modifying factors which cannot be controlled, as well as to genetic modifiers of uncontrollable type like susceptibility to disease (which has a stunting action) or adaptability to laboratory breeding. Therefore, experimentation with this character cannot reveal any more significant racial differences than those already detectable by inspection of many individuals.

The smallest forms are those from southern Europe. A medium size characterizes the North European forms, as well as those from Turkestan. This is also true of the races from Hokkaido and northernmost Honshiu. Going southwest we find all northern Japanese forms large, and those from the border region and West Japan extremely large (see fig. 8). (In this group, in addition, the males are nearer in size to the females than in other races.) In Kyushiu size decreases again, and in Korea it is again medium.

It is surprising that body size and velocity of development, the latter of which we discussed as an adaptational character, are not correlated completely; they are inherited independently; e.g., southern European and northernmost Japanese forms have the same rate of development, but one is a very small and the other a large race. But in a general way the distribution of the races according to their size is very similar to the distribution in regard to velocity of development. There is no easily visible relation between size and environment which would stamp size at once as a character adapted to the relative surroundings. But the general parallelism with the distribution of the character velocity of development, which certainly is adaptational, suggests that size differences might be the visible result of physiological differences (metabolic rate or the like) which are adaptational and therefore must be attuned to environmental conditions.

6. *Larval pattern*

A very conspicuous racial character, which however would in most cases escape a taxonomic investigation, governs the markings of the caterpillar. There is present on the cater-

Fig. 8. *Lymantria dispar* ♂♂. Above tall and dark race found near Lake Biwa, below small, light Hokkaido race. (From Goldschmidt.)

pillar's back a basic pattern of light markings which, however, does not become visible in some races because it is overlaid with a dark pigment. In addition, the basic pattern is more or less different in its details. Geographic races differ in both respects. With regard to the cuticular pigment we can distinguish three main groups of forms. (*a*) The first shows the basic pattern through all instars without formation of overlying pigment. This type is found in southwestern Japan and in Korea. (*b*) In the second group the basic pattern is slowly covered and encroached upon by dark pigment which increases during development. This type is at home in northeastern Japan. (*c*) In the third group the dark pigment completely covers the basic pattern. This is the type found all over the Eurasian continent, but with a certain irregular fluctuation into the lower grades of the second type. Within these main types many subtypes can be distinguished, partly by different degrees of the pigmentation process, partly by specific features of the basic pattern. Thus, Hokkaido contains a special type with a very bright pattern in early stages and subsequent pigmentation. In Honshiu the darkest type of the group (*b*) is found in the extreme north, and this brightens step by step going southwest until the brightest type is reached in western Japan. There are many individual features characterizing definite regions. Thus, the Hokkaido pattern is unique, in the Gifu region an orange pigment is found, in southern Europe the longitudinal elements of the pattern are increased, etc. The series of increasing overlaid pigmentation is genetically based on a series of multiple allelmorphs. The pattern differences have not been analyzed in detail, but judging from the accumulated raw data a multiple-factor inheritance is superimposed upon the multiple-allelic series, a situation which would be difficult to unravel in terms of Mendelian formulae. There is no adaptational value of this trait visible though some metabolic situation (like: quick development—much pigment) might be hidden behind the visible differences and constitute an adaptational trait.

Fig. 9. First row: ♂♂ of *Lymantria dispar,* gray, light brown, brown, chocolate. Gray typical for Europe, the browns for Japan; the three multiple alleles for brown are distributed irregularly. Second row; Mendelian recombinations for color genes typical for the Hokkaido race but occasionally found elsewhere. Third row: Mutant females, not characterizing definite races. (From Goldschmidt.)

7. *Imaginal pattern*

I have already mentioned that this most conspicuous taxonomic character is of small importance for distinguishing subspecies. Though innumerable small differences are found, it is not possible to assign them to only one racial group. The number of races which can be distinguished thus, exactly and without error, is rather small. A more whitish wing in the females and a definite gray in the males characterize the Eurasian form; and cream females and brownish males, the Japanese forms. Both types may occur in Korea. The wings of the Hokkaido males show in most cases a rather distinctive light center. In central Japan, around Lake Biwa, females are gray, sometimes rather dark, and males are chocolate brown (see fig. 9). Further, definite mutant types occur in certain groups. Such mutants have been described for northernmost Honshiu, for central Japan, and for Korea. But occasionally forms of one type occur among the others. The reason probably is that most of the color types of the males are based on simple Mendelian differences which might occur everywhere by mutation. But the differences among the females seem to be better racial characteristics, though only a few such types can actually be distinguished. It is, however, possible that in this case the use of statistical methods might reveal a difference of the type as represented above for coccinellids; namely, percentage differences of recognizable forms among the different populations. That the population is so constituted does not become apparent when the breeding method is employed but it ought to be checked after the differences have been analyzed. This was not done in the present case.

May I interrupt the description at this point to describe an interesting test which was made to find out the relations between a taxonomic analysis and a physiological-genetical one as performed in this case. The diagnosis of a racial series (rassenkreis) by taxonomists is sometimes a clear-cut procedure. There are subspecies which are easily and unequivocally distinguished by visible characters, and any specialist could select a definite race from a mixture laid before him. But this seems to be a rather rare situation. In the majority of cases the procedure is a different one, as was aptly described by one of the protagonists of the rassenkreis concept (Rensch, 1938): Rensch studied land snails of the genus *Cyclophorus* from the Sunda Islands. Many species had been described and named. In studying a large collection, Rensch encountered difficulties. Many individuals could be assigned to two different species, and a decision was impossible. Therefore the presence of a rassenkreis was expected, and hence the geographic distribution of each "species" was checked. It turned out that, e.g., the form *borneensis* was found only in Borneo, and the forms *perdix* and *zollingeri* only in Java. But in Sumatra five species occurred. Thereupon all former determinations and descriptions were discarded, and the material was arranged geographically, for individual islands. Now the rassenkreis became apparent. Each of the islands contained a definite form, more or less variable but apt to be considered a definite geographic race. Thus a rassenkreis was established: *C. perdix perdix*, Java and Bali; *C. p. tuba*, Sumatra, Malacca; *C. p. borneensis*, Borneo. The difficulties had been based upon wrong determination of extreme variants which were overlapping. Many other "species" turned out to be only individual variants within a definite group (this applied to the other two of the original five "species").

There is no doubt that a similar procedure is usually followed when considerable material is available. There is, further, no doubt that the situation as described for *Lymantria dispar* is of exactly this kind. Here we found that only a few races could be distinguished by taxonomic study of the imago alone, though many more appear if physiological and larval characters are added. Within these major subspecies from northern Europe, southern Europe, Hokkaido, Japan, and the Japan borderline region a considerable number of variants may be distinguished in the imago, many of which are confined to a single locality, while others appear in different localities. These variants are mostly mutants and their type and occurrence have been described in detail

(Goldschmidt, 1933). Also, in this species taxonomists have described a number of subspecies and named them. Some of these, like the Hokkaido form, correspond to actually existing units, while others are based on a few chance individuals and have no basis in actual facts. To clear up this situation I made the following test with a taxonomist, specialist for this group, a very keen observer as it turned out, who agreed to collaborate in this test: Mr. X received material from all the different major racial groups. From each he received a complete brood from a pure inbred strain. But he was not told that the—say two hundred—individuals were brothers and sisters. The whole material was marked by numbers and the origin unknown to Mr. X. He was asked to determine the subspecies without knowledge of the origin of the material. He secured type specimens and others of the subspecies described in the literature and by means of comparison determined my material. The result was exactly as expected; it will be reported in some detail, because many writers on evolution, especially geneticists, have very little knowledge of the way the all-important lowest systematic categories are arrived at.

(1) Race from Hokkaido. The broods contained the different mutant wing colors found in this region. Most of the individuals were correctly assigned to Hokkaido. But aberrant colorations (mutants and recombinations—see fig. 9) created difficulties; they received wrong assignments.

(2) Races from Japan proper. One hundred fifty different broods from all localities studied were given. All but one were correctly assigned to Japan. This one, from northernmost Honshiu, is very typical. Here one of the mutant characters usually found only in Hokkaido was present, and therefore the conclusion was that the males came from Germany, females from North Africa.

Within this large Japanese group no subgroups could be discerned by Mr. X, but again, the mutant types, which actually do not characterize one region alone, were assigned to different subspecies previously described by collectors from chance specimens. Wherever individuals happened to be smaller or showed one of the light wing alleles, which are really insignificant, they were assigned to Europe or Turkestan.

Let us parallel these last results with facts previously reported. In older taxonomic descriptions definite forms had been assigned to definite regions as characteristic of them, a result which was actually based on incomplete taxonomic work. Let us assume that the regions in question were better known and large collections had been made. Then probably the different regions would have been characterized by different percentages of the different forms (as in the *Cepaea* or *Coccinella* example), and this percentage again may have been a chance result, if siblings of a segregating brood had been caught by chance, or the collection made during one season only. Even from a large collection it would have been difficult to derive a correct description, and even under the most favorable circumstances most of the facts regarding the typical seriation of subspecies within Japan could not have come to light.

Another of the errors already mentioned is interesting. One of the broods of northernmost Japan containing males with light wings, a mutant discussed above, was assigned to Europe. It actually resembles the Hokkaido race. But it had been tested genetically and was typically northern Japanese in all decisive characters. If these individuals had been collected, the taxonomist would have been forced to draw the erroneous conclusion that the Hokkaido race occasionally passes the Tsugaru Strait and is found side by side with the Japanese form.

To return to our test:

(3) Races from Korea. Out of twenty broods, seven were assigned to the Vladivostok region, which is only a slight error. But seven were assigned to Central Japan, and four to Germany. As a matter of fact, we know that wing colors appear in Korea which are also found in those localities. The wing colors then led the excellent taxonomist astray. He had, of course, no means of finding out that all these differently colored individuals were completely characterized as a definite group by larval and physiological characters.

(4) Manchuria. The forms were partly attributed to Amur, which is practically correct, and partly to Germany on the basis of the same criteria as mentioned before.

(5) Turkestan. This is the only group in which the taxonomist completely failed, and I cannot see how he could have succeeded, as distinctive imaginal features are lacking.

(6) Mediterranean forms. The majority were correctly diagnosed, though a few were assigned to the Amur region, an error clearly attributable to chance variations in size and color.

(7) Northern Europe. Of three German broods one was assigned to Germany, one to Amur, and one to the Mediterranean, again on the basis of characters as before.

This test is very interesting. Mr. X showed a keenness which I could hardly match in distinguishing actually distinguishable forms. Where he failed certainly everybody would have failed and would be bound to fail because the visible characters are simply not characteristic for the subspecies in question. In a general way the test shows further that Rensch's method of taxonomic procedure, as described before, is correct. With that type of work available beforehand, Mr. X would probably have avoided most of the errors he made. But one more thing is brought out by this test; namely, that the taxonomic study of a rassenkreis even under the best of circumstances still gives a very incomplete amount of information regarding the actual conditions in nature, because the student is forced to work with differential characters which might not be the really important ones. This does not mean much to the taxonomist, but very much indeed to the evolutionist.

We now return after this detour to the *Lymantria dispar* races. There are a few additional characters which distinguish races, only one of which will be mentioned in conclusion.

8. Chromosomes

All geographic races have a haploid set of thirty-one chromosomes, a number which is found in numerous species of Lepidoptera (see below). But the size of the chromosomes differs in the different races of the gypsy moth. Generally speaking, the mass of chromatin is correlated with the potency of the sex races. The weak races have the largest

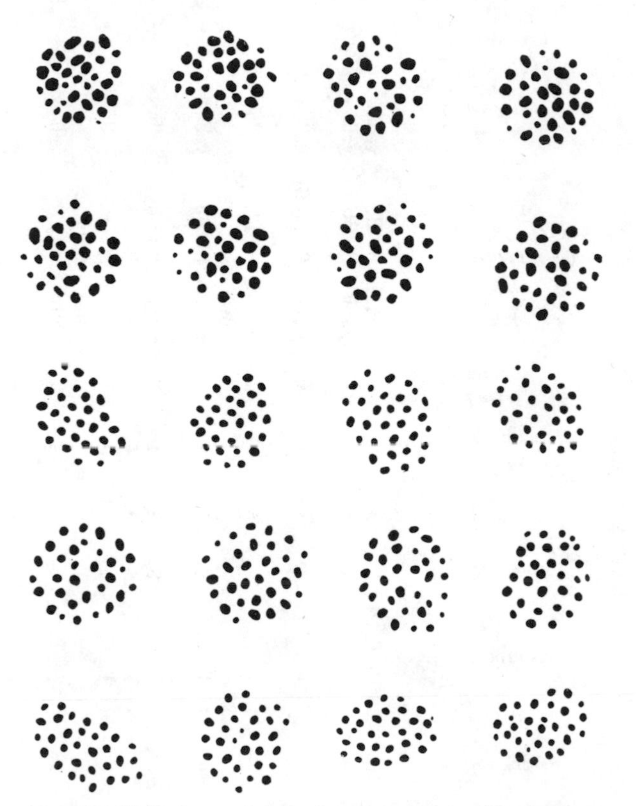

FIG. 10. Metaphase plates of spermatocyte chromosomes of *Lymantria dispar* of a central and a western Japanese race to show the size differences. (From Goldschmidt.)

chromosomes, and the strong races the smallest ones, with all transitions in between (fig. 10). Of course, this character fluctuates so greatly that minor differences could hardly be located. An interpretation of the fact is not available.

The facts thus far reported permit us to draw a general

TABLE 4

(EXPLANATION IN TEXT.)

	Racial group	*Sex races*	*Larval time*	*Molting genes*	*Diapause*	*Growth and size*	*Mass of chromatin*	*Larval pigment*	*Female wing color*	*Genes for male wing color present*	*Abdominal wool*	*Sensitivity to infectious diseases*	
Japanese Isles from N.E. to S.W.	Hokkaido	1	2	T_1 T_2	1	3	?	1(6)	grayish	Ah SF	brown		1(6) means first very light, later very dark
	Northernmost Japan	10	2	T_1 T_2	5	3	1	5	grayish	Bv Bc	yellow		
	Northern Japan	9	3	T_2	4	4	1	4	grayish	Bc Bb Bh	yellow		
	Mountain region Central Japan	8	2–3	T_2	4	4	2	4	grayish	Bc Bb Bh	yellow		
	Border region Central Japan	7	4	T_2 T_3	3	5	2	3	grayish brown	Bc Bb	yellow		
	Eastern part W. Japan	6	3	T_2 T_3	2	5	3	2	grayish	Bb Bh	yellow	++	
	Western part W. Japan	5	3	T_2 T_3	3	4	3	1	grayish	Bb Bh	yellow	++	
	Kyushiu	4	3	T_1 T_2 T_3	3	3	4	1	grayish	Bb Bh	yellow	+	
Korea		3	2	T_1 T_2	1	2–3	5	1	grayish white	grayish Bh, S	yellow	+	
Manchuria		4	?	?	7	?	?	6	white	gray	yellow	+	? no sufficiently large series
N. Europe		2	1	T_1 T_2	1	2	6	7	white	gray	black		
Mediterranean		2–3	3	T_1 T_2	6	1	6	6	white	gray	yellow		
Turkestan		4	3	T_2 T_3	1	3	5	6	white	gray	yellow		

picture of the geographic variation of this animal, a picture which will suggest definite generalizations. To make it easier to look over the main facts at a glance, we have assembled them in the following tables. Table 4 shows the specific seriation of the variants for all individual traits within the geographic range. The upper part of the table contains the series from eastern Asia starting in the northeast in Hokkaido and going southwest. Then follow the other regions. Where different grades of a character can be clearly distinguished, a numerical grade has been assigned, and the largest number always means the highest grade; e.g., ten for the sex races the strongest race, four for larval time the longest time, etc. Thus the vertical rows show the geographic variation for each individual character, and the horizontal rows the composition of each group with regard to the individual characters.

Looking over this table, we see that by the use of taxonomic methods involving only imaginal characters we could distinguish:

(1) Three main sections, those respectively with yellow, brown, and black abdominal wool.

(2) Among these, four subsections for female color (white, gray-white, gray, grayish brown).

(3) Size groups 1–5.

The combination of these characters would permit us to name the following subspecies (we never used such names because they tell only a part of the story, but they might as well be introduced here leaving it to a professional taxonomist to give such names as the rules require.

(1) *dispar:* small; females whitish, males gray, abdominal wool black. Northern Europe.

(2) *mediterranea:* very small; abdominal wool yellow. Southern Europe.

(3) *bocharae:* like (1), and larger; yellow abdominal wool. Turkestan.

(4) *hokkaidoensis:* small; brown abdominal wool, females gray, males very light. Island of Hokkaido.

(5) *japonica:* females gray, males different types of brown, large size. All Japan, except borderline.

(6) *obscura:* females dark-brownish gray, males brown or chocolate, very large-sized. Borderline in the Lake Biwa and Gifu region.

(7) *chosenensis:* color and size of females like (5), of males and sometimes females like European races. Korea.

But on the basis of some of the characters invisible in the imago, further subgroups may be distinguished. For example, in the "subspecies" *japonica* as defined above, the different types from northernmost, northern, and western Japan and Kyushiu may be distinguished. In the border region, as mentioned before, different groups are distinguishable on the west coast, on the western shore of Lake Biwa, eastern shore, Gifu, and Nagoya regions. And even within these subgroups (and also among the North Europeans), minor differences were observed in some characters. This makes it probable that specific forms are found in very small areas. The lower systematic categories within this species, if studied with more refined methods than ordinarily used, then appear as indicated in the diagram on page 61.

If only the subspecies as named here are considered, the different forms can hardly be arranged into a chain of forms with quantitative differences increasing with their geographic distance. If, however, the analysis is pushed to the point represented in table 4, a perfect geographic order appears, which is typical though different for the individual traits. Take, for example, the first column with the sex races and their perfectly orderly arrangement in eastern Asia, or the diapause with another, different regularity. It is just this set of facts by which it could be demonstrated that the seriation parallels climatic series in nature and by which it could be proven that the genetic differences are actually adaptations of the life cycle of the animal to the seasonal cycle in nature.

A few more points are recognizable. Each race is distinguished from the others by a series of hereditary traits which may vary independently over the geographic areas.

One character may be different in two adjacent areas, another common to two or three areas. One character may show an increase from one end of the geographic series to the other, another may first increase and then decrease as the specific types of adaptation for which they stand severally may require. Major breaks in the seriation are usually

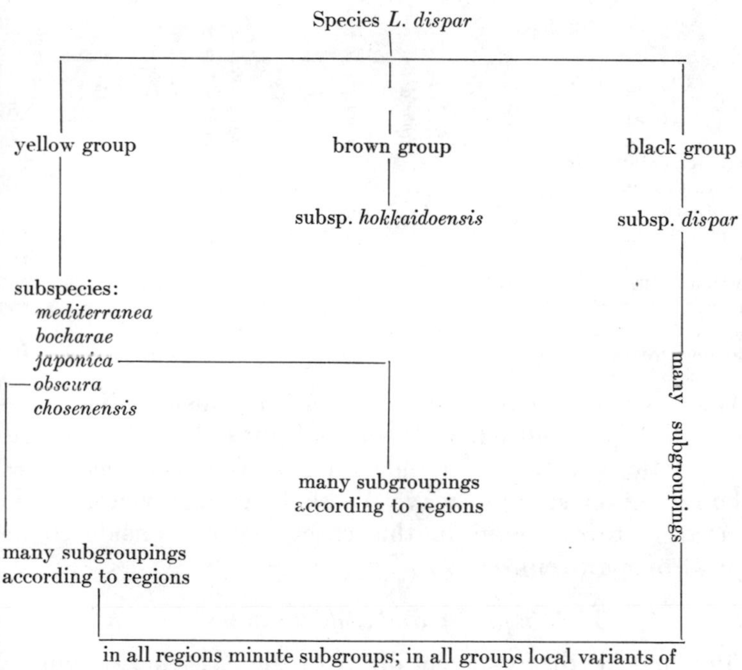

caused by complete isolation by a major barrier like the Tsugaru Strait, which separates Hokkaido, geologically belonging to Amur, from Honshiu, geologically belonging to Korea. Where quantitative traits are involved, there is a transgressing variability, and where two racial groups meet, hybrids will be found.

It is, finally, of interest that all the genetic differences which have been analyzed are based severally on most of the known types of heredity. But it is rather rare that simple

Mendelian differences exist, if they occur at all. Table 5 shows the genetic types found for some of the characters.

TABLE 5

TYPES OF HEREDITY FOR THE DIFFERENT RACIAL CHARACTERS OF *L. DISPAR*

Character	*multiple alleles*	*di- or tri-hybridism*	*multiple factors*	*cytoplasmic influence*	*sex-linked inheritance*	*sex-controlled inheritance*	*modifying genes noticed*	*adaptational value*
female sex determiner				+			+	+
male sex determiner	+				+			+
time of larval growth	+(?)		+(?)	+				+
instars	+			+				?
diapause			+					+
growth and size	+(?)		+(?)	+				?
markings of larvae	+		+				+	?
wing color	+	+				+	+	−
abdominal wool			+			+		?

We have represented this case with a considerable number of details, because here the actual physiological, ecological, genetic, and taxonomic meaning of a rassenkreis comes out in a rather striking way. We shall see now whether the general features found in this case may be considered as typical of a rassenkreis.

b. Subspecies and still lower units

The first feature as exemplified in the case of *Lymantria* is that the individual geographic races or subspecies are not of equal value. One race may differ more from all others (e.g., Hokkaido in *Lymantria*) than these among themselves. A series of races may form a definite subgroup within the species. Within the subspecies a more or less large number of typical minor forms may be distinguished, and among these again differences may be found which may go down to the individual family, sib, or colony. Actually, the latter phenomenon has frequently been observed in taxonomic work, and it seems to be visible wherever large collections are studied. Some taxonomists have therefore introduced further categories and use a quaternary instead of ternary nomenclature. As we have seen, there is no end to such subdivisions, especially if physiological characters are added to the morphological ones. For practical taxonomic purposes such subdivisions are not in general use, but if the facts are to be used as a basis for evolutionary discussion, it is safe to keep in mind that the existence of named subspecies is only a part of the whole story.

These results, which became so clearly visible in the *Lymantria* material, also turned out in the same way in another zoological material which was studied taxonomically and genetically: in Sumner's work on deer mice (*Peromyscus*), to which we shall have to refer repeatedly (Sumner, 1915–32, review in 1932). Thus, an exact statistical study of the subspecies *gambelii* in different parts of its range in California showed that the *gambelii* of La Jolla is not identical with that from Berkeley or with that from Calistoga. Other examples of the same type are found in this work. The same conclusions may also be derived from Anderson's work on Iris (1928, 1936), where, however, only a few subspecies were tested. But within these the individual colonies stood out as rather well-defined and in themselves constant subgroups, which, if isolated, might be called separate subspecies. A rather extreme case of subdivision has been presented by Kinsey (1929, 1936). He worked upon an immense number of gall wasps of the genus *Cynips* all over its North and Central American range, and his work included not only the usual taxonomic characters but also, in part, the life cycle and the galls. This made it possible for him to distinguish many more subgroups than are ordinarily accessible to the taxonomist, each of which groups is said to inhabit a definite geographic area. The result is strikingly similar to the one described for *Lymantria*, though the terminology is different. What would be called a subspecies by others he calls a subgenus, and further subdivides it into complexes and species. Thus he comes to a quaternary nomenclature which in the usual nomenclature would be species

—group of subspecies—subspecies. Aside from the nomenclature, this reveals, of course, the actual finding of the same type of subunits within the species as described before. We shall return to this work below. I do not doubt that any completely known material of a species with a considerable range of distribution would show the same groupings and that some still better-analyzed species might be split into even more types of subgroups, provided that visible differences are involved. As an example I might mention Crampton's work on *Partula* (1932) where a whole nomenclature of subsubspecific variation is used (gens, cohort, socius, etc.). We refrain from mentioning other material, which is to be found in numerous modern monographs in different fields: insects, mollusks, birds, mammals. An excellent discussion of these and similar facts is found in Remane (1927) and in the other general works quoted. See also Philiptschenko (1923). Nor shall we enter upon a discussion of nomenclature; i.e., which forms should be named or not, which category ought to be called a subspecies, etc. This is a completely utilitarian problem which the taxonomists must solve as they please. But when it comes to a discussion of evolution it must be kept in mind that, whatever the nomenclature, the species may be subdivided to a final limit set separately for each case, and sometimes extending as far as the individual colony. No such subdivision may, therefore, be called an elementary species, or the like. And if we discuss later the problem of subspecies as incipient species, we have to keep in mind that the term subspecies includes the whole gamut of eventual subsubspecies, etc., down to the individual family or colony.

There can be no doubt that the differences which the authors indicate with regard to the number of possible subgroupings in different cases are frequently due to the relative amount of material which has been analyzed. But in other cases it is actually the species itself which is more or less prone to form smaller units. In Rensch's publications (*loc. cit.*) a large number of groups have been analyzed to find out how many species of a genus or a family form more or less considerable rassenkreise. He finds that in each such group a minority of single species remain which have formed no subspecies. Though some of these will disappear with better knowledge, others will be left. (We shall later discuss such an example, the swallowtail *Papilio machaon*, where one species has developed numerous geographical races, but the next related species none.) The same author (Rensch) has drawn attention to the fact that the more sessile species are apt to contain many subspecies, whereas vagrant forms have fewer or no subspecies. An appropriate example which is quoted by Rensch is found among the Carabid beetles where the nonflying *Carabus* species contains very numerous subspecies, whereas the flying *Calosoma* are without subspecies. We shall return to this point when discussing adaptation.

c. Racial chains (clines)

A point which we noticed in the *Lymantria* case is the orderly seriation of distinctive characters within a rassenkreis. In some of the early work in this field, racial chains were described in mollusks (Sarrasin, 1899; Plate, 1907), with an orderly sequence in the change of the characters over the area of distribution. The rassenkreis concept, however, showed that rassenkreise might also stretch in different directions over a large area, which would result in a more or less checkerboard-like arrangement of the races, which again would preclude a linear arrangement of the geographically varying characters. In our paradigm the Japanese forms were not connected with the North or South European forms by a chain of intermediates. But within the Japanese Isles we found a very typical chain or gradient for definite subspecific characters stretching from northeast to southwest. It is obvious that this chain formation has in this case a relation to environment as the direction of the chain coincides with a similar chain in climatic conditions. J. Huxley (1938) has recently proposed calling a chain or gradient of differences arranged in a definite direction a "cline." We might then say that the cline of the subspecific characters parallels a climatic cline. Such subspecific clines are actually a frequent feature of geographic races and may even be a very characteristic feature. One might say generally that wherever subspecies which stretch over a considerable area have been described, and wherever this area shows a typical cline with regard to temperature or moisture or seasonal cycle or the like, a corresponding cline with regard to at least some of the characters which distinguish subspecies can be found. Innumerable data exist to show that the length of something increases in subspecies in a direction from north to south, or that a color brightens in the same way, or that a shape changes in a definite direction. Examples of this are abundant in all groups of animals, and many have been studied by statistical methods. A very considerable importance is attached to these facts, because they are in closest connection with the problem of adaptation, as will be discussed later.

One of the most remarkable features of these clines of subspecific characters is that they frequently occur in the form of a continuous fluctuation of characters, so that two distant forms are connected by a continuous series of intermediate conditions. The idea has frequently been expressed that these intermediates might be hybrids, but this can be easily ruled out in cases which have been analyzed genetically. In some cases intermediates might be nonheritable modifications by environmental action. But this is again ruled out where breeding under identical conditions can be performed. Some taxonomists, therefore, take refuge in Lamarckian ideas, making environment responsible for the production of the cline (for example, Rensch). The existence of the cline has, in fact, nothing to do with environmental action, nor does it require a special hereditary process. It is, rather, the consequence of ordinary genetic changes in conjunction with selection, as will be discussed later. I may add that such clines have also been found in plants. Langlet (1937), for example, has described such a continuous cline for the pine tree. In Iris, Anderson (1928) finds that only one subspecific trait, size, is arranged in an orderly cline following temperature, whereas other traits are arranged in an irregular way.

d. Independent variation of individual traits

A third group of facts seen in our paradigm was that individual traits based upon different hereditary factors vary independently over the cline of subspecies. One trait might be common to more than one subspecies, another be different in each. Also, the direction of the cline may be different for different traits. In discussing this point, it must first be emphasized that the geographic races, studied taxonomically, are distinguished by a series of different traits. In most cases only morphological traits are used, and these may be few or very numerous in different materials. Size, color, proportions are predominant features. But also ecological and psychological features, like breeding habits, song, or migration in birds, are sometimes distinguishable. Rarer are physiological characters such as specificity of gall-producing secretions in gall wasps, or annual and perennial habits or relations to the chemistry of the soil in plants. The most elaborate individual study of the quantity of subspecific differences has been made by Zarapkin (1934) on subspecies of the beetle *Carabus granulatus.* He measured all possible organs and all available numerical differences, involving altogether 116 different traits. The statistical treatment of this material showed that most of the differences involved resulted in different proportions of the organs, or of proportions of the general pattern. A good example of a well-studied rassenkreis with detailed analysis of the differential features and numerous illustrations is given by Endrödy (1938) for the beetle *Oryctes nasicornis.* I mention only these two studies, as their results are typical of innumerable others.

Where such different traits are studied statistically, the independent geographical variation of individual traits is also frequently observed, just as in *Lymantria.* An example which has been analyzed statistically is found in Alpatov's work on the geographical variation of the honeybee (1929, and earlier papers). Here the absolute size of the body, the

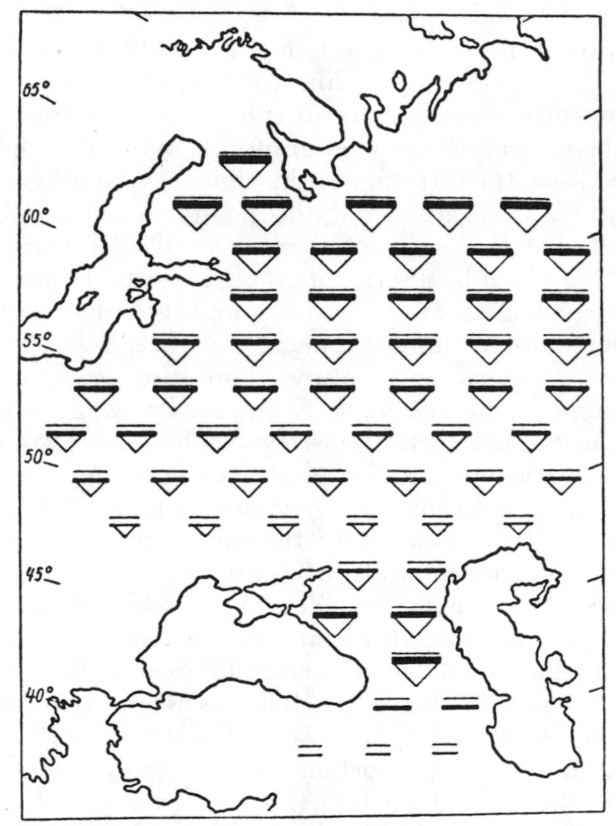

Fig. 11. Diagrammatic map of European Russia. The size of each group of signs represents the size of the bees. The width of the middle bar corresponds to the coloration of the abdomen. The relative size of V-shaped sign represents the relative size of the wax-gland surfaces. Owing to the lack of space, North Caucasus is covered with signs corresponding to a more northern or mountain type of bee. (From Alpatov.))

relative size of the wax gland, the color of the abdomen, the length of the tongue, and the relative length of the hind legs show a beautiful cline going from north to south in the plains of Russia. But in the Caucasus this cline is interrupted for some of the characters; namely, size of body, coloration, and size of wax gland, which again resemble the condition in

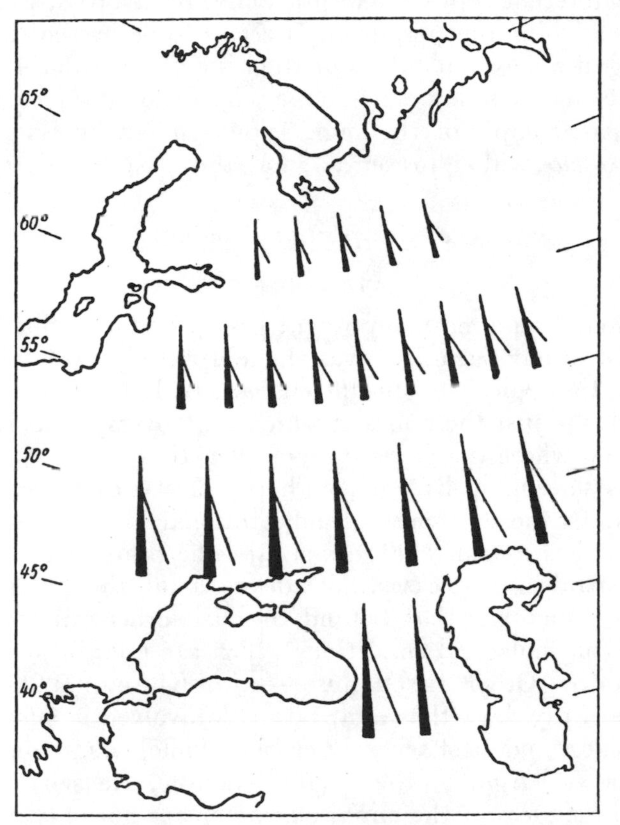

Fig. 12. Diagrammatic map of European Russia. The vertical bars with broader bases represent the tongue length of bees. The lines attached to the right side of the bars show the proportion of legs to body size. (After Alpatov.)

northern forms. Figures 11 and 12 illustrate this situation diagrammatically, and might be compared with table 4 for *Lymantria*.

There are, further, many instances to be found in Kinsey's monographs on *Cynips* where galls, wings, and hypopygial spines vary independently along one geographic cline. The other phenomenon found in *Lymantria*, the difference of direction of the clines for different individual traits where different types of adaptation were involved, probably also occurs in other material. It seems to be present in Kinsey's gall wasps, but tables of the type of our table 4, which permits us to follow such clines, have not been compiled. Another example of the same type, studied by Sumner in *Peromyscus*, will be presented in a later chapter.

e. Adaptational value

AA. GENERAL

A fourth and most important group of facts which came to light in our paradigm was the adaptational value of certain of the typical subspecific characters. It is not surprising that this is just the point at which controversy is at its hottest, and where the views of some geneticists and some taxonomists are most divergent. There is first a divergence with regard to the existence of adaptational characters. Some taxonomists assume that the subspecific differences as such are neutral in most cases, not adaptational, though they are willing to assume that behind the visible neutral traits invisible physiological conditions which are adaptational may be hidden. Other taxonomists—e.g., Kinsey, 1936—most emphatically deny that adaptation is involved in subspecific differences and most severely criticize biologists who mention adaptation. Again, others (for example, Rensch) assume that differences of the clinal character are not only adaptational but the result of a direct influence of the environment upon the hereditary constitution. The latter need not be discussed. No geneticist would be willing to accept this Lamarckian view, though Sumner (1932) cautiously indicates that he cannot help looking at the facts this way. There is actually no need whatsoever to accept such a view, even if it were not completely discredited, as the known facts, genetic and otherwise, permit us to account completely for such adaptations without direct environmental action, as will be discussed below. As a matter of fact Rensch has recently forsaken his former Lamarckian outlook.

But there are also taxonomists who express viewpoints which are completely in harmony with those developed here, at least as far as subspecific differentiation is concerned. Grinnell (1928), one of the leaders of modern taxonomy, has given clear expression to the facts under discussion. Special value is given to his opinions by his mastery of the fauna of the Pacific Coast of North America, an area which furnishes innumerable examples for all aspects of our problem. Grinnell believed in the idea that geographic races are incipient species. Leaving this point out of our discussion at present, we may quote his paragraphs on subspecific variation and adaptation (without accepting every single point):

"No matter how heritable variations in individuals may arise, no evolution in the phylogenetic sense can have taken place until said variations have been subjected to the drastic process of trial for survival. This endurance test is imposed by environments. And the critical factor for divergence of stocks under differing environments is isolation. Otherwise there is swamping, with resulting uniformity of populations, instead of divergence. Of course, the amount and rapidity of effect by environment is immediately limited by the conservatism of the organism—the animal will stand only so much ecologic pressure. Its inheritance prescribes a certain limit of modifiability; but counting that in—then, with a more or less segregated population, whose variations are of the inherited sort, the Darwinian factor of selection comes into play. These heritable variations of selectional value are of small compass, certainly not of large amount, as the old 'mutation' concept had it.

"By the action of selection a population is able to accommodate itself to conditions as they change; it becomes less liable to outright extinction should conditions change abruptly. Animal adaptation, so-called, is merely the demonstration of a capacity to survive under conditions at the moment existing—just that! And animals do just as little adaptation as they can and 'get by.' Inertia is a characteristic of the organism. The direction of such modifications as are acquired is determined by the course of environmental history.

"The accumulating experience of the field naturalist is bringing conviction that the incipient species in nature, the subspecies, owes its origin to a process, on a vast scale, of trial, discard, and preservation, of individuals, and of groups of individuals comprising populations, which populations from generation to generation are thereby rendered more nearly adjusted to such environments as they can endure at all. But environments themselves never stabilize; they are changing, proliferating, evolving continually. A balanced state of perfect adaptation of the organism can never be attained, but only continually approached, such approach being forced, under penalty of extinction.

"It seems to me, then, that the problem of the origin of species ought to be dissociated largely from the problems of inheritance. The problem of speciation would seem to lie much more nearly to the provinces of the geographer and climatologist than to that of the geneticist. The studies of the systematist, if he be also a field zoologist, in his definition of minor species and of subspecies, and of the geographer, may be looked to, accordingly, if properly correlated, to bring an improved understanding of the conditions, methods, and results of evolution, more especially as regards the higher vertebrate types."

There can be no doubt that the characters by which subspecies in animals, and to a smaller extent in plants, may be distinguished taxonomically, look, as a rule, neutral, which means that they do not show any obvious adaptation to environment. If races are involved which are more or less isolated (see below) and not members of a continuous cline, the neutral-looking type of the distinguishing characters will actually often be nonadaptive. If, however, clines are involved; i.e., continuous variation of the subspecific character parallel to some continuous feature of the environment, like temperature, the suspicion arises that a physiological adaptation is hidden behind the apparently neutral trait. To return to our paradigm, nobody could claim that the pigmentation of caterpillars which are usually found in large numbers on the same tree exposed to all kinds of illumination and which are eaten only by a few birds and beetles, could be of any adaptational value. (Actually, the real enemies of *Lymantria* are bacteria, viruses, and larvae feeding on the eggs.) But if we find that the clines of this trait closely parallel those of another physiological character which is clearly adaptational, we are led to suspect that the pigmentation is the visible consequence of a metabolic peculiarity which is adaptational. Or, looked at from the side of the environment: We found a major line of demarcation between the geographic races of *L. dispar* in what we called the borderline region in Japan. Looking at the meteorological tables, we could not discover any hiatus in the cline from northeast to southwest at this line. Let me now suppose that we had first found the data regarding soil differences in Japan (see above, p. 46) for which a line of demarcation coincides with the subspecific borderline. We would have suspected that the subspecific differences between the forms east and west of this line were adaptations to the different soils. But there is hardly any correlation visible between soil conditions and the vital needs of an omnivorous animal feeding on many trees and, therefore, an adaptational relation between soil and subspecies would be denied. But now it is found by climatologists that the soil conditions are indicators of definite seasonal cycles, which are not evident from inspection of meteorological tables. At once the adaptation becomes evident, the adaptation of the life cycle to the seasonal cycle.

It is very remarkable that in Sumner's work on *Peromyscus* a case has been described which very closely resembles the one just discussed in *Lymantria*, though in this case not a whole cline of subspecies is involved but only two. The very pale subspecies *albifrons* is found near the Gulf Coast of Florida, and another subspecies, *polionotus*, occurs inland. Within a stretch of only forty miles the decisive hereditary characters, especially the extension of the colored area of the pelage and the intensity of its color, change in a slow gradient from the *albifrons* type to an intermediate type found in a small intermediary zone, and then slowly, though

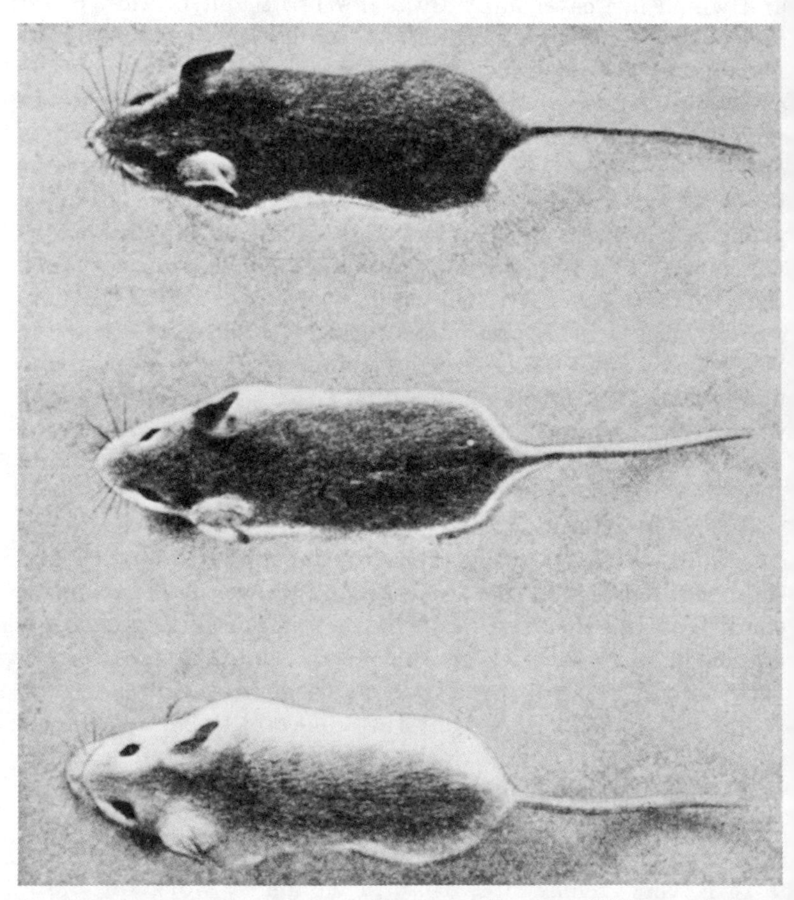

Fig. 13. The forms *Peromyscus polionotus polionotus, albifrons,* and *leucocephalus* on white background. (From Sumner.)

at this point a little more abruptly, grade into the *polionotus* type (see figs. 13–15). It may be added that near the coast on an island of pure white quartz sand a still paler form, *leucocephalus,* exists. In looking for possible adaptational significance, Sumner points out that the white sand to which *albifrons* and *leucocephalus* might be an adaptation is limited to a narrow zone close to the sea. There is no pronounced

Fig. 14. Map of portions of Florida and Alabama showing stations where the subspecies *polionotus* and *albifrons* occur. (From Sumner.)

gradient of soil color inland. ". . . it is true, the sandy soil is paler in comparison with that encountered farther north, this difference depending upon a difference in the respective geological formations. Furthermore these soil types succeed one another rather abruptly. . . ." Sumner cannot see any relation to adaptation here and advances an explanation

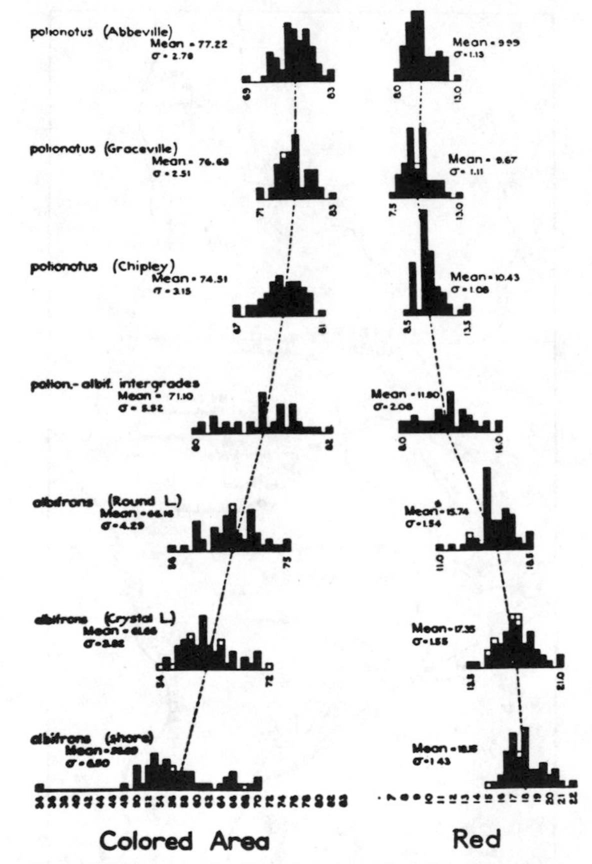

Fig. 15. Values of colored area and red in the pelage of *Peromyscus* at each of seven stations (see map, fig. 14). The broken lines connect the means. The grades represent the color in standard measurements. (From Sumner.)

based upon population pressure to account for the diversity between *albifrons* and *polionotus* in spite of lack of a proper barrier between them. If we remember our last example, we realize that a soil difference may be an indicator of an actual climatic difference, though it might be difficult to demonstrate the relation in a given case. In addition, there is no reason why the color of the sand must be the factor of adaptation. An indirect physiological adaptation to moisture or other soil conditions might become visible as a coat color. Thus, this cline might be after all an ordinary case of climatic or ecological cline, to which is added a new ecotype in the white sands. We shall return to the same set of facts soon in a discussion of parallel clines.

It is not possible to mention all cases in which a direct or indirect adaptational value of subspecific clines has been surmised. Only two more examples will be mentioned, because the basic facts of subspecific differentiation are in many respects parallel to those found in *Lymantria* and are also based on exact statistical study. Alpatcv, in a series of papers (see his review, 1929), studied the geographical races of the honeybee. The typical racial conditions involve a number of rather minute morphological differences like length of tongue and number of hooks on wings, which we discussed above (see figs. 11, 12). Alpatov now points out "that the southern bees are obliged to have a longer tongue, not only because of a lower nectar level, but also because of a probable difference in the composition of the whole nectar-secreting flora. It has been reported by many bee-keepers that the southern, and particularly the Caucasian bees, can fly longer distances gathering nectar, and it is probable that their wings are in consequence more developed and have a larger number of hooks. The smaller size of the wax glands is probably connected with the condition that the bees in the south have perhaps less need to work upon the reinforcements of their nests. Hence the differences in the tongues, the wings and the wax glands (also probably in the first joint of the tarsus of the last pair of legs) may be considered as adaptations to different biological ends."

In a very different group of animals an excellent and exhaustive study has been made by Miller (1931) on the American shrikes *(Lanius)*, birds which in some species have formed large series of subspecies. A statistical study of external and skeletal features, together with a close scrutiny of ecology, propagation, and instincts, has led Miller to the conclusion that the majority of the subspecific features are clearly adaptational. He finds variations in coloration correlated with climate; characters of size and ratio of wing and tail which are correlated with powers of flight, as regards either migration or openness of the habitat; characters of the bill and feet, which are probably correlated with small differences of behavior in feeding or with the floral environment or with the general bulk of the animal; i.e., adaptations to various perching and feeding conditions. In addition to these subspecific characters, which are most probably of an adaptive nature, there are others which are considered as neutral and therefore called palingenetic. Altogether the species in question is not extremely restricted to very specific ecological niches and therefore the adaptive traits are not as marked as in other cases. That they can be located nevertheless makes the case instructive.

These examples demonstrate that what apparently is nonadaptational may turn out to be strictly adaptational if only the proper environmental factor and the proper physiological process can be located. Therefore I am inclined to consider all subspecific characters which vary in a cline parallel to a geographical, climatological, or other environmental cline (salinity, moisture, soil composition, insolation, etc.) as, at least indirectly, adaptational.

This statement must not be misunderstood, however. It applies only to cases in which parallel gradients of clearly adaptive and of not visibly adaptive traits are found and in which they are both to be based upon independent genetic traits. Very frequently different races show an obviously adaptive gradient of one character and in addition a visible hereditary difference which is not clinal. The latter is probably a fortuitous mutation without any adaptational significance. A nice example of this kind has been analyzed by Melchers (1932). He studied two Alpine races of *Hutchinsia (alpina* and *brevicaulis)* with strictly separated distribution. The decisive factor is the lime content of the soil, as physiological experimentation showed sensitivity to lower concentrations of lime of the form growing on soil rich in lime. With this adaptive trait based upon genetic factors a difference in the shape of the petals of the flower is combined. This is also genetic and inherited independently, but the distribution of this character does not coincide with that of the other and is not clinal.

BB. ECOLOGICAL RACES

There are two sets of facts, found in many taxonomic and other studies, which favor the conclusions reached in the last chapter. The first group of facts related to the existence of ecological races and their relation to the subspecific variation. The most extensive data on this subject have been accumulated in plants. In plant taxonomy the rassenkreis principle has not been employed as extensively as in animals, though R. von Wettstein (1898) had made use of it. We shall later have to mention some more recent work which parallels completely the situation in animals. But in plants the ecological relations between type and environment are usually more conspicuous than in animals and have led to extensive analytical work. As far as the genetic and evolutionary side is involved, the foremost investigator is Turesson (1922 ff.) whose work in many respects closely parallels mine on *Lymantria*. Plant forms frequently may be modified by external conditions in a much more extreme way than animals, as has been shown in the classic work of Nägeli, Kerner, Klebs, Bonnier, Massart, *et al.*, who found the phenotype of aquatic forms, dune forms, alpine forms, etc., in certain cases controlled by the action of external agencies. From such facts the erroneous conclusion was drawn that these adaptational types are actually produced in nature as well by the direct action of the environment. Nowadays we know (as will be discussed in detail later) that by action of modifying influences many if not all hereditary forms of a species, the mutations, may be reproduced as nonhereditary phenotypes, so-called phenocopies. The successful copying of a mutant type by an experimentally produced identical, nonheritable phenotype does not give any information about the origin of the hereditary type, though it does give highly important information regarding the potentialities of the organism (see later). If such an ecological type, however, is based on genetic differences, it will retain its features after transplantation to different conditions. Evolutionary study is, of course, mainly concerned with those adaptational types which are based upon genetic differences. It is these with which Turesson experimented. He bred the different habitat types of different plant species side by side, thus separating the hereditary differences from the nonheritable modifications, just as I did in *Lymantria*. By this means he found within one species sand types, dune types, cliff types, etc., all hereditarily different with regard to these types. These he called ecotypes. It is obvious that these ecotypes are in part identical with what we have thus far termed geographic races, and in part not identical. This is simply due to the type of ecological adaptation involved. If the adaptive trait adapts the subspecies to definite environmental conditions which change geographically, i.e., in the majority of cases involving climatic conditions, the resulting geographic races are simultaneously ecotypes in Turesson's terminology. Turesson actually calls them climatic ecotypes and gives examples of the same type as found in all rassenkreis work.

If geographic races exist the features of which are not adaptational, directly or indirectly, they would be subspecies without being ecotypes. But in addition to the typical geographic variation and subspecies formation, usually arranged in a cline, hereditary adaptational subgroups may be formed within the species or within the geographic races which adapt the type to an environmental condition which does not change with a geographical cline. If, in our paradigm, *L. dispar* had not been omnivorous, as it actually is, but had been adapted to definite food plants, as is the case with many other Lepidoptera, a diversification according to this physiological feature might have arisen. There might have been ecotypes feeding only on oak, others on pine, etc., and these ecotypes might have occurred within the different geographic races. Similarly in the plants studied by Turesson, a dune ecotype might occur in Scandinavian, French, African subspecies, if such exist. And if a plant species is found only in a small geographic area, excluding typical geographical races, only ecotypes might be encountered within the species. We might even visualize a species found only within a single square mile somewhere, but broken up

Fig. 16. Ecotypes of *Primula acaulis;* left, from Trieste; right, from Scotland. (From Turesson.)

into ecotypes if gross- or microclimatic differences are present in this area (dunes, swamps, etc., or open spaces, underbrush, etc., as microclimatic sections). Thus we see that the ecotype, both in plants and animals, may coincide with the adaptive geographic race or may subdivide this into smaller genetic units for the sake of specific adaptations. These relations have not always been realized in discussions of this problem. In Turesson's papers many botanical examples of all these types and their combinations may be found. There is also a very instructive set of photographs in Turesson, 1926. In figures 16 and 17 we illustrate two such examples of ecotypes. Figure 17 shows three ecotypes of *Geum montanum*, forming a cline in adaptation to different verti-

FIG. 17. Ecotypes of *Geum montanum* from (*a*) 2,800 m., (*b*) 2,400 m., (*c*) 2,000 m. altitude in Tyrol.

cal levels in the Alps; figure 16 shows two ecotypes of *Primula acaulis* in which the time of flowering is adapted to two different climates.

CC. PARALLELISM OF SUBSPECIFIC CLINES

The second group of facts relating to the adaptational value of subspecific differentiation is that connected with the parallelism of adaptational variation in different species. This feature is rather obvious when ecotypes are involved. Many plant species, for example, have developed alpine or halophytic types showing definite common features which are obviously of an adaptational nature. In animals the most obvious examples are desert–cave–deep sea–polar forms. Turesson (*loc. cit.*) has made a special study of such ecotypes, the results of which are of importance. They show that such parallel ecotypes are hereditary and therefore demonstrate the formation of genetic subgroups in adaptation to definite habitats, which has occurred in the same way within very different species. Thus he gives a large list of species (Turesson, 1925) all of which have developed within the proper habitat definite and similar growth types which he characterizes as the types *campestris* (field t.), *arenarius* (sand t.), *salinus* (shore t.), *alpinus* and *subalpinus*. In animals, where the subspecies as characterized by taxonomists do not as clearly indicate the type of adaptation which might be hidden behind apparently neutral characters, it is more difficult to find such simple ecotypes, of course with the exclusion of nonhereditary modification, though they exist; e.g., in mollusks (soil and shells). The parallelism of adaptational hereditary variation therefore is found more in a parallelism of subspecific traits within a geographic range, the adaptational value of which has to be proven in any single instance. Facts of this nature were known before the rassenkreis concept was applied, and they were laid down in such much discussed rules as Bergmann's, Allen's, and Gloger's rules, to the effect that body size or pigmentation increases in a definite geographic direction. (For details see Hesse's book [Hesse-Allee, 1937].) In our present discussion the important point is that such features are observed within different rassenkreise occupying a similar area. Rensch (1924) has made such studies in relation to Bergmann's rule, which says that size increases with decrease in temperature. He finds examples for this in rassenkreise of different species. Regarding wing length of birds he even thinks that he can specify an increase of 1 to 2 per cent per 1° C. decrease in temperature. At this stage in our discussion I wish to emphasize an important point to which we have already alluded. Rensch (formerly), as well as some other taxonomists, take it for granted that such phenomena as are expressed in Bergmann's rule demonstrate that the hereditary differences are produced by the action of the surroundings, and he cites as proof such parallelism as is discussed here. We shall later show the fallacy of this argumentation. Here I want to draw attention to the fact that such rules are observable when the adaptive character hidden behind the phenotype, size in this case, actually adapts the subspecies exclusively to such climatic conditions; e.g., temperature, as are apt to form simple geographic clines. If adaptation externally expressed in size is in fact adaptation to another climatic factor which does not change with temperature, a very different result will follow. We have already seen such a nonconforming example in the size variations of *L. dispar* in Japan, where the series (which, by the way, runs inversely to Bergmann's rule) changes direction west of the borderline region. Here the reason was clear; the adaptation was not one to temperature but to the time available for larval feeding, a time which in the North was limited by the length of the vegetative period of the food plants, but in the South by the extreme summer heat. A rule like Bergmann's, then, may apply to those animals among which, according to Bergmann's classic concept, temperature adaptation means adaptation of the size of the body surface to loss of heat. But with different types of adaptation different results will be found which do not disprove such a "rule" but simply are on a different level. This, of course, again emphasizes the precariousness of conclusions based exclusively on taxonomic work without the check of physiological experimentation. Much superfluous discussion would have been eliminated had the situation been properly understood.

In accordance with this argumentation the facts indeed sometimes do, and sometimes do not, agree with the different "rules." Unfortunately, most of the facts are based only upon taxonomic information, and therefore the specific adaptational value can hardly be ascertained. Rensch (1936) has collected a considerable amount of material from his own work and from the literature in order to check quantitatively the application of such rules to members of recognized rassenkreise. Bergmann's rule that in warm-blooded animals the races of warm regions are smaller than those of cold regions was tested with birds and mammals. He found exceptions in 8 per cent of the cases in Palearctic birds, 12.5 per cent in birds of the Sunda Islands, 26 per cent in North American birds, 40 per cent in West European mammals, 19 per cent in American mammals. Allen's rule that in a rassenkreis of warm-blooded animals protruding parts of the body (tail, legs, ears, etc.) are shorter in colder climates was also tested, with similar results: tail length of central European mammals, 14 per cent exceptions; wing length of American birds, 20 per cent exceptions; relative length of beak, 10 per cent exceptions; and so on. Gloger's rule that among warm-blooded animals those living in warmer and moister climates develop more melanin pigment, whereas forms in dry, hot climates have more yellow and red pigment, was similarly tested, with the same percentage of fit. To these old "rules" Rensch added a number of new ones: (1) Within rassenkreise of birds, races in cooler regions have narrower and more pointed wings, which are supposed to be more efficient mechanically. (2) Within a rassenkreis of mammals those inhabiting warmer regions carry shorter hair. (3) Within a rassenkreis of birds, races in cooler regions lay more eggs in one batch than do those of warmer climates. (4) Within a rassenkreis of mammals those living in cooler climates produce larger litters. (5) In races of migrating birds within a rassenkreis stretching into the tropics the tropical forms are nonmigratory. For all these "rules" about the same percentage of fit was found as before. In all these cases we know that these or comparable traits are hereditary. The adaptive character can be inferred, though only rarely proven.

In the groups of facts which we have just discussed, the adaptational value of the parallel features of variation in different species occupying the same region was assumed on a more or less vague basis. There is one similar group of facts which contains somewhat more direct evidence; namely, the variation of color in mammals on different soils (see above, p. 74). A comprehensive study of the problem, including a review of former work, has been recently presented by Dice and Blossom (1937). (In this connection special mention should be made of the previous work of Benson [1933], who studied the morphological basis of coloration.) The following groups of facts are included in these studies.

There is first the subspecific (or subsubspecific) variation of color of pelage with the color of the soil. By physical measurements of both, Dice could establish a high degree of correlation, so that he does not doubt that the color is actually adaptive; i.e., a concealing color. This set of facts is especially clear in desert regions. A second group of facts involves specific features; namely, the presence of rock or lava beds, isolated or not isolated within a desert area. These beds are frequently inhabited by dark races, the color of which closely coincides with that of the respective rocks. The third group of facts applies to colors of subspecies in different life belts, desert, forest, etc. Here again the colors correspond to that of the soil, which latter is controlled by the amount of humus. The fourth group of facts proves that the foregoing features occur in a perfectly parallel way in many different mammals of different species, genera, and families, and therefore might also be called a rule. As it may be regarded as certain that these color races are genetically different, they demonstrate again the adaptive value of sub- and subsubspecific differentiation, though the discussion as to what is actually the selective factor is not yet

:losed (concealing coloration versus unknown physiological
:eatures).

Reinig (1938, 1939) has recently discussed the facts con-
ained in this chapter, as well as in some others, from what
ıe believes to be a new point of view. He starts (see below)
vith a group of facts which tend to show that geographic
aces have frequently originated by migration from a series
of centers, which coincide with faunal refuges during the Ice
Age. In these refuges not only old species but also an ac-
umulation of species are found. Within the species a con-
iderable genetic variation is supposed to occur, the reason
eing that within the refuges more manifold conditions
vhich would permit the survival of mutations and their re-
ombinations are extant. Vavilov (1922, 1928) had already
ointed out (see also Schiemann, 1939) that "gene centers"
xist for cultivated plants in which large numbers of races
re found. With an increase of distance from these centers
he number of races decreases. Vavilov had explained this—
s probably most geneticists would do—by selection of more
pecialized mutants in conditions more and more different
rom the original ones, an explanation which is also borne
ut by the facts in *Lymantria*. This explanation certainly
overs adaptive traits only, as discussed before. Reinig, how-
ver, thinks that nonadaptive traits will also exhibit the
ame phenomenon, which he calls (rather unfortunately)
llele diminution. If the migration from a center does not
ccur populationwise but by dispersal of individuals, the
robability is that the individuals carry fewer mutant char-
cters than are contained in the population. The number of
nutants will therefore decrease toward the periphery, a
rocess which he calls elimination. Reinig points out, then,
hat the facts underlying the "rules" which we just discussed
o not agree with a selectionist explanation. He thinks that
he clines in question, especially body size (Bergmann's
ule), follow the lines of expansion of a species from the
entral area. Therefore the process of "elimination"; i.e.,
npoverishment with respect to mutants, would be the under-
ying cause. He thinks in this connection especially of poly-

meric genes (multiple factors), the elimination of which would account for a clinal decrease in size, which is frequently observed. I do not think that these deliberations help our understanding of the situation. They are not in agreement with the genetic facts. They forget that mutation and subsequent recombination seem to occur freely in wild populations. If no selection were involved, the migrating individuals would start new populations, which soon would be as polymorphic as the old ones. There is no such thing as an elimination of genes (read: mutants) as long as the process of mutation is not stopped, or unless selection sets in. But the taxonomic facts also do not agree. There may be conforming cases of clines coincident with the direction of occupation of new areas. But there are also cases in which the situation is different. We may point to the case of size in *Lymantria*, which has a cline decreasing from central Japan northward and southward, with no indication that central Japan was a refuge and southern and northern Japan areas of later migration. In this case the adaptational character of the cline could actually be demonstrated. I therefore prefer the standpoint which has been developed in the preceding chapters.

Finally, the relation of the problem of domestication to the facts under discussion ought to be at least mentioned. The evolution of cultivated plants from wild ones is certainly a case of microevolution combined with migration and selection. Vavilov's theory of gene centers applies to this situation, which obviously closely parallels that of natural subspecific variation, adding only man as one of the selective agencies. The facts demonstrate that the centers of origin of the races are simultaneously the primary (Vavilov) or secondary (Schiemann) centers of the wild ancestors, and in some cases the mutants which probably started the process of domestication are also present in the wild populations. The very interesting details (see also the discussions by Turesson, 1932, and Schiemann, 1939) agree with the general conception developed here, to which, however, the facts of polyploidy and species crosses, which are a special feature of evolution in plants, have to be added. These short remarks may suffice.

We may finally return once more to the Lamarckian interpretation which has been applied to some of the cases of subspecific adaptation (Rensch, Sumner). There are certainly many cases known in which the possibility exists that a trait found typically in a definite environment, and there in different species, is nothing but a nonhereditary modification. But there can be no doubt that in the majority of cases real hereditary differences are involved, as is proved by all the experiments reported. In such cases sometimes a typical Lamarckian pitfall can be demonstrated. Grinnell (1928) mentions the behavior of birds in the peculiar climate of a region in Lower California. Here is found a "humid desert"; i.e., a region of meagre rainfall but high atmospheric humidity. In this region very different birds, like flycatchers, finches, woodpeckers, show parallel subspecies with deepened coloration, smaller bill, different proportions. It has been known for a long time (see below) that birds raised in humid conditions (Beebe, 1907) show such a type of nonheritable variation. As the adaptive value of such features is not directly visible (though hypotheses may be formed), the Lamarckian will point to the situation as in favor of his view. This is, however, not justified. We shall see later that almost all simple hereditary variations may be copied under experimental conditions (phenocopies, Goldschmidt) by nonheritable modifications. This is a consequence of definite potentialities of development, facts of great evolutionary significance outside of the Lamarckian concept, which will be discussed in a later chapter.

DD. SPECIAL ADAPTATIONS

All the facts discussed in the foregoing paragraphs then point out that such geographically varying hereditary characters as are arranged in a continuous cline will in the majority of cases have an adaptational nature, though this might not always be easy to discern, because the character in question is of a physiological nature which may or may not be visible behind the morphological features. Therefore considerable importance is attached to those cases in which the adaptational value of subspecific forms is beyond question or where subspecies are actually characterized by physiological differences—one might also say, where the subspecies is identic with the ecotype. In our paradigm we found and discussed such cases, the most convincing of which was the distribution of the different lengths of the diapause in the races of *L. dispar*. The number of such examples is, however, not large, though numerous facts such as those contained in the "rules" just discussed strongly point in this direction. Actual data closely resembling those given for *Lymantria* have been furnished by Brown (1929) for daphnids. Weismann (1876–79) had previously found that different species of daphnids (species and geographic races are not distinguished here) show a different resistance to high temperature; i.e., a definite lethal temperature. This feature parallels very closely the climatic conditions within the areas inhabited by the different forms. The common species, according to Brown, may be divided into three groups according to their distribution: widespread, southern, and northern. The first group is subdivided again into those having a summer maximum of individuals and those having spring and fall maxima. Examples of these different types were reared under closely comparable conditions at constant temperatures and the time of development from the first young instar to the first adult instar was measured, as was the temperature coefficient Q^{10} for different temperatures. It was found that the temperature coefficient for the interval 20°–30° was typically different, and some groups were distinguished: (*a*) *Moina macrocopa* and *Pseudosida bidentata* with a Q^{10} of 2 to 2.4; (*b*) *Simocephalus* and certain varieties of *D. pulex* with $Q^{10} = 1.7$ to 1.8; (*c*) *Daphnia longispina* with $Q^{10} = 1.5$; (*d*) *Daphnia magna* and *pulex* with 1.3 and 1.2, respectively. These groups correspond rather closely to the relative positions of the lethal temperatures of the same species and thus also to their climatic habitat. All these differences were found to be constant over long periods under dentical conditions. It is obvious that these results are of xactly the same type as those reported for growth rates in ur paradigm (where, however, no different temperatures vere used). To what extent some of the species are actually eographic races (the *Daphnia* species and types) cannot e determined.

Here is another case which has been analyzed experi- nentally and clearly relates to geographic races. Flanders (1931) studied the temperature relations of races of *Trichogramma minutum* (parasitic wasp). In this form, olor races exist which show a definite geographical group- ng. The differences are clear if the form is reared at a lefinite temperature, though they may disappear at other emperatures. (This point belongs to a later discussion.) Temperature experiments with these races showed that the ength of the life cycle is influenced differentially in the dif- erent races. There is actually a close parallel to the *Lymantria* case, as each race has its specific optimal tempera- ure. To mention only one paradigm: "The different stages of the yellow race varied the least in their reaction to sudden hanges in temperature except on the seventh day. . . . This may be an adaptation to a northern climate, tending o prevent emergence during the fluctuating temperatures of early spring when no hosts are available."

A case of the same type as in *Lymantria* possibly exists n *Cynips*, as Kinsey points to differences in the time of atching in different subspecies (though denying even the xistence of adaptive subspecific traits). The foregoing esults are also in agreement with the facts concerning the listribution of geographic races (frequently called species n older literature). It has been emphasized innumerable imes that certain temperatures are limiting values for the urvival and spread of certain "species." To mention only ne characteristic example: The well-known pest *Trialeu- rodes vaporariorum* (Homoptera), of tropical origin, is ound in our climate exclusively in hothouses. Innumerable xamples of this type are found in the literature on applied ntomology, and the facts have been frequently discussed in a way very similar to that followed here, though species and subspecies are usually not distinguished properly. (Discussion and experimental analysis of the last-mentioned case are found in Weber, 1931; Bodenheimer's writings contain much material; an extensive review of experimental facts is given by Uvarov (1931). Outside of applied entomology numerous data are found in Hesse-Allee, 1937, and for plants in Lundegaard, 1930. As a very instructive analysis by taxonomic and experimental methods I mention a recent paper on terrestrial Isopods by M. A. Miller (1938). All these facts taken together demonstrate that adaptation to temperature or climate or other conditions is actually accomplished by hereditary traits, which are typical of different races of a widespread rassenkreis, though there is not much material available (except *Lymantria*) in which an entire rassenkreis was tested for other types of ecological adaptation paralleling those in plants (soil, salinity, etc.). (See also Hesse-Allee, 1937.)

Some interesting data are derived from the recently discovered geographical races (see, however, below) of the mosquito *Anopheles maculipennis* (Martini, Hackett, *et al.*), races which are of importance in connection with problems of malaria prevention. They are named *elutus, labranchiae, maculipennis, messeae, atroparvus*. They differ in their association with man and animals (a physiological difference), whence their practical importance. The morphological differences are rather small; they deal with the bristles on the last segment of the larva, form of spines of the hypopygium, length of wings, color and pattern of eggs. In addition, there are differences of optimal temperature and salinity in the larval surroundings. These races might be called ecotypes, but there is certainly a geographic association also, though all the details are not clear (see below). A typical rassenkreis of the clinal type obviously is not present and also may not be expected in an animal which associates with man and beast. According to Missirolli, Hackett, and Martini (1933), the race *messeae* spreads from the North Sea to central Italy and generally prefers (i.e., in the larval stage) the soft water of lakes, and rivers, and the stagnant water of the plains. The race *maculipennis* is very adaptable and found everywhere. The race *atroparvus* is found in lowlands, along the seashore, and in swamps. The race *labranchiae* characterizes the brackish waters of the swamp area in central and southern Italy, and the islands. The race *elutus* is also a southern race inhabiting brackish water, though it lives in fresh water in Palestine. It spreads into Asia Minor and is able to withstand a high degree of salinity. The imago, according to Martini and Teubner (1933), is very sensitive to changes in temperature and moisture, as are also other related species of Diptera. Under experimental conditions these mosquitoes chose a definite optimal condition. This was different for the different races: during hibernation *messeae* chose 6°–8°, *atroparvus* 10°–13°. Both races also have a different optimal temperature for egg production. Another physiological difference is the behavior during hibernation, which is complete for *messeae* and *maculipennis*, incomplete for *labranchiae* and *atroparvus*. These few facts show that physiological characters are found as distinguishing features of ecotypes and geographical races, when these are investigated. The behavior in experiments (choice of definite temperatures and moistures) showed that the races are adapted to conditions typical for the natural environment (see also Hundertmark, 1938). It ought to be added that, according to the most recent experiments of Bates (1939), the taxonomic status of the Anopheles races has become dubious. Bates crossed the races and found all grades of fertility in different crosses, from complete sterility to more or less fertility, suggesting that actually species are involved, ecospecies instead of ecological races. (A study of the salivary chromosomes of the hybrids would be most important.)

In another group of insects a systematic study of adaptational physiological characters has been made by Krumbiegel (1932). He studied the geographic races of *Carabus nemoralis*, distinguished by small morphological characters, with regard to physiological characters which might be adaptive. He worked out the optimum temperature and found it to vary in close parallelism with the temperature of the area inhabited by the race. The reaction of the races to light varied in the same way, indicating an increasing tendency to diurnal life paralleling the increase of temperature of the area. The extremes which were found are just as extreme as those present between different species of different habits. Table 6 contains a few data; namely, average

TABLE 6

CARABUS NEMORALIS

T, average of locality	*Optimum T of local race*	*Shock T of race*
8.7° C.	26.1° C.	39.5° C.
7.8	27.1	39.8
8.4	27.9	43.1
9.3	29.3	45.1
9.2	29.4	46.8
10.3	29.4	46.4
13.7	29.7	49.4

temperature of the locality (annual mean), temperature optimum for the respective local races, and shock temperature of the same.

Finally, an example taken from the plant kingdom will be given. J. Clausen and collaborators have made an extensive study of Californian plants with all available methods, of which only preliminary reports are thus far available (Clausen, Keck, and Hiesey, 1939). Some of the species investigated have developed races or ecotypes each physiologically adjusted to different environments, as, for example, to coastal, montane, or alpine conditions. Such races can usually be distinguished by their morphology as well as by their reactions to different environments. Within each ecotype individual variations of genetic nature, which are often associated with minor habitat differences, further enrich the diversity within the species. This is, indeed, the same situation which we found in *Lymantria*. Regarding the geographical situation, it is stated that in the region of the Pacific Slope species of wide distribution are often differentiated into four major races or ecotypes corresponding to different climatic belts of this area; namely, coastal, lower montane, subalpine, alpine. In addition, a maritime and a type specific for the "Great Basin" may be found. Here, then, we have a clear coincidence of subspecies and climatic conditions, and the adaptational nature of the differential traits is patent. Clausen and collaborators state (just as we emphasized it in our discussion of other material): "Evidently the conditions in California are so varied that four to six major changes in hereditary set-up are required if a species is to occupy the entire area from west to east." In this case, also, extensive experimentation with transplants has clearly separated the modificability (norm of reaction) under new conditions from the hereditary differences, thus demonstrating the adaptational nature of the latter. In this case, also, morphological differences run parallel to the physiological, adaptive traits.

f. Genetic analysis

In our experiments on *Lymantria* the hereditary basis not only of the subspecific traits but also of the characters of still lower subgroups was demonstrated. There can be no doubt that this applies to all other rassenkreise, though sometimes also a nonheritable modification might have been erroneously introduced into taxonomic work. Many individual crosses have been made between two different geographic races, thus demonstrating their differences to be hereditary. This applies to many groups of animals and plants. But only in a few cases, as in our paradigm, has the genetics of a series of members of a rassenkreis been worked out. Therefore a special importance is to be attached to such analyses. We have repeatedly mentioned Sumner's (1915–32) work on the deer mouse *Perómyscus*, which, carried out simultaneously with my own on *Lymantria*, established the hereditary nature of the members of a rassenkreis for mammals in a way which is strictly comparable to my demonstration for insects. In both cases the subspecies were bred in a common environment for many generations, and they kept their distinguishing features. In both cases a genetic analysis of the individual differential traits was made. In *Peromyscus,* just as in *Lymantria,* a number of individual mutations of different types was found, partly in different subspecies, partly in specific localities. These mutations are inherited in a simple or nearly simple Mendelian way and do not bear any relation to the geographical cline of subspecies. We can refer to our former discussion of this phenomenon. Many species of *Peromyscus* are known, some of which live side by side, and each of which has formed subspecies over its geographic range. In *P. maniculatus,* which ranges from Labrador to Alaska in the north and to Yucatan in the south, Osgood (1909) has described thirty-five subspecies inhabiting specific areas. We pointed out before that the observer working with statistical methods may distinguish within these subspecies lower groups, down to a single local population, which differ significantly from each other but which are not named as subspecies. Sumner calls such groups simply "race from Berkeley," just as I did in *Lymantria.* I emphasize again this complete parallelism in order to show that the phenomena observed in *Lymantria* are actually typical for geographic variation. The differences between the subspecies again closely parallel the *Lymantria* data. Some subspecies are different in many different characters without overlapping and therefore may be easily recognized. In other cases the difference may be confined to fewer and overlapping characters, which require a statistical treatment. The characters which make up the subspecific differences as well as those of a still lower category are found in every organ or part which has been carefully observed or measured: namely, size of body, length of tail, feet and ears, skull, pelvis, femur; number of caudal vertebrae, width or length of dorsal stripe; area of colored portion of pelage; shade and color; proportion of numbers of different types of hair; length of hair; detailed features of the hair—like length of dark tip, character of pigment; depth of melanic pigment in different external and internal parts; also a few not very clear physiological characters like propensity to fattening.

There is another complete parallel with *Lymantria.* We saw that the individual traits making up the racial differences may vary independently over the geographic range. Therefore, in some instances there is a series of changes in different characters parallel to the geographic series of subspecies or, expressed differently, a high correlation between the seriation of the races and that of the individual traits. In other characters the direction of variation is independent of the geographic order of the races—no correlation. In some series of races which could be studied in such detail, Sumner found the same situation when comparing interracial correlation for different characters. The details are less easily demonstrable than in our table 4 because highly fluctuating characters are involved, yielding results only if coefficients of correlation are calculated. But in some cases, e.g., the previously mentioned races *polionotus, albifrons*, etc., such facts could be established (if I have correctly interpreted the rather involved data). Length of body parts showed "erratic" correlations; they varied independently between races, whereas the different pigmental characters of the pelage were correlated; i.e., varied in a parallel way.

Regarding the genetic differences between the subspecific characters only one type of inheritance seems to be represented; namely, multiple factors. As the differential features were all of a quantitative type, this is according to expectation. It is surprising, however, that a very large number of multiple factors was obviously involved in all the different traits, though the phenotypic differences were rather small. (Actually it was first assumed by Sumner that no Mendelian inheritance was discernible at all, an assumption which was corrected later.) A racial difference due to a single Mendelian gene was never observed, which again agrees with *Lymantria*, if we do not include multiple-allelomorphic series, which have not been discovered in *Peromyscus*, but are frequent in *Lymantria.*

It is of importance to note that the situation in the plant kingdom does not seem to be essentially different at this level, though it would be difficult to quote much material which has been analyzed as closely as the examples taken from the animal kingdom. The reason is partly that rassenkreis studies have not yet been as popular among plant taxonomists as among animal taxonomists, and, further, that plant material has been more favorable for the study upon the species level and therefore has been most frequently analyzed with distinct species, as will be discussed later.

As an example we may mention the rassenkreis of the snapdragon section (probably = species) *Antirrhinastrum*, which has been described as a series of different species, which however are obviously subspecies. Within this species, it seems, the situation is exactly as described for the other examples, if Baur's (1932) unfortunately rather general description is interpreted in the light of other facts.[10] There is a series of larger groups (named as species instead of subspecies) with a definite geographic range and considerable morphological and physiological differences. But, just as in all other cases described, some of the subspecies are well defined, while others are connected by intergradations. These subspecies may be arranged in groups. Within these subspecies again subgroups are found, local sibs, down to individual colonies, each rather uniform but altogether forming a series within the subspecies. In addition, special ecotypes like desert or alpine forms are found. These subspecies and sibs were bred side by side and remained constant, just as we saw in the other examples. Most of these forms are completely fertile *inter se*, and in F_2 the hybrids segregate for the typical differences. No details are available, but the short report indicates that here, as in the other cases, multiple factors with small individual effects are predominant. But a series of twenty-five multiple alleles for flower color is also mentioned.

Another excellent example has recently been analyzed by Babcock and Cave (1938), in the species *Crepis foetida* and here exact data are available. The rassenkreis in question

10. E. Baur was unfortunately prevented by his untimely death from publishing more exact data.

spreads over the whole Mediterranean area into Asia, and consists of seven subspecies which can be clearly distinguished and have been named. Again, they remained constant if bred side by side. They are fertile *inter se* and have the same chromosome complement. The differences between the subspecies are partly quantitative or morphological, partly qualitative or physiological, as table 7 indicates.

TABLE 7

DIFFERENCES BETWEEN GEOGRAPHIC RACES OF *CREPIS FOETIDA*

Columns 1–7 are the subspecies. Fvt = *vulgaris typica* (Spain); Fvg = *vulgaris glandulosa* (Sicily); Fvi = *vulgaris interrupta* (Cyprus); Fvf = *vulgaris fallax* (Syria); Fr = *rhoeadifolia* (Tiflis); Fct = *commutata typica* (Crete); Fcl = *commutata lesboa* (Lesbos); T = *Crepis Thompsonii;* E = *Crepis Erithreensis.*

From Babcock-Cave

Character	*Fvt*	*Fvg*	*Fvi*	*Fvf*	*Fr*	*Fct*	*Fcl*	*T*	*E*
Caudical leaves	P	P	P	P	P	P	P	D	d
Lobes or teeth	A	A	R	A	A	A	A	A	A
Lobes or teeth	E	D	E	D	D	P	P	E	E
Central axis	L	L	S	S	L	S	S	S	I
Glands on stem	+	+						+	+
Glands, inner bracts	+	+		+	±	+	±	+	+
Diameter open head, mm	25	32	32	37	33	37	28	21	15
Bract ratio	2	2	1.7–2	1.7–2	1.7	1.7	2	2	2.5
Style branches, mm	2.5	3	3.8	4	4.5	2.8	2.9	1.6	2
Pappus length, mm	6.5	7	4	5.5	5.5	3.5	4.5	5	5
Outer achenes, beak	S	S	S	S		S	S		L
Paleae						+	+		
Ligule color	Y	Y	Y	Y	Y	Y	Y	C	Le
Red ligule teeth								+	+
Leaf spots								+	
Life cycle	6	5.7	5	4.5	5	3.5	4.5	3	3.1
Flower-light relation	Cl	O	O	O	O	O	O	Cl	Cl
Self-compatibility	high	high		low	low			high	high

Abbreviations: A, acute; C, cream; Cl, flowerheads closing midday; D, dentata; d, shallowly toothed; E, entire; I, intermediate; L, long; Le, lemon; O, open in sunlight; P, pinnately lobed; S, short; Y, yellow.

(Two more "species" are added, to which we shall return later.)

This table shows the same rules which we noticed in the former examples, especially those regarding individual variation of the traits. It might be added that small differences in chromosome size could also be found. The genetic results may be reviewed in table 8.

It is remarkable that here an apparent monohybrid ratio was found rather frequently, though sometimes not in all crosses. This is surprising, as in all other cases analyzed simple Mendelian behavior was rather rare. But there is

TABLE 8

GENETIC RESULTS OF CROSSING SUBSPECIES OF *CREPIS FOETIDA*

From Babcock-Cave

Character	*Monohybrid*	*Dihybrid*	*Duplicate*	*Multiple*	*Modifiers*
Self-incompatibility			+	+	
Leaf shape				+	
Early caudical leaves	+				
Stem lengths				+	
Ligule color	+	+		+	+
Paleae on receptacle			+		+
Glandular hairs			+?		
Nodding heads	+				
Color of style	+				+
Achenes	+				
Anthocyanin spots	+				+

a possibility that multiple-allelic series, which were not easily detectable, may have actually been involved here. There is another feature to this case which rather separates it from our former examples. According to definition as well as to actual data, the different subspecies of a rassenkreis occupy different areas overlapping only at the borders where hybrids may be found. Some of the areas of these subspecies of *Crepis*, however, overlap very largely, and actually hybrids occur wherever two forms occur together. One might conclude from this that the distinguishing characters in this case are hardly adaptational, or only adaptive to such climatic or other differences as obtain at the extreme ends of the range, where the forms are really separated. Conditions in nature can certainly not be described according to a single nonflexible scheme.

Looking over the results of genetic analysis of the subspecies within a rassenkreis, we cannot fail to come to general conclusions regarding the genetic differences between these lowest-named taxonomic units. Actually these conclusions have been reached in an identical way and independently by all geneticists who have worked with such material: Babcock, Baur, Clausen, East, Goldschmidt, Sumner. It is clear that the standard type of mutation which is used in analytical work plays no role, or only a very limited role, in microevolution. The decisive differences, which must have arisen by mutation, are based on groups of extremely small but additive deviations, as revealed by multiple-factor or multiple-allelic differences. These differences accumulate, beginning with differences between colonies of such a minor order that they can hardly be described, though each investigator knows them, and aggregating into the easily distinguishable quantitative differences separating actual subspecies.

The genetic picture, *within* the species, then agrees with Darwin's ideas, formulated in the recent genetic era as the occurrence and accumulation of micromutations, though occasionally also a larger mutational deviation may be added to the process of diversification.

g. The evolutionary aspect

We now have to assay these facts in terms of microevolution within the species. It is at this point that taxonomists and geneticists are apt to disagree. Taxonomists are frequently inclined to assume that the close parallelism between external conditions (climate) and subspecific differentiation can be understood only if these genetic differences are produced by the respective environmental factors. Rensch, for example, points out that mutation must go in all directions and that, therefore, an orderly seriation of types cannot be produced by such mutations. He thinks that the existence of series of parallel variation (Bergmann's rule, etc.) could not be understood on the basis of selection of chance mutations. He further thinks that selection for minimal differences, in size, for example, is out of the question. We have already seen that Sumner inclines, though with considerable reserve, toward such a standpoint. The geneticists, on the other hand, are convinced that small deviations by micromutation are present in each population, thus forming a basis for eventual selective action by such selective agencies as climate. This selection might act in two different ways: either by destroying through a secular change of external conditions such mutant combinations in the population as are not preadapted to the new conditions; thus the preadapted members of the population, i.e., those fitted by chance hereditary mutant combinations for life under changed conditions, will survive. Or there is another way of action: the preadapted types of a population can migrate into a new environmental niche and reproduce there. In any case, a certain amount of isolation is the decisive factor.

The question then arises as to whether the difficulties which the taxonomist finds actually exist. (I shall not stress the well-known fact that heritable effects of the environment with a purposive response of the germ plasm to environment have never been proven and are considered as actually impossible on the basis of our present genetic knowledge.) As there is no doubt about the presence of innumerable micromutations within a population, the problem narrows down to the problem of gradated series of characters, their parallel variation in different forms, and the selective value of the smallest steps. Let us consider the last point first. We have repeatedly emphasized that many of the visible traits which vary subspecifically in orderly series paralleling an environmental cline are not adaptational in themselves but the visible results of physiological differences, which are adaptational. As an example we pointed to size in *Lymantria*, which is not adaptational but obviously a consequence of the highly adaptational velocity of larval differentiation. Physiological traits of this type may have an extraordinary selective value even for very small differences. Selection works not only through the average differences, say, in temperature, but even more so through the extreme differences. For example, a race of moths whose hereditary hatching time or feeding time does not completely coincide with environmental conditions may survive in an environment to which it is not well adapted in this respect, for many generations of average years, but is immediately wiped out by one extreme year. These adjustments are obviously very exact and attuned to very small differences. Biological literature is full of data showing how after extreme winters forms which had migrated into a region without belonging there by preadaptation have disappeared. In this connection I have repeatedly pointed in former papers to the following facts. *Lymantria dispar* has been introduced into the United States and has settled there and multiplied immensely. The same species has been repeatedly introduced into England but has never established itself. We do not know the reason, but we do not doubt that the lack of some preadaptational trait is responsible, possibly one concerned with the mild and moist climate of the British Isles, which does not provide the low temperature required during the diapause. (The claim that the species was formerly indigenous to England but died out later is not well substantiated.)

In this connection we may also mention the following facts. In southwestern Japan definite races of the gypsy moth occur, with the same population density as is found in other Japanese races, judging from my experience in collecting. When bred under laboratory conditions, in Japan as well as in Europe, these races invariably perish from infectious diseases, usually in the first generation. They are obviously adapted to a very narrow range of environmental conditions.

Further data in our work with *Lymantria* show in a more detailed way how small such limiting climatic features may be. *L. dispar* never spread north into regions with very short summers, the minimum length of the vegetative period and the maximum speed of larval development being the limiting factors. In Europe the species does not go farther north than Stockholm, and in Russia it is absent from the northern provinces. In eastern Asia the northernmost range is Hokkaido with five summer months, whose average temperature is above 10°C. In Sakhalin only four such months exist, and the species does not occur. At the southern limit, the limiting features are extreme summer heat and warm winters. The normal life cycle of the species requires cooling off of the eggs in winter. They may be hatched after artificial cooling for a short time. If kept without cooling they will hatch only after long delay, and then in small numbers and in much weakened condition. This hereditary physiological condition excludes a spread farther south. Actually the form occurs in southernmost Japan but does not spread into the adjacent Ryukyu Islands, where the coldest winter month has an average temperature of 14° C.

It is possible to confront the facts just analyzed with rather similar ones studied from a Lamarckian point of view. J. W. H. Harrison (1920), one of the few geneticists who entertained (at least formerly) Lamarckian concepts, made a very remarkable study of the differentiation of the geometrid moth *Oporabia autumnata* into subspecies and still lower units. He recognized only one subspecies, *filigrammaria*, which differs from *autumnata* in many individual characters. They all remain constant if the moths are bred under identical conditions. The chromosome number is also different in this case (38 and 37 = n). Both these subspecies (they ought to be written *O. a. autumnata* and *O. a. filigrammaria*) form many local races in a rather parallel fashion. The history of this subspecific differentiation is reported in the following way. It is assumed that the original species fed on birch in Mid-Tertiary and was at home in the circumboreal continent. During the glacial period the species was separated into two groups, one in southeastern Europe and another one west to the present British Isles. The latter became the present subspecies *filigrammaria*. In the course of this development it became adapted to a new food plant, a low-growing evergreen shrub, *Calluna*, growing in the short glacial summers. This led to a condition which closely parallels the one analyzed in *Lymantria* "Aestivating as it did in the pupal condition, and its emergence as imago in later summer being correlated with the development of the pupae in response to an exposure to a period of diminishing daily temperatures, it is clear that the individuals derived from early-hatching ova and from more rapidly feeding larvae would tend to be exposed earliest to this optimum state of falling temperature, and thus would emerge earlier. On the contrary, individuals exhibiting retarding development would tend to emerge so late that the rapid appearance of early winters would destroy them before they secure the perpetuation of their species. In this fashion, by natural selection, a race provided with early-hatching ova and pupae would be built up. *Calluna* not being a deciduous shrub would by that very fact favor early emergence and assist this development" (Harrison). Though Harrison realizes that the structural differences of the subspecies must have arisen by mutation, he nevertheless prefers a Lamarckian explanation for the evolution just described. It is not necessary to discuss this any further, as a more probable explanation has already been given. I want to add only one point, which is sometimes not clearly recognized in discussions of preadaptations and their migration into empty niches of environment. If the whole environment changes by catastrophic or secular processes (Ice Age, cooling of climate), the new condition amounts to a new niche. Preadapted groups of individuals within a variable population survive and alone populate the zone with new environmental conditions. Preadaptation then covers horizontal as well as vertical occupation of new niches, present horizontally (in space) or produced vertically (in time).

These facts and their interpretation are also in harmony with the facts of plant ecology. Turesson (1932) has pointed out over and over again that preadaptation based on manifold mutant combinations within a population is responsible for the existence of well-adapted ecotypes in definite localities. He speaks directly of "the plant species as an indicator of climate" and gives numerous examples of the limits of distribution of Scandinavian plants, which bear out this conception. Additional facts may be derived from plant breeding, for instance, wheat, where races adapted to life in new climates are selected; i.e., combinations of genetic characters preadapting the race to new environments are sought for and experimentally recombined. The reciprocal of these facts is found in those cases in which one species can form subspecies adapted to extreme conditions (Arctic, alpine) and another is unable to do so. For example, a subspecies in Scandinavia has been able to produce an alpine form, but the subalpine species in central Europe has not succeeded in so doing. Obviously, the preadaptational mutant combinations were present in one subspecies and absent in the other. Many examples of this kind are found in Turesson's papers. J. Clausen's work, leading to the same conclusions, has been reported above.

If these points are granted—and I think that there has been unanimity in this regard among geneticists since Cuénot and Davenport first formulated the conception of preadaptation—the existence of graded series of races parallel to a climatic cline is easily explained without recourse to Lamarckian doctrine. Given a selection of race A from a population containing the necessary micromutations and preadapted to the environment α, and similarly B to β, etc., it follows that an arrangement of α, β, etc., in a cline must be accompanied by a parallel cline, A, B., etc. From the point of view of the geneticist there is no difficulty whatsoever in explaining such situations.

The basis of the foregoing discussion is the assumption that the characters of geographic races, especially those arranged in clines, are usually adaptational. The authors who have done experimental work with such rassenkreise, as well as many taxonomists, agree on this point. I need mention only Rensch, J. Clausen, Turesson, Goldschmidt, who have emphasized this point in their writings. But this does not mean that all subspecific varieties are adaptational or that a subspecies might not be formed without adaptation being involved. The latter might be expected when subspecies are considerably isolated. In this case major differences have arisen in time, independent of adaptation, and have been preserved by isolation and lack of negative selection value. When such isolated types, none being members of a cline, are found, it is always useful to inquire into the geological history of the area. It will then frequently be found that an early isolation explains the condition. I studied such a case connected with the sudden interruption of the typical cline of Japanese races of *L. dispar* at the Strait of Tsugaru separating northern Japan from Hokkaido (see Goldschmidt, 1932a). The geological history of the region showed that Hokkaido was separated from the present Japanese Isles before these were torn from the mainland. Similar situations were discussed by Anderson for subspecies of *Iris* in connection with the glacial period, and by Babcock for *Crepis* on Mediterranean islands.

A third point which to taxonomists appears difficult to understand without direct environmental action is the existence of parallel clinal variation of the same type in different species. We discussed these facts before as the rules of Bergmann and others. It is obvious that if a physiological adaptive trait underlies the visible neutral traits like size and pigmentation, it will be frequently expressed in the same visible trait, which therefore appears in similar clines in different species. There is no need for directed mutation due to environmental influence to explain such facts, which fit most naturally into the general picture which we have drawn.

All these discussions lead to one general conclusion, which I have repeatedly drawn in former work and which will later turn out to be of decisive importance. The formation of a rassenkreis of subspecies (including the still lower categories) is the method by which a species adapts itself to different local conditions within the area which it is able to inhabit. This adaptation, strictly within the limits of the species, is produced by micromutation in different directions, involving all known types of Mendelian inheritance of manifold morphological and physiological traits. Selection of preadaptational combinations accounts for everything else.

5. LIMITING FEATURES OF SUBSPECIFIC VARIATION

We used in the last paragraph the phrase, "strictly within the limits of the species," indicating that nothing in the facts thus far studied leads beyond the confines of the Linnean species. This limitation, which later will become the basis of further conclusions regarding microevolution, requires that special attention be directed toward such features of the rassenkreis as might be considered to be leading toward or even beyond the confines of the Linnean species. Some of these features will be considered now as a special group of facts found in rassenkreis studies.

Geographic variation may be of a checkerboard type; i. e., the subspecies spread in all directions over a considerable area which is then covered by a mosaiclike grouping of these subspecies. They do not form a continuous series, though such might be present in individual stretches of the whole area. This type, which seems to be possible only in large continental expanses like North America and Eurasia, is apparently less frequent than the gradated type of continuous clines. Sumner describes the *Peromyscus* material (on the basis of Osgood's taxonomic work) as generally of the mosaic type, though Osgood as well as Sumner indicates additional typical clines. It is conceivable that the type of distribution in question is only an apparent one, which would be dissolved into a more complicated pattern of clines if studied from this viewpoint. (We exclude from this statement certain peculiarities of insular faunae, which will be discussed soon.) One might draw this conclusion from the very elaborate data of Kinsey for the genus *Cynips*. Here an apparently irregular distribution over the area of North America has been dissolved into a complicated branching and rebranching system of clines of different size, value, and direction (see below, p. 161).

As we mentioned before, the type of arrangement of geographic races which we just called a checkerboard type (see *Peromyscus*) has recently been given a different in-

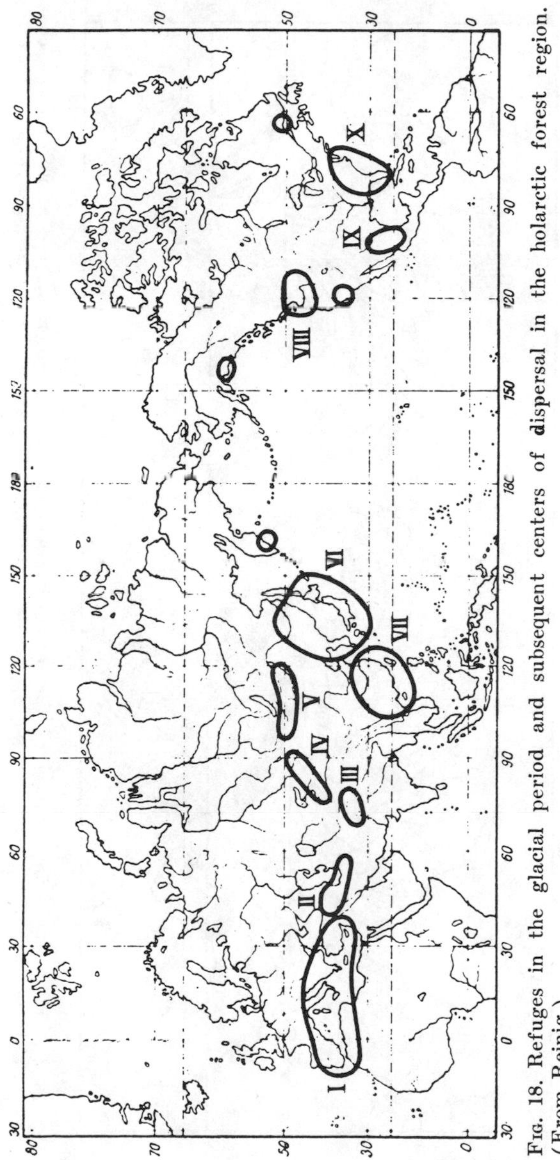

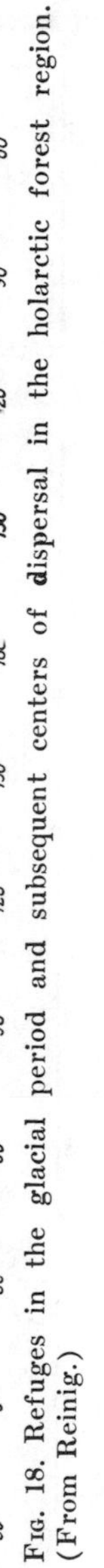
Fig. 18. Refuges in the glacial period and subsequent centers of dispersal in the holarctic forest region. (From Reinig.)

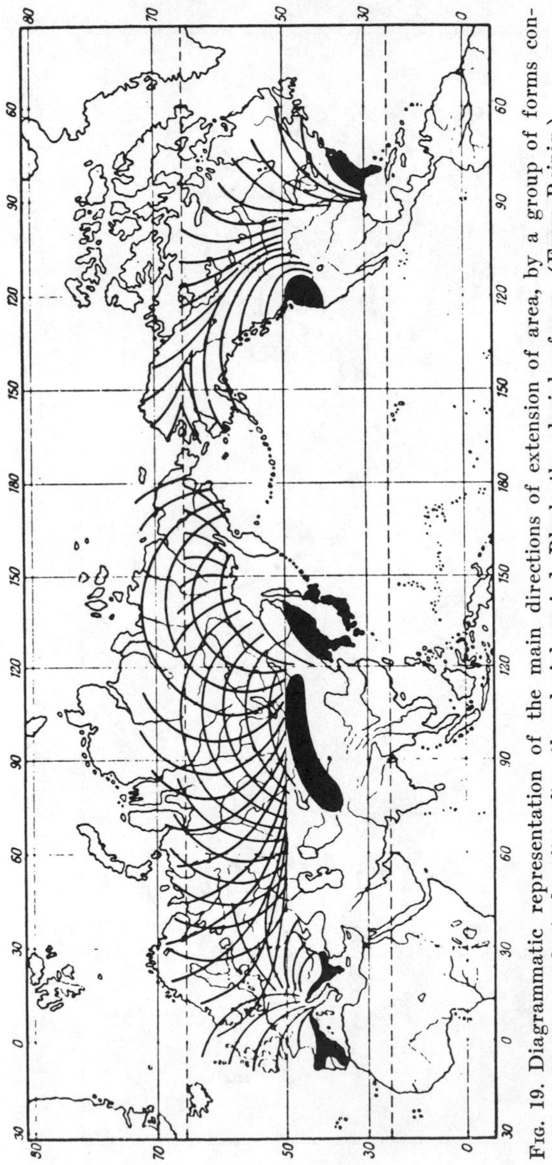

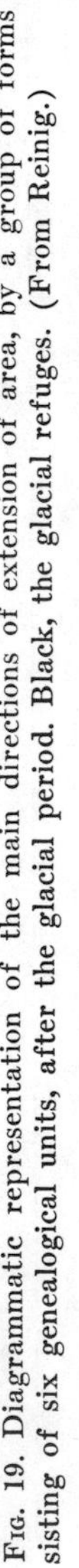

FIG. 19. Diagrammatic representation of the main directions of extension of area, by a group of forms consisting of six genealogical units, after the glacial period. Black, the glacial refuges. (From Reinig.)

terpretation. Reinig (*loc. cit.*) has come to the conclusion that in such cases of continental subspecies formation a series of centers of dispersal can be recognized which coincide with faunal refuges existing during the Ice Age. We have already reported and criticized the conclusions drawn from the facts observed in connection with this viewpoint. For the problem under discussion the facts which are represented in figure 18 (after Reinig) are of interest. This figure shows the centers of species accumulation in the forest habitats of the holarctic region; ten major and a few minor centers are distinguished, which are said to coincide with the glacial refuges. Figure 19 represents Reinig's view as to how clines of subspecies are spreading from these centers. If this is the case, as it seems to be in some forms, clines will frequently intersect and very different forms may be adjacent to each other, where clines from different centers meet. (We shall return to the facts later when discussing so-called artenkreise.) This type of subspecific distribution seems to be represented in the swallowtail *Papilio machaon*, according to Eller (1939). The group (subgenus) consists of four distinct species, one of which, *machaon*, has an immense holarctic distribution and has formed numerous (over sixty) distinguishable subspecies with specific habitats. A few of these are pictured in figure 20. Twelve of these races are said to be confined to the glacial refuges whence clines of secondary races are spreading in different directions, corresponding to the views of Reinig (fig. 21). There is no reason to assume that such a distribution, caused by geological features, follows any other laws than do the simple clinal distributions discussed before. (It might be interesting to consider Kinsey's data on *Cynips*, with their complicated system of branching clines, in the light of Reinig's conception.)

From the standpoint of microevolution all the types of subspecific arrangement described are of the same significance. Certainly the more frequently found type of geographic variation is the one in which the subspecies are arranged in a continuous cline or a branching combination

Fig. 20. Races of *Papilio machaon* L. from Eurasia and North America. Left vertical row: *asterias,* North America; *taliensis,* Yunnan; *aliascus,* Alaska; *kamdschadalus,* Kamtschatka. Right row: *hippocrates,* Japan; *oregonius,* British Columbia; *stabilis,* Costa Rica; *siccimensis,* Sikkim, India. (After Eller, from Pagast.)

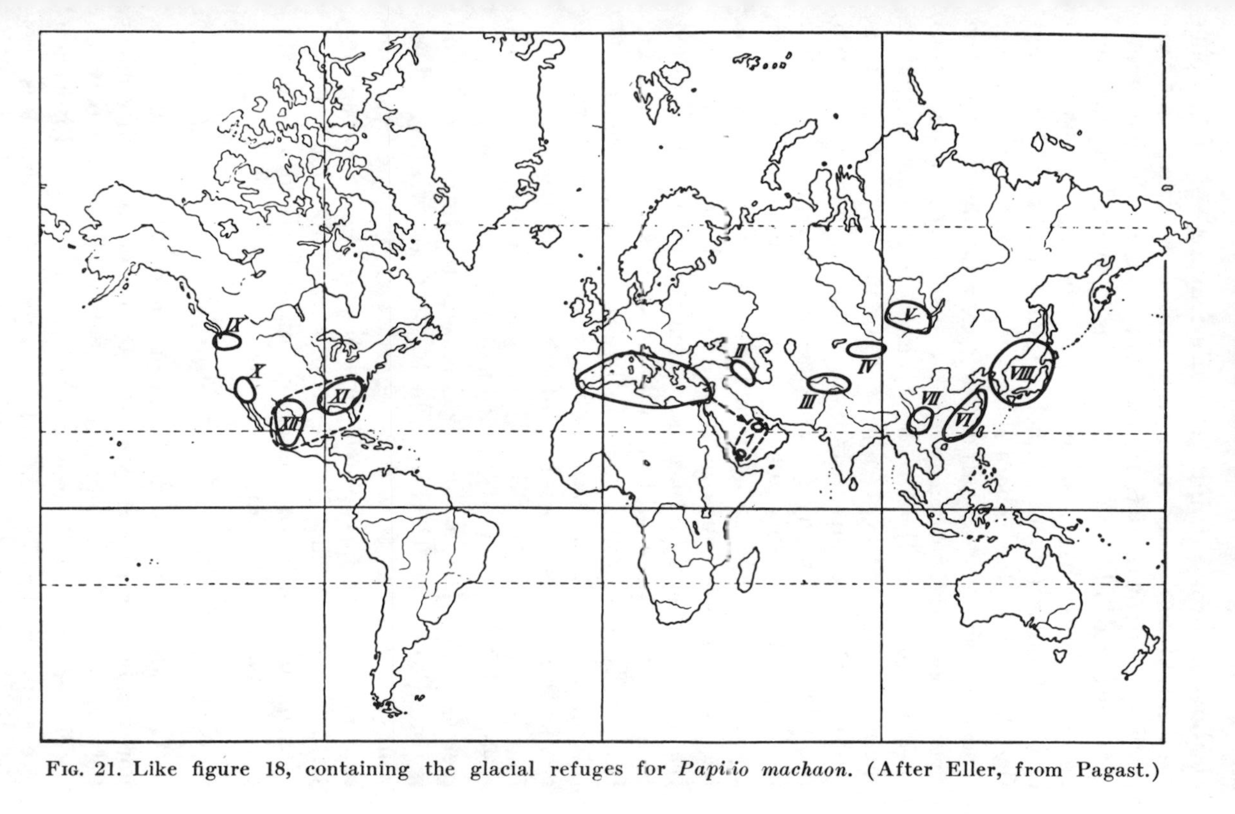

Fig. 21. Like figure 18, containing the glacial refuges for *Papi.io machaon.* (After Eller, from Pagast.)

of clines following definite climatic features, which in some cases might be decisive in a very indirect way; e.g., by controlling the growth of food plants of animals. (The presence of specialized additional ecotypes is neglected in this connection.) In such cases, then, a rassenkreis will have two extreme points (if not branched), and it is to be expected that the two end-members show the extremes of difference. The question arises whether such extreme members of the rassenkreis may be considered in some cases as nearer to specific difference than to subspecific, thus indicating that these extreme members represent incipient species. This is actually the opinion of the majority of taxonomists and probably also of many geneticists who come in contact with these problems. This idea—an application of the strict Darwinian concept to modern taxonomic conceptions—has been especially emphasized by Rensch, who claims that extreme members of a rassenkreis may not be distinguished from species, and would actually be species if by migrating backwards they would come to live in the same area as the intermediate subspecies. Kinsey goes still farther when he claims that the end-members of such a cline are so different that they might be called genera according to ordinary taxonomic standards. The student who reads these statements without personal experience with the form in question is bound to be skeptical when he learns that a feature (short wings) which was formerly regarded as of generic significance occurs as a seasonal variation within a subspecies. Before we can discuss the merits of such conclusions we must be acquainted with such facts as might be favorable to the taxonomists' contention.

A. Fertility

The oldest conception of species, and the one which has not yet been replaced, states that the decisive differences between species of animals are sterility of hybrids between species and the more or less complete physiological isolation preventing hybridization. (In plants the situation is less simple, though J. Clausen adheres strictly to the fertility test for the concept of a good species.) This definition is also involved in the modern taxonomic conceptions, which consider as species forms inhabiting the same area without interbreeding. The development of intersterility is, therefore, to be regarded as a decisive step in the isolation of species. We do not intend to discuss here the general problem of hybrid sterility. A very clear discussion from the genetical viewpoint can be found in a paper by Stern (1936). We shall discuss only such facts as appear to be pertinent to the problem of incipient species. The members of a rassenkreis are by definition fertile *inter se*, and wherever genetic tests have been made, even the most extreme members of a series are completely fertile *inter se* (see the material studied by Babcock, Clausen, Goldschmidt, Sumner). But there are certain facts which indicate that a transitional state may be encountered and these facts are, naturally, used most prominently by defenders of the idea of continuity between subspecies and species. These facts may be grouped into the following categories: (*a*) lowered fertility between subspecies; (*b*) noninterbreeding subspecies within the same area; (*c*) impaired interbreeding due to differences in morphological or physiological details; (*d*) impaired fertility due to chromosomal differences.

a. Lowered fertility between subspecies

In nature only neighboring subspecies will interbreed in the zone of contact. In experimental analysis all types may be tested and, as mentioned before, are fertile *inter se*. This situation is found to occur wherever it has been genetically tested (*Lymantria, Crepis*) and it has also been assumed for many cases not tested by genetic experimentation. But there are a few facts available which may be interpreted as a kind of incipient intersterility. We may recall that the sex races in *L. dispar* differed as to the potency of the sex genes. These differences were discovered when crosses between different races led to sexually abnormal individuals. The hybrid combinations resulted in an upset of the proper balance of the sex-determining factors, with the result that the hybrids in certain combinations became intersexual (details in Goldschmidt, 1932a). As all the higher grades of intersexes are sterile, we find a subspecific difference resulting in sterile hybrids. As a rule, this happens between races otherwise rather far distant from each other; e.g., central Japanese versus Eurasian races. The question is whether this type of sterility is comparable to typical interspecific sterility. In *L. dispar* the crosses which give intersexual offspring do this in one direction only. The reciprocal crosses, however, are completely fertile. In addition, mutations which result in intersexuality without racial crossing occur within a single race (see Goldschmidt, 1934, for details and literature). This looks as if this type of sterility were on a very different level and not comparable to specific sterility. The other case in animals in which a comparable series of sex races exists, the frogs (Pflüger, R. Hertwig, Witschi, [*loc. cit.*]) lends itself to the same conclusion. Here racial crosses produce changes in the sexual balance which, however, lead in the end to normal and fertile individuals, the expression of the unbalance being only an embryological feature.

There is another case reported by Standfuss but not analyzed. He found that two geographic races of the hawk moth *Smerinthus populi* produced sexually abnormal offspring (intersexual?) after crossing. The only comparable case known in plants (Goldschmidt's interpretation of Oehlkers' work, 1938a) relates to different "species" of the South African form *Streptocarpus*. The crosses were probably made with subspecies of a rassenkreis, but no information on this point is available.

On the other hand, it cannot be denied that an upsetting of the sexual balance might also be produced in crosses of species. But wherever such a situation could be analyzed it was found to be the result of a disturbance of the whole chromosome mechanism (e.g., triploid intersexes in species backcrosses of Lepidoptera). Another case in fishes will be discussed in a later chapter. We conclude, then, that partial sterility in interracial crosses due to a disturbance of the sexual balance; i.e., to a considerable quantitative difference of certain physiological processes, may not be considered as a genuine type of hybrid sterility. Wherever it may be found within an otherwise homogeneous race as well as between subspecies and species, it appears to be a special physiological feature which can hardly be regarded as typical for the direction in which specific intersterility is established.

b. Noninterbreeding subspecies within the same area

Considerable importance has been attached by taxonomists to a phenomenon which might be termed the convergence of a rassenkreis. Either by way of migration or as a result of geological events, the end-points of a rassenkreis may secondarily converge, and as a result the two most distant members of a cline will be brought into the same area together. Here are a few examples from different groups of animals. There is first an example in our gypsy-moth material of how such a situation may come to pass. The present families of Lepidoptera go back to the Tertiary, the same period in which the Japanese islands were formed. The present climatic conditions also were initiated toward the end of the Tertiary. *L. dispar*, which must be derived from the mainly tropical forms of this family, therefore has produced the climatic races only since that time. According to Arldt (1910), the coast of the Asiatic continent was found in the Miocene east of present-day Japan. Toward the end of this period Sakhalin was broken off and the northern part of the Japanese Sea formed. In the Lower Pliocene, Hokkaido was separated from Sakhalin, and in Middle Pliocene from Japan, by the formation of the Tsugaru Strait. Only in the Upper Pliocene was Japan separated from Korea by the Tsushima Strait, and only in the Lower Diluvium was the land bridge connecting Japan with southern China severed. Though we are completely ignorant as to whether the original form of *Lymantria* was already present when all this happened, or migrated there from the south, two things are clear: first, that the Hokkaido form, originally continuous with the Asiatic continent, was first separated from the rest of Japan by the Tsugaru Strait; second, that the series of races stretching from Korea through all Japan to northernmost Japan must have differentiated *in situ* or else migrated into their present area when Japan plus Korea was a long tongue-like peninsula attached to the eastern Asiatic continent. Whatever the details may have been, here a situation obtains which makes it possible for a racial chain to be bent into a circle and for the closing of the ring—the circle Hokkaido—Amur—Manchuria—Korea—Kyushiu—Honshiu closing from both sides at the Strait of Tsugaru—to bring the two most extreme races near together. In this case they cannot enter the same area, as the intervening Tsugaru Strait prevents this. If they could meet, however, they would still be fertile *inter se* (races Hokkaido and northernmost Japan) as the experiments show, though they represent actually the most extreme sex races. Another question is whether they would interbreed in nature. Failure to do so is not necessarily identical with intersterility.

The first example of a convergent rassenkreis (or part of one) to be given is mentioned by Osgood (1909) for *Peromyscus*. The subspecies *arcticus* and *algidus* of *P. maniculatus* occur together in the Upper Yukon Valley (Alaska) and "apparently maintain themselves distinct." *Arcticus* ranges southward and eastward and intergrades with *oreas; algidus* follows the coast, being contiguous to *hylaeus*, which is followed by *macrorhinus*, which again intergrades with *oreas*. In the detailed description, however, the phrase is found: "About the upper waters of the Lewes River, of the Yukon drainage, *arcticus* is found in company with *algidus*, and apparently distinct from it, though elsewhere the two are connected." This can hardly be called a convincing statement for the author's claim, repeated in subsequent literature, that a natural event which would remove the intermediates between the two subspecies and the form *oreas* would leave two good species living in the same area.

An entomological example of the same type is the following described by Forbes (1928). The nymphalid butterfly *Junonia lavinia* contains three well-marked series of geographical forms, the North, Central, and South American

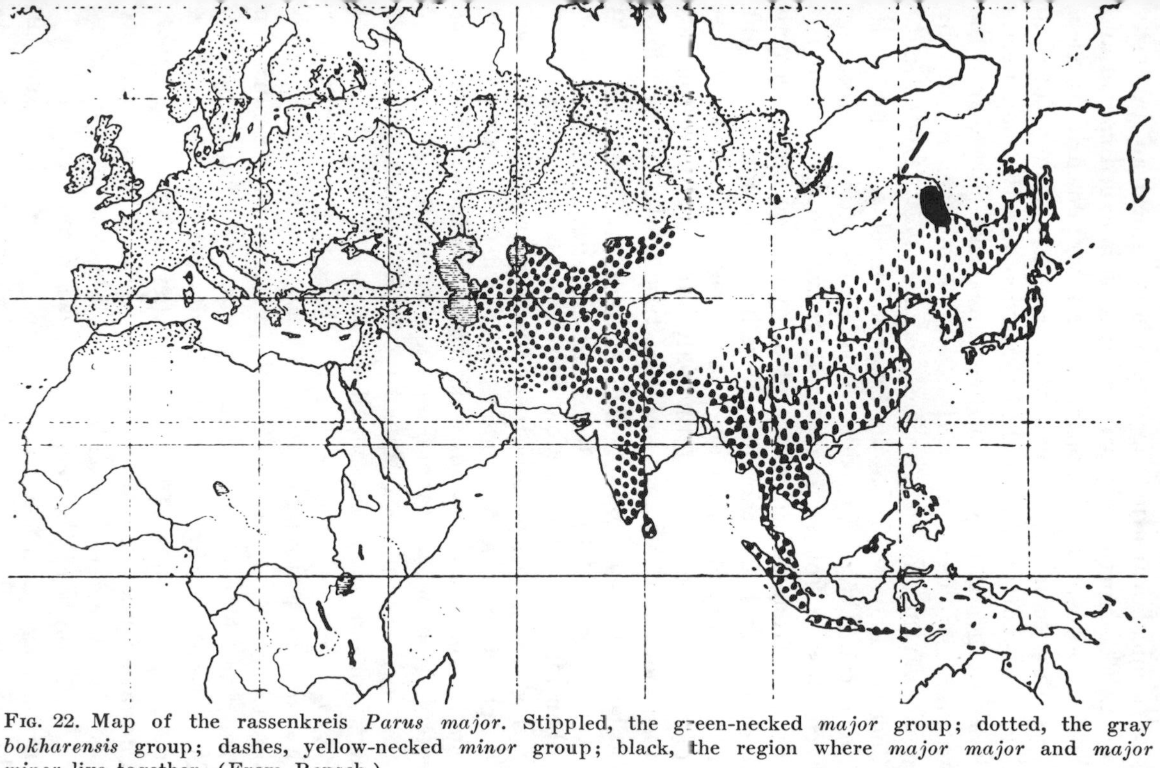

FIG. 22. Map of the rassenkreis *Parus major*. Stippled, the green-necked *major* group; dotted, the gray *bokharensis* group; dashes, yellow-necked *minor* group; black, the region where *major major* and *major minor* live together. (From Rensch.)

types. According to Forbes, this is the most striking case of geographical variation in the whole American butterfly fauna. The North American subspecies is called *coenia* and shows subsubspecific variation in its range down to Mexico. Here the second subspecies *zonalis* starts, going down to Peru and northern Brazil. In Mexico the two subspecies show intergradations. Both these subspecies spread into the Antilles, and, according to Forbes, *coenia* reached Cuba by coming from North America, and *zonalis* from South America. Thus in Cuba the two ends meet again and both forms are said to live side by side without interbreeding. According to Forbes, they "have become so distinct on the way that from the Cuban point of view they would seem to be real species. In other words, we have here species caught in the very act of formation." If we look coolly at the facts, however, we see that intermediate forms between the two subspecies are not found in Cuba, though they occur where the latter meet in Mexico. Unfortunately, we are not told more. Do these two forms actually fly together in definite Cuban localities? Is their biological and ecological behavior similar enough to permit them to interbreed? Would they be able to mate and produce fertile offspring if brought together, as I strongly suspect? Again, I cannot see that the sweeping conclusions mentioned are warranted.

A third example is frequently quoted in European literature (for example, Rensch, Stresemann), to wit, the case of *Parus major* (a titmouse). This little bird has formed an Eurasian rassenkreis, *major*, *bokharensis*, and *minor*, as indicated in the map (fig. 22). *Major* spreads as a northern form from Europe across Siberia into northern East Asia. In the south *major* gradates into the Persian *bokharensis*, and this, in turn, in India and southern China, into *minor*. The latter reaches north through China into Amur. Here it meets an eastern branch of the Siberian *major* and again two end-points of the series are shown converging. According to Stegmann, these two subspecies live side by side in this region without hybridizing. The case closely resembles the one in *Junonia* and the question marks are the same. We do not know what keeps them apart; but it might be a very small physiological or biological difference of the same order as individual micromutational differences, whether geographical or local. But this is not what real species differences consist of. As long as the impossibility of mating or the production of sterile hybrids is not proven, I cannot see in these cases more than an interesting ecological feature. Rensch, by the way, makes the most of this example in order to prove that this is the way in which species are primarily formed, but he remarks in another place (1933, p. 338): "If in the Amur-Region the green necked . . . *Parus major* (*major*) behaves like a species towards the *P. major (minor)* living in the same area, it is obviously due only to the small difference in size, color, and voice. But it is very probable that both forms could be bred in captivity. . . ."

It might be added that among animals cases are known in which races inhabiting a common range do not interbreed for some ecological reason; e.g., the *Anopheles* races, as mentioned above. In plants the same situation is known to be due to simple mutational differences of a sterility factor (Melchers, 1939). In none of these cases is a convergence of a racial cline present which would warrant such conclusions as those of Rensch.

At this point of our discussion there ought to be mentioned a rather informative example which is rarely discussed in the light of our problem; namely, the subspecific differentiation of the human race. Though it is possible that different species of the genus *Homo* have existed and have disappeared again, nobody can fairly claim that present mankind belongs to more than one species. Let us suppose that a giant collector from Mars visited the earth, made a collection of human beings, and returned to work them up in his Martian museum. He would most certainly come to the conclusion, in applying usual taxonomic standards, that he had found a new family, Hominidae, and within this a number of very distinct genera, like the white, the black, the brown, the yellow man. Within these genera he would distinguish species or ecospecies, replacing each other geographically. For example, he would identify in the black genus the species *Bantu*, *Bushman*, *Hottentot*, *Pygmy*, *Australian*. Within some species with a rather large geographical range he would find geographical races; e.g., the different tribes of Negroes across the center of the African continent. If the collection were large enough he would meet with isolated subspecies, with very different insular forms, with subsubspecies down to small hordes, with differential specific traits. (Regarding the latter point, not generally known, I think, I might mention a personal observation among the semisavage head-hunting tribes of Formosa [1927a]. I noticed that in two different small tribes of this Malaylike group the men within the tribe resembled each other to such an extent in certain features of the face that they might have been picked out of a crowd as brothers. The genetical basis, homozygosity by inbreeding, is obvious.) In short, his description would closely compare with innumerable other taxonomical studies, and it would also be perfectly correct, as far as information goes. But the next collector might have better chances to observe his specimens and he would find difficulties. He might reach the same conclusion as have recent students of insular faunas (Galapagos finches, Hawaiian drepanids) (see below), that from a taxonomic point of view all the forms might also be assigned to a single species, though the morphological and ecological differences between Negrito and Swede, Papuan and Eskimo, Hottentot and Chinese are quantitatively just as large as are those between different so-called genera; e.g., of gall wasps. The next Martian visitor might be a geneticist who would notice that all these forms, if given a chance, interbreed and produce fertile offspring. He would notice that this also applies to cases in which differences in the structure of the genitals exist (the Hottentot–Boer hybrids), and he would state with perfect confidence that only a single species, with many sub- and subsubspecies, exists. Now, there can be no doubt that many of the isolated human subspecies or end-members of a series are as different from each other as are extreme subspecies in animals. There is no doubt that some subspecies, like those in animals, have been isolated for a very long time. There is no doubt that the time available for subspecific differentiation has been about the same as that which is assumed for the cases in animals and plants. There may also be detected at some points the presence, due to migration, of two races which are interfertile but which do not produce hybrids on account of psychological isolation. Such an occurrence would be a special feature without any evolutionary significance. We conclude, then, that if the subspecies is an incipient species, this must also be the case for the major human races. I wonder whether anybody would be willing to accept such a conclusion!

I should like to add a few more points to this example, although they will anticipate a later discussion. In an oral discussion of the problem which I recently had with one of the leading experts, F. Weidenreich, he informed me that the recent discoveries, both for Pithecanthropus and Sinanthropus, indicate that at the early level of human evolution represented by these forms the subdivision into the main future races had become visible. If this is the case, it follows that the human species first subdivided into geographic subspecies, and that the following evolution—actually the major part of human evolution after the first separation of man from apes—occurred within these subspecies by the formation of subsubspecies, etc. by mutation, selection, and hybridization. If we view the doings of our first Martian taxonomist in this light he acted just as numerous modern taxonomists do: he describes man as what we shall soon discuss as a genus geographicum, with many species in separate habitats, and these again subdivided into subspecies, etc. Here we know that the Martian taxonomist is wrong and that he has studied only a single species. We shall later extend this conclusion to comparable cases. (Meanwhile Weidenreich has published his views, 1939.)

c. Impaired interbreeding on a morphological basis

The foregoing facts demonstrate that interbreeding between different geographic races may be impaired in special cases, though there is no reason to assume that an actual isolation through impossibility to produce fertile offspring or any offspring at all was present in these cases. But it has been claimed that actual differences may occur which make interbreeding physiologically impossible. Rensch mentions extreme size differences in races of beetles which would make interbreeding impossible if these races should meet in the same area. I am inclined to be rather skeptical as regards this argument. (Think of the Pekinese and the St. Bernard dog; is each of these an incipient species? Dachshund and St. Bernard have been crossed.) For many years I tried in vain to cross larger northern Japanese females of *L. dispar* with small Hokkaido males, though the reciprocal cross was easy to perform. But later the same cross succeeded easily, for unknown reasons. Impossibility of mating, then, cannot be argued but must be demonstrated. More interesting are the cases where the genital armature is involved. (See also the Hottentot–Boer case, above, and Baelz's description of differences between Japanese and white women, which, however, do not prevent normal fertility.) It is known that in many insects differences in the structure of the genital armature are very characteristic features distinguishing different species (see below). Such differences, however, also occur between geographic races (Jordan, 1905, 1927; Drosihn, 1933; Franz, 1929). Jordan, who did very extensive work in this field on Lepidoptera, came to the conclusion that geographic races may be identical with regard to genital armature but different in other characters, while other races may differ both in somatic traits and in genital armature. In rare cases only the genital armature may be different. In a special case studied about half of the subspecies showed differences in the genital armature. There is, of course, no reason why this character should not vary between the geographic races, as any other trait does. This variation would be of special significance only if the differences found among species were such that copulation would be made impossible between different species. Only then might the racial differences be considered a step toward specific ones. But we know that some species have quite different genital armature, whereas others show small differences or none at all. In addition, the differences do not prevent successful mating, as the innumerable species hybrids produced in Lepidoptera (and even some considered to be generic) demonstrate. There is no fact available to indicate that the differences in genital armature found in geographic races are such that physiological isolation is affected, or even would be affected by further variation in the same direction. In discussions of this subject, it is frequently forgotten that actually the differences do not involve any major features but are of a more or less ornamental type, with a few teeth or spines or processes here and there but not at physiologically decisive points. Frequently, according to Jordan, the armature of the races is identical for females but different for males, which of course shows that no physiological importance in the direction of sexual segregation can be attached to this variation. There is even a case mentioned by Jordan in which two seasonal forms (modifications) of the same form differ in genital armature. It must be emphasized (as many evolutionists who discuss these problems have never dissected such an armature) that the racial differences in genital armature involve exclusively organs which do not behave like key and hole in the two sexes. Copulation in Lepidoptera proceeds, as one is apt to forget, by the insertion of the penis into the bursa and not into the female sex-aperture. The rest of the genital armature serves only as claspers, or, in the female, is concerned with egg laying, not with copulation. These decisive facts are generally overlooked when this material is presented as a demonstration of transitions toward sexual isolation.

A corresponding situation found in beetles (Jeannel, Franz) does not lend itself to different conclusions. Thus Franz emphasizes that extreme differences in the form of the tip of the penis are found among the males of species of the beetle *Orinocarabus*. But the females do not vary correspondingly, which excludes any importance of this variation for the problem of isolation of species. Between different subspecies, also, definite differences are found, and here a very important point has been noted by Franz. For example, there are typical quantitative differences present in the so-called vaginal apophyses of the females. The males of the same races also differ in the structure of the preputial sac. But actually these two parts do not act like key and hole in the act of copulation, and their variation does not seem to be of any physiological importance. Franz discusses the possibility of both features being simply a consequence of different degrees of chitinization, heritable in the respective races. It is needless to say that here, as in Lepidoptera, species may frequently cross, thus demonstrating how little even the larger differences in genital armature between species mean physiologically. Within a single rassenkreis differences in armature may be small or nonexistent, or they may be more considerable. In the latter case all transitions exist within the series. This indicates that these features show the same type of variation as do all other variable characters within a rassenkreis without leading to physiological isolation, because only unimportant details are involved.

d. Impaired fertility due to chromosomal differences

It would be of great importance to know whether chromosomal differences between the geographical races of a rassenkreis exist which could act in the direction of physiological isolation by causing the production of inviable gametes in the hybrid. We have already mentioned that such rassenkreise as have been studied cytologically (*Lymantria:* Goldschmidt; *Crepis:* Babcock) show the same chromosome number in all races, though differences in chromosome size have been observed. The hybrids have a perfectly normal chromosomal behavior. In *Peromyscus,* Cross (1938) found forty-eight chromosomes (a typical mammalian number) in all species but one; namely, *eremicus,* with fifty-eight chromosomes. But among the subspecies of *maniculatus* there is one, *hollesteri,* with fifty-two chromosomes. It is assumed that here a fragmentation of chromosomes is in progress. Other cases involving small chromosomal differences between distinguishable or undefined races will be discussed later, as their significance lies in a very different field. But there is a phenomenon which occasionally produces isolation through chromosomal differences; namely, polyploidy. Thus, Turesson (1931; detailed cytology by Levan, 1935) found a Siberian race of *Allium schoenoprasum* with a tetraploid chromosome number and giant size, a so-called autotetraploid. Cases of this type are not rare in plants. But we know that tetraploidy is one of the "mutations" frequently found in plant species, mutations which are probably combined with physiological features of a presumably preadaptive type. (This will be discussed later.) Though a tetraploid is at least partly isolated from a diploid (the triploid hybrid producing many unviable combinations), it can hardly be claimed that the occasional existence of a tetraploid as a geographic race is to be considered as a sign that subspecific differences are leading into specific ones at the end of a series. A tetraploid race or ecotype is just one special type of subspecific variation.

In animals, where polyploidy is rather rare, its significance for geographic variation seems to be still smaller, as polyploid forms might occur without definite relation to distribution and even side by side with the normal ones. All cases which have been described show a relation between chromosome number and propagation. In *Artemia salina* (see Gross, 1932) diploid bisexual races exist, and, in addition, diploid, tetraploid, and octoploid parthenogenetic races, which also show morphological differences. In the parthenogenetic psychid moth *Solenobia triquetrella*, according to Seiler (1938), bisexual diploid races exist. Though they have been obtained from different places, nothing is known as to whether a subspecific or geographic or ecotypical differentiation is involved. Since a cross between diploid and tetraploid races in these animals produces sterile intersexes, one might call this an isolation due to chromosomal difference. In this connection the case of *Ascaris megalocephala univalens* and *bivalens* may also be mentioned. These races are found side by side and differ only in chromosome number. Recently races with six and eight chromosomes, found by Li (1936) in Mongolian horses, have also been added to the 2- and 4-chromosome races. The chromosomal races are apparently otherwise identical. However, the case is too unique to be of general significance.

A comparable case, though not involving polyploidy, is that of *Trialeurodes vaporariarum* as described by Schrader (1926). In an American race, parthenogenetic eggs are haploid and produce males; in the English race the diploid number is restored and females are produced. A very similar situation also occurs in plants (Chara, Ernst, 1918). But the rarity of such cases precludes any evolutionary significance, except for the presence of a more or less freakish type of microevolution of the nature of a blind alley within the confines of a species.

B. Isolation

We do not intend to discuss here the importance of isolation for selection. A masterly discussion of this problem is found in Dobzhansky's book (1937). The problem with which we are concerned here is to find out whether the extreme members of a rassenkreis are incipient species. Whenever this problem is discussed by taxonomists and such geneticists as have worked in this field, statements may be found of the tenor: if by any geological or other events this or that subspecies were to be isolated, it would actually have to be considered a different species. We shall later discover the rules or rather, lack of rules, for the distinction between subspecies and species in border cases. Here we have to see whether, if we stick to actual facts, isolation within a rassenkreis leads to specific differentiation or not. We have seen that the average rassenkreis is a continuous one with all transitional conditions between different subspecies. Even when the rassenkreis cannot be arranged into a simple cline, as; e.g., in the checkerboard distributional type illustrated by *Peromyscus*, or in the polycentric type of Reinig, individual subspecies or groups present typical clines with all intergradations. But there are also rassenkreise existing, especially where oceanic islands are involved, in which major geographic barriers separate the individual subspecies, thus producing a considerable amount of isolation. As an example which shows both types of variation we may again take the *Lymantria* case, where a continuous cline occurred from the East Asiatic mainland via Korea to northernmost Honshiu. But at the Tsugaru Strait this was interrupted, and in the island of Hokkaido, isolated since the Tertiary, a subspecies was found which all in all was more different from the others than any known race. Actually, Matsumura had called this race a species. We know in this case that the Hokkaido subspecies produces fertile offspring with all other races and is not genetically different in any special way. In *Peromyscus* there are similar examples of different subspecies isolated on the Pacific islands near the coast of California next to a typical rassenkreis. Our problem now is whether isolation of members of a rassenkreis enables them to differ in such a way that these subspecies have greater chances to start new evolutionary lines toward higher systematic categories. It is interesting to see what Rensch, who is a strong supporter of the idea that subspecies are incipient species and who in addition has much firsthand knowledge of insular rassenkreise, has to say on this problem. He writes (1934): "Some taxonomists consider two vicarious forms, separated by a barrier of distribution, as geographic races if the distinguishing characters are continuous as a consequence of convergence of the extreme variants. They consider them as species if these transitions are missing. This distinction, however, is gratuitous, as two such forms are sometimes distinguished only by a very minor, though constant, trait. Thus the bird *Oreosterops superciliaris* Hart. lives only in the small Sunda Islands Sumbawa and Flores. In Flores the superciliar band is always pale sulphur-yellow (race *superciliaris* Hart.); in Sumbawa it is brilliantly golden (race *hartertiana* Rensch; in addition the throat is yellower and the size somewhat smaller). There are no transitions known. But it would be gratuitous to call these forms therefore two species, as they are otherwise completely alike and as their differences are much smaller than those of many other different geographic races, linked by transitional forms." This shows that isolation of the members does not in itself substantially change the condition of a rassenkreis.

We can illustrate the point under discussion with another example taken from the rassenkreis of *Papilio machaon*, which was discussed above. Among the races are a few which are restricted to rather isolated areas, partly of insular nature; e.g., Arabia, Kamtschatka, Japan, Newfoundland (see fig. 20). Eller considers these as incipient species, though there is no extreme deviation of these races from some of the nonisolated ones visible. The difference becomes still smaller if we notice that within one and the same race spring and summer forms exist (seasonal dimorphism) which exhibit differences of similar degree. The production of species by this type of isolation, occurring within a rassenkreis, has still to be demonstrated. We may also point to a case which was discussed above, the racial group of *Crepis* forms. We mentioned that on account of their isolation some of the forms were treated as distinct species by Babcock and collaborators. But we pointed out that the genetical facts did not reveal any notable difference upon which to base such a distinction. It is just this type of fact which is frequently used to bolster the claim that specific characters Mendelize like racial characters (see below). If the data are considered within the whole body of facts as discussed here, we feel obliged to exercise considerable caution.

We have discussed quite a number of facts relating to the spreading of subspecies over large areas, facts which were frequently used in favor of the conception that extreme subspecies are incipient species. But such a conclusion could not always be drawn. There is the much discussed, accepted, or criticized age-and-area hypothesis of Willis (1923) derived from a huge body of facts on plant distribution. Willis starts from the observation that a rare species is frequently confined to a very small area. He tries to show that adaptation to very specific conditions; i.e., a unique ecotype, is out of the question, and also that in many cases an explanation by the assumption of relics—species from former epochs (e.g., the glacial age relics in many groups of animals and plants) —is ruled out by the facts. He therefore assumes that the area occupied by a group of allied species (which certainly would include the subspecies in modern nomenclature) depends, *ceteris paribus*, upon the age of the species in the place in question. Thus the area occupied is a function of the age of a species. A species of very limited or isolated range, with which we started this discussion, is therefore a very young species and not a very old relic, as frequently assumed. If this is correct, such species cannot have evolved gradually but must have appeared suddenly. Without taking a definite stand on this hypothesis, not having the necessary mastery of plant geography, I mention it in order to show that the facts under discussion do not *necessarily* lead to the strict neo-Darwinian explanation. We shall return later to the same subject.

In discussing such a body of facts, it has to be kept in mind that the eventual results of isolation within a rassenkreis may be of two very different types. If it is accepted that the subspecific characters in cases of typical clines are adaptational, patently or cryptically, the lack of intermediates in the case of isolation might mean nothing but the lack of intermediate environments, requiring transitional adaptational traits. If, however, nonadaptational; i.e., fortuitous, traits are involved, the situation is a different one. The differences would have nothing whatever to do with typical geographic variation, which is orderly, but would be the result of chance mutations building up in different directions for no other reason than the presence of chance initial differences. In the latter case the result would be a completely haphazard, disorderly arrangement of the separate subspecific forms and of course discontinuity between adjacent forms.

Good examples of this type of subspecific differentiation may be found among insular birds. One such example is represented in figure 23 (after Murphy, 1938). In the Marquesas the flycatchers of the genus *Pomarea* have formed subspecies (partly called species by the ornithologists), though "they obviously comprise a single formenkreis." Their distinguishing features are plumage differences in the nature of secondary sex characters. In the map the so-called species live in the islands surrounded by the solid lines, the subspecies within the broken lines. In the diagrams of plumage pattern white represents white, black is black, and

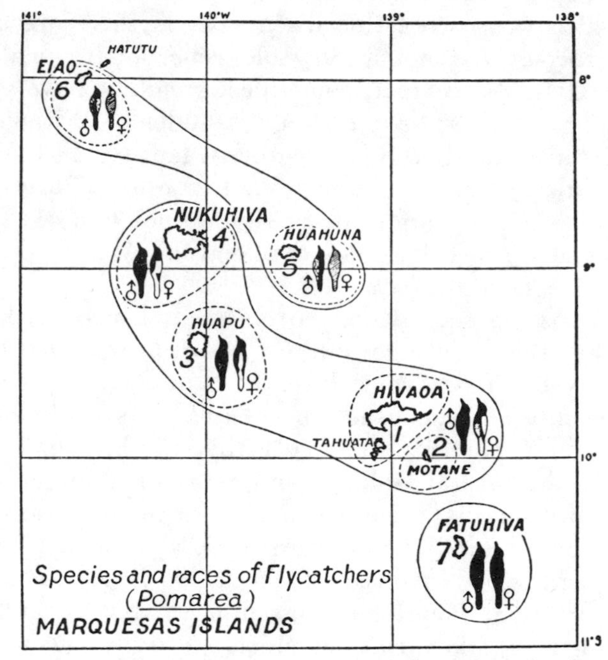

FIG. 23. Map of the rassenkreis of flycatchers in the Marquesas. Explanation in text. (From Murphy.)

ruled lines brown. The different types of pattern and their relation to sex dimorphism are evident from the sketches. Here we have, then, again a discontinuous, not clinal, obviously nonadaptational diversification within a rassenkreis. It appears as a result of playful chance. But why should any of these forms be an incipient species? Why should this type of variation, in principle not different from the polymorphisms treated above, be anything but a special case of microevolution? (For much material and conclusions opposed to those drawn here see the numerous ornithological papers by E. Mayr 1924–39.)

One more example of this very irregular type of geographic variation combined with isolation, and also formation of clines, ought to be mentioned because it has been frequently discussed from an evolutionary standpoint. The significance of the case in question ought to be made apparent by a realization of its actual relation to the normal and most frequently found type of subspecific geographic variation, the orderly one. I refer to the polymorphic variation of some land snails in tropical islands. This intensely interesting group of facts can hardly be called completely transparent and therefore has found very different interpretations.

The classic example, which was used by its discoverer, Gulick (1905) for elaborate evolutionary speculations, some of which strike one as quite modern, is the case of the Achatinellidae of Hawaii. Gulick's son has published a very good review of the facts, adding the results of more recent taxonomic work (Pilsbry, Cook) which we may use as a basis for evaluation of the data (A. Gulick, 1932). In doing so we may call the forms which replace each other in different localities subspecies, though they have been described as species. But it will be seen that the facts point to subspecific variability within a large species, which in the taxonomical work is referred to as a subgenus. Gulick found (as is also found in so much typical rassenkreis work) that the most divergent forms within each species are very different from each other. But it is very difficult to describe the limitations of all such subspecies. There is an immense variation in size, shape, texture, and color. It is possible to select from these types groups of a certain uniformity and local distribution, but they are bridged by all intermediates. Order is obtained, however, by the study of geographical relations (see p. 53 for Rensch's procedure for studying a rassenkreis!). "In general, if any string of colonies of related snails are situated in a row, the intermediate colonies will have shells that average intermediate to those between which they are located. If an intermediate colony is only a mile or two distant from the neighboring stations on either side, it will carry various forms identical with shells in each of the other locations and there might be one or two that range through all three collecting stations. Mixed with these will be clear-cut forms that are peculiar to single localities. Some of the forms will be definite with but little tendency to show intergrades. Others have many intermediates that connect them by infinitesimal shadings into other forms, perhaps of the same locality, perhaps of the next one adjoining. Colonies that are more remote differ after the same method, but to a greater degree" (Gulick). In the typical *Achatinella* series in Oahu a string of such colonies was studied all around the island. In any series of closely related forms, the degree of similarity between the populations was roughly measurable by the number of miles that separated them. Only for remote colonies (5–15 miles) could each individual be reliably assigned to its respective colony.

Up to this point the picture of this type of variation very closely resembles the one which we drew before for all typical rassenkreise. There is only one additional feature. Together with the "cline" of different forms parallel to a geographical cline we find the phenomenon of local polymorphism which we discussed on page 23, 26 for other land snails. Within each colony Mendelian recombinations of such more or less fortuitous characters as color, banding, sinistrality are present in typical numbers. These polymorphic recombinations occur in a great many different snails in a perfectly parallel way (e.g., *Helix*, *Amphidromus*, *Achatinella*, *Partula*) and are simply local occurrences. If this phenomenon, typical for shells of snails, but also encountered in other cases, is discounted, a typical cline of geographical races remains. The really distinguishing feature is the occurrence of a gradient within an extremely small area. According to Gulick, all these forms are ecologically alike, all feeding upon the same food, living on definite trees which are found only in the same type of tropical rain-forest. No climatic or other difference which might form a cline, paralleling the racial cline, can be found for miles. A geographical adaptation is therefore ruled out. (We accept this statement for argument's sake, but add that considerable differences might be found if we made microclimatic studies which, in applied entomology, have recently turned out to be very significant.) Therefore another factor presented itself to Gulick as the most potent one: isolation. (This is why these facts are discussed in this chapter.) These snails usually do not leave their tree and do not migrate at all. The deep ravines in which the food trees grow are separated from each other and from those on the other side of the island by high bare ridges which would in any case prevent migration. One might therefore compare the populations to those on very small islands. Isolation of varying populations has thus made possible the subspecific differences. "Heredity never stays still; every generation blossoms into multitudes of little novelties, that immediately add themselves into the hereditary sum total of their racial stock, and become the starting point for ever more novel departures. By the law of statistical probabilities the various novelties will turn up very unevenly distributed in the different colonies and the new bents which they initiate will tend to make the different local races become continually more and more unlike each other" (Gulick).

This latter conclusion could be accepted if the facts had shown that the isolated colonies vary in a haphazard way. But actually there are two types of variation: the more or less haphazard (though not completely so) recombinations of the polymorphic Mendelian characters in the different colonies, and the orderly cline of different forms in a definite direction for which isolation cannot account. But even if we accept isolation as decisive, it has only resulted in a copy of geographic subspecies formation within a small area, where it is more conspicuous than it would otherwise be. Isolation, if accepted as the decisive factor (not considering the special features of local polymorphism), has not led to any larger deviations than the typical subspecific ones, linked by all intermediates, and the material in question therefore does not help us to understand diversification beyond the limits of the species.

Since Gulick's work was done, a similar case, the *Partulae* of Tahiti, has been studied most thoroughly by Crampton (1916, 1925, 1932). Here we have at least some intimations as to the genetic situation, as in the viviparous forms young from the brood pouch may be compared with the mother. If we take this information together with our knowledge of similar characters in *Helix* (Lang, 1906, 1911), there can be no doubt that the distinguishing qualitative traits are based on simple Mendelian recombinations and that the quantitative traits are inherited in some way. The *Partula* material otherwise closely resembles the *Achatinella* shells. That the composition of the individual colonies (based on local polymorphism) varies in time, as actually found, is not surprising, and will be found in any European *Helix* colony revisited from time to time. That the subspecies spread in recent times from their original area, as found by Crampton, is another interesting detail concerning population problems. Again no relation between environment and subspecific differentiation was found, though for many species on different islands a typical subspecies is described for each valley or area within the distributional area of the species. We shall return below to the same material in another connection.

The facts discussed in this chapter thus show that there is no reason to conclude that isolation of subgroups within a species leads to the formation of categories other than those formed by ordinary continuous geographic variation. The subspecific variation as obtained by isolation may be less orderly than otherwise, and in some cases may even result in somewhat wider gaps between two adjacent forms; but there is no reason, at least as far as the factual material goes, to suppose that isolation makes subspecies develop into species. The conclusion is the same as that derived from our former discussion. Isolation or no isolation, the subspecies are diversifications within the species, but there is no reason to regard them as incipient species.

There is, I think, in the whole idea of subspecies as incipient species a psychological element. It is taken for granted that species are evolved from each other by a slow accumulation of small individual steps (by means of selection, of course). If, therefore, a subspecific series is found to exhibit different degrees of small differences, the situation must indicate the presence of exactly the evolutionary process which is postulated. If, nevertheless, the individual rassenkreise remain separated by large gaps, and if the most extreme members are still only ordinary subspecies, the preconceived idea forces the neo-Darwinist to look for the most impossible explanations to fill the gaps. One of these which always works is the time-honored phylogenetic idea that the existing gaps were formerly filled by missing links. In other words, the subspecies are incipient species because a strictly Darwinian view requires such an interpretation, and because it is taken for granted that no other possibility exists.

The adherents of such a view derive much comfort from the results of population mathematics, especially Wright's calculations (1931), showing that small isolated groups have the greatest chance of accumulating mutants, even without favorable selection. I do not want to create the impression that I underrate the mathematical study of selection problems, as found in the brilliant work of Fisher, Haldane, Volterra, Wright. Actually, I had tried to work out a special case of selection (nun moth, Goldschmidt, 1920b) with insufficient mathematical equipment before Haldane furnished the proper formulae, and therefore I am fully aware of the importance of this now-popular branch of evolutionary research. But it is necessary to remember an old remark of Johannsen in his criticism of Galtonian biometry; namely, that biology must be studied *with* mathematics but not *as* mathematics. This means that the most brilliant mathematical treatment is in vain if the biological rating of the material is not correct (see Pearson and Mendelism). I am of the opinion that this criticism applies also to the mathematical study of evolution. This study takes it for granted that evolution proceeds by slow accumulation of micromutations through selection, and that the rate of mutation of evolutionary importance is comparable to that of laboratory mutations, which latter are certainly a motley mixture of different processes of dubious evolutionary significance. If, however, evolution does not proceed according to the neo-Darwinian scheme, its mathematical study turns out to be based on wrong premises.

In our present discussion of isolation and the incipient species, it is the contention that small isolated populations have the greatest evolutionary chances from the standpoint of population mathematics. This contention must fall to the ground simultaneously with the neo-Darwinian concept. But it might also be pointed out that the mathematical conclusion does not agree with many biological facts. Anybody who has seen the regal primrose grow in a single crater of Java, or collected *Apus* and *Limnadia* in their rare and isolated haunts, or has studied the occurrence of innumerable so-called relics, is impressed by their uniformity and their obvious position at the end of an evolutionary blind alley, in spite of isolation in small populations, in addition to generalized, primitive features (Phyllopoda, Anaspides) most suitable for evolution. On the other hand, large isolated populations frequently show most extreme variation. I once observed a population of a *Helix* species in Paestum, Italy, which was so dense that the plants were hardly visible under the innumerable snails. The variation among the snails (of the well-known Mendelian type) was immense, and certainly could not have been greater. There is no factual basis for the assumption that such a Mendelian polymorphism leads beyond the existence of whatever recombinations are possible. Another set of facts which clearly does not agree with the mathematical theorem is found in Vavilov's gene centers, assumed also by Reinig (see discussion on p. 87). Whatever the theoretical interpretation may be, the facts show a small area containing a multitude of species side by side, and numerous mutants within the species. By dispersal of these mutants rassenkreise may be formed, but nothing indicates that species are produced in these centers by isolation and accumulation of mutations.

The contents of this chapter, as well as all the data presented thus far and to be presented below, show that the neo-

Darwinian conception, which works perfectly within the limits of the species, encounters difficulties and is not sustained by the actual facts when the step from species to species has to be explained. Selection will certainly be involved also in the accomplishment of this decisive step, but we shall see that selection in nature probably has much easier work than that required by the neo-Darwinian idea of slow accumulation of micromutations.

6. THE SPECIES

Our discussions up to this point have shown microevolution at work within the confines of the species, diversifying the primary form either by adapting the species genetically to diverse conditions of the environment within the area suitable for occupation; i.e., by subspecific, geographic subdivision, or by a diversification which is more haphazard and nonadaptational, occurring in the form of mutations, local polymorphism, and polymorphism enhanced by isolation. In all cases the diversification could be subdivided almost without limit down to differences between individual colonies, showing that taxonomic subunits could be multiplied if it would serve a purpose. Wherever known, this diversification was based on the different types of Mendelian differences, implying origin by accumulation of micromutations. It further turned out that the subgroups, wherever tested, were completely fertile *inter se*, though this would not exclude an occasional lack of actual interbreeding which might be on the same biological level as; e.g., noninterbreeding between Brahmin and Pariah.

Darwin's classic concept of the origin of species, which, as we saw, is the one to which modern biologists have largely returned—we spoke of neo-Darwinism—is found in the following phrases from the *Origin of Species* (Chapter II): "Certainly no clear line of demarcation has as yet been drawn between species and subspecies—that is, the forms which in the opinion of some naturalists come very near to, but do not quite arrive at, the rank of species: or, again, between subspecies and well marked varieties or between lesser varieties and individual differences. These differences blend into each other by an insensible series; and a series that impresses the mind with the idea of an actual passage.

"Hence I look at individual differences, though of small interest to the systematist, as of the highest importance for us, as being the first steps towards such slight varieties as are barely thought worth recording in works on natural history. And I look at varieties which are in any degree more distinct and permanent as steps towards more strongly marked and permanent varieties; and at the latter as leading to subspecies, and then to species. A well marked variety may therefore be called an incipient species."

All these facts have become apparent in our previous discussion, where the modern factual additions to the classic conception were recorded as microevolution *within* the species. We now come to a consideration of the next step in evolution, as set forth in the words of Darwin: "Certainly no clear line of demarcation has as yet been drawn between species and subspecies." Do subspecies actually merge into species as gradually as one subspecies grades into another one? In other words, are subspecies incipient species and is specific differentiation, as well as that of higher categories, a continuation of microevolution, based upon the same principles of accumulation of small mutations, adaptational or otherwise?

Darwin's term, "incipient species," has been frequently used in our discussion. I am not sure that the many authors who use this term stop to think what is actually meant by it. Incipient species must mean that any variation, large or small, within a species has the potentiality of becoming a new species, and, further, that this probability increases with the accumulation of different traits and is therefore greatest in extreme subspecies. If this is true, it follows that subspecific differentiation is a *necessary, obligatory* step toward species formation. This, in turn, means that the differences between two closely related species must be a continuation of the series of differences between subspecies, as we found subspecific differentiation not to be haphazard but orderly. And since this orderly behavior of subspecific differences is found to parallel geographical or ecological clines, the decisive step from subspecies to species must occur only at the extreme points of the range of the species. Localized species, not forming clines of subspecies, are therefore excluded from further evolution. There is no possibility of other interpretations within the concept of incipient species. Rensch is one of the few who recognized this clearly and actually postulated (see below) that the new species are formed at the extreme end of a subspecific cline and later return to the point of origin to live side by side with the old species.

But geneticists who use the concept of incipient species do it in a different way. They think that a subspecies will be isolated and then have a chance and even greater probability (see above, S. Wright) of producing new mutations, which accumulate until the specific difference is reached. It is usually overlooked that such a conception does not require at all the existence of incipient species. Any isolated group within a population, whether already different from the rest or not, will have the same chance for evolution as any other (provided an equal rate of mutation) if the genetical premises are correct and if the direction toward the new species is not bound to coincide with the direction of subspecific differentiation. The only apparent advantage of a subspecies over any ordinary mutants would be that a few mutations have already been accumulated to start with on the path toward the species. How little that would mean for evolution becomes visible if we remember the numerous species which have needed all the time since the Late Tertiary to produce their subspecies. The difficulty caused by adaptive subspecific traits will soon be discussed. The Darwinian incipient species makes sense, therefore, only if the track leading to specific differences is a continuation of the subspecific clines. Otherwise any isolated population would potentially be an incipient species, and the rassenkreis might at best be called only a model of specific differentiation (i.e., from the point of view of neo-Darwinism).

A. The Good Species

We do not intend to discuss definitions, and we do not feel entitled to tell the taxonomists what they ought to call a species. But if we want to analyze the all-decisive step from within the confines of the species to the next higher category, which is generally called the species, we must know what kind of taxonomic distinction we are discussing. Let us see, therefore, how some modern taxonomists look at this decisive point.

It is of course well known that in older literature the members of a rassenkreis are listed as species; in some recent work this system is still followed, particularly in plants. There is even a quite recent and very elaborate piece of taxonomic work, Kinsey's on *Cynips (loc. cit.)*, in which there is found not only a return to the older method, but even a step beyond that, for he calls all clearly recognizable forms species. These, however, are mainly problems of taxonomic technique. We are not interested here in the names given to the categories. The different viewpoints have been amply discussed by numerous authors. Their respective merits may be weighed and compared by a study of such comprehensive treatments as those published by Berg (1926), Du Rietz (1930), Lotsy (1916), Remane (1927), Robson (1928), Semenov-Tianschansky (1910), and many others. Our problem is to find out whether the lower categories which we treated as subspecies or geographic races, according to the rassenkreis concept, show a continuous intergradation with members of another rassenkreis or species. If subspecies are considered to be incipient species which only need isolation to become species, such a merging of one rassenkreis into another must be observable, unless we suppose that the links are bound to be missing. If, however, subspecies are nothing but an intraspecific diversification which adapts the species, at least in the majority of cases, to definite conditions within its area of distribution, the limit between two species or rassenkreise ought to be in the nature of a hiatus, an unbridged cleft.

It is remarkable that many modern taxonomists who have worked with rassenkreise seem to be inclined to take the latter viewpoint, in spite of their usual neo-Darwinian leaning. As a matter of fact, it is one of the primary tenets of the rassenkreis theory that forms which replace each other geographically and are able to interbreed freely are members of a rassenkreis. But forms which live in the same area (forming their own rassenkreis or not, as may be) and which do not interbreed are called species. (The important border cases will be discussed in the next chapter.) Kleinschmidt (1897; English translation, 1930), the father of the formenkreis concept, held that every species was separated from every other by a gap. His intensive studies as an ornithologist of considerable authority, as well as his study of other groups, demonstrated to him that this was the rule. (He was even led to deny an evolution from one species into another, but here the preacher might have influenced the ornithologist.) He categorically denied that subspecies are incipient species. His modern followers in the rassenkreis theory do not agree with the latter conclusion (see below). But they nevertheless regard species as entities which, barring certain border cases to be discussed later, have a separate existence and do not grade into each other. (This statement is of course independent of the terminology; i.e., whether an author calls the absolutely distinct types genera or subgenera or species.) Turesson says (1922): "Thus while the belief that the Linnean species of the present genetically represent complicated products of recombined Mendelian factors, or genotype compounds, has been strengthened, few would maintain that the problems connected with the formation of the Linnean species are exhausted by this demonstration. Most of these species are, *as every earnest inquirer will find*, in their natural areas of distribution rather circumscribed products, which do not live in any extensive connubium with congeners of other species. *The bridgeless gaps*[11] found between species of the same genus, the final molding of the Linnean species, then remain to be explained." Not

11. Italics mine.

very different is J. Clausen's (1937) statement: "It has been shown, first of all, that species really do exist as natural biological entities. Each is fitted to live in the environment in which it is found, as a key fits a lock." Still more explicit is E. Anderson's (1936) statement: "The conclusion was reached that closely related though these Irises might be, variation within either species was of quite another order of magnitude from the hiatus between them. . . . The variation *within* could never be compounded into the variation *between. The two species were made of two different materials.*"[11a] This latter statement, however, relates only to a single case. Crampton *(loc. cit.)*, though a neo-Darwinian, also describes the facts relating to the *Partula* of Tahiti in the same sense. He emphasizes that the species differences, as opposed to the subspecific variation, are clear-cut. Different species may be found browsing on the same tree without any question arising regarding their specific diversity. Further statements of experienced taxonomists presenting exactly the same conclusions as those now reported will be mentioned below.

The above quotations illustrate the experiences of biologists working both experimentally and morphologically with the lowest taxonomic groups. My own results with the material I was analyzing are exactly the same (see 1932, 1933). The two nearest relatives of *Lymantria dispar* are the species *mathura* and *monacha*. The latter inhabits the same area as *dispar*, the Palearctic zone, and the two species may be found side by side in Europe as well as in Japan, though their life habits are somewhat different. *Mathura* is a subtropical form stretching from India through southern China up to northern China and Japan, where all three species may be found side by side. *Mathura* and *dispar* are rather similar in habit and may be found laying eggs on the same tree, as I once observed in Shantung Province of northern China. I do not doubt that *monacha* might also be present by chance on the same tree with the others, though its ecology is somewhat different. Taxonomically there is no doubt

11a Italics mine.

that the three species are members of the same genus. But it is not difficult to show that between them exists the "bridgeless gap" which we are discussing. This applies to practically any morphological, physiological, and ecological character which has been studied (much unpublished work

. *Lymantria dispar* (top row), *mathura,* Japanese subspecies (second row), *ha* (third row); left, female; right, male. (Original)

by the author). The difference of wing pattern and sexual dimorphism may be easily seen in figure 24. The genital armature is different (see fig. 25), and so are the shape, size, hairiness, and color of the body. The caterpillars are utterly different in pattern of markings, color, and hair (fig. 26). The pupal case is extremely different in color, texture of

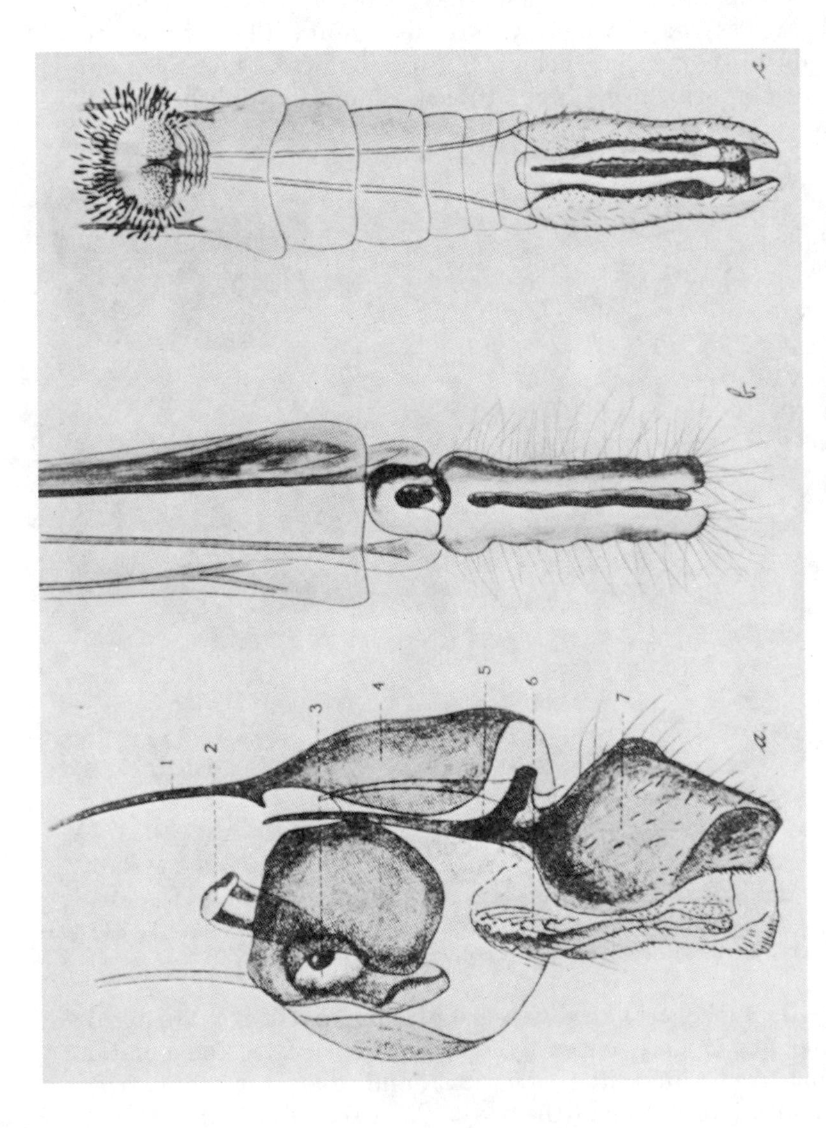

Fig. 25a. Genital armature of females (*a,b,c*) and males (*d,e,f*) of *Lymantria dispar* (*a,d*). *L.*

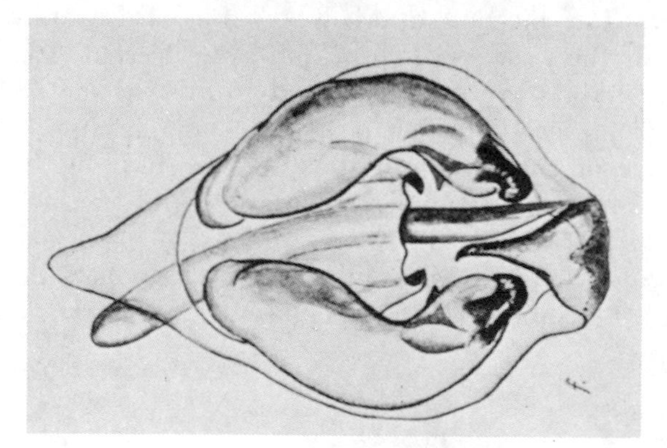

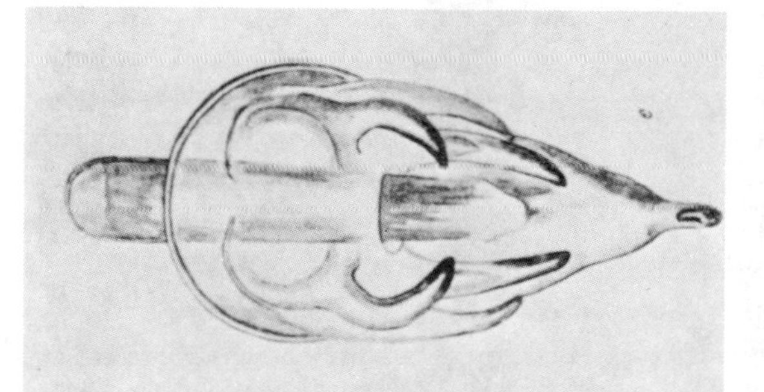

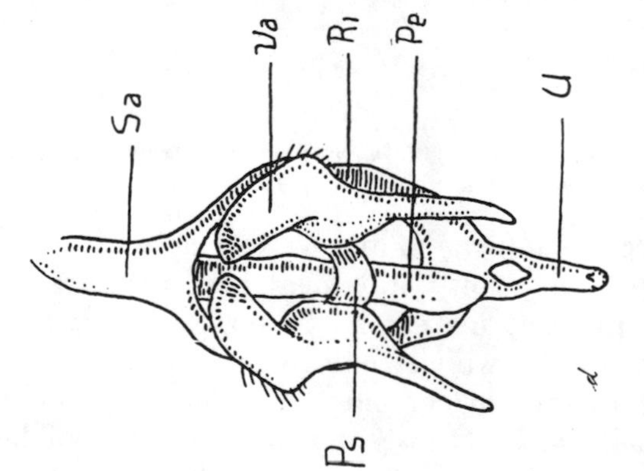

FIG. 25b. Genital armature of females (*a,b,c*) and males (*d,e,f*) of *Lymantria dispar* (*a,d*), *L. mathura* (*b,e*), *L. monacha* (*c,f*). (Original.)

chitin, hair. The method of attachment of the pupae and the details of the process of pupation are different. Ecology and feeding habits are different, and so are the methods of dispersal. Completely different are the egg-laying habits, the structure of the female ovipositor, the hair on the abdomen, the function of the cement glands, and the instincts. Finally, there is also a difference of sex chromosomes between *dispar* and *monacha*, as my former students Seiler and Haniel (1921) found. (The chromosomes of *mathura* have

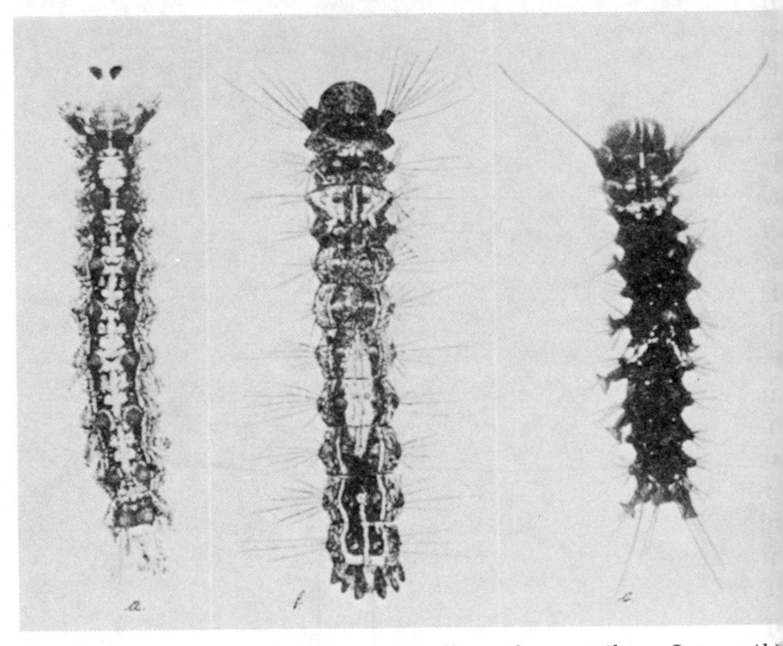

Fig. 26. Caterpillars of (*a*) *Lymantria dispar* from southern Japan, (*b*) *L. monacha* (from Goldschmidt), (*c*) *L. mathura* (original).

not yet been studied.) Thus we can actually apply to this description Anderson's statement that the three species are made of different materials. Nevertheless, *dispar* and *mathura* are species which replace each other geographically with a small area in common. Both have formed subspecies within their range (those of *mathura* have not been as yet properly described, but many are known). In *dispar* we know all the trends of subspecific variation, but these are on a level very different from that of the specific differences. For example, where patterns are involved (caterpillar, wings) the subspecific variation may be described as a plus or minus change within a particular pattern. But between the species even the basic patterns are different.

Let us now recall the discussion on page 140 regarding the meaning of the idea of incipient species. If the geographic races are incipient species or, to put it more clearly, if the road to species formation is bound to be traced via subspecific variation, differences between closely related species, especially those replacing each other in adjacent areas, ought to be *on the same genetic level* as subspecific differences. This means that the differences are such that they can be *conceived* of as having arisen by further accumulation *of the type of differences* which are found in subspecific differentiation. An earlier statement of this conclusion (Goldschmidt, 1933) has been misinterpreted (by Rensch and J. Huxley) as meaning that closely related species ought to be different in exactly the same features as subspecies, if derived from these. This is not necessarily meant. What is meant is that the level of differences; i.e., their order of magnitude, must be the same, though the species might be characterized by a more extreme departure. If we look at different rassenkreise in Lepidoptera, butterflies or moths, the subspecific differences are always of the same order of magnitude; i.e., shifts in a few directions within the general pattern of organization of the species. If these are the incipient species, the closely related species must be different in the same way, though on a larger scale. But actually the specific differences are on a completely different level of organization, the result of completely different materials, as Anderson expressed it. Take as examples of subspecific differentiation those described in detail in former chapters. Compare with those, to use only one character, the difference in the process of egg laying between the three *Lymantria* species. In *dispar* there is a blunt abdomen, without ovipositor, the instinct to lay the eggs in a cluster of definite shape, the instinct to lay on the surface of trees, boards, rocks, the instinct to cover the batch with a thick sponge of rubbed-off abdominal hair, and in addition all the morphological features which make all this possible. In *monacha* there is a long ovipositor with all the necessary muscles and innervation, which can be bent and inserted in clefts of bark. The eggs are deposited in loose clusters kept together by a little cement. In *mathura* the ovipositor is still more flexible, enabling the moth to stick the eggs below the bark. They are held together by a very hard white cement in which the eggs are completely embedded. In both cases no abdominal hair is present and the instinct to rub off the hair is missing. It is clear that such complicated differences are on a quite different evolutionary level from the simple subspecific differences and cannot be conceived of as merely a quantitative departure from the type of difference found in subspecies. Here we actually have the "bridgeless gap" of Turesson, the species "made of entirely different material" of Anderson, the completely different reaction systems (see below), the different order of magnitude of differences, the different evolutionary, morphological, genetical level.

I should point in this connection to another example whose specific biological features, however, do not permit us to generalize. Two well-known species of parasitic nematodes, *Ascaris megalocephala* from the horse and *lumbricoides* from the pig, are adapted to practically the same conditions of life within a mammalian intestine. Taxonomically they are distinguished by a few minor differences in size and in the features of the anterior end. But in studies of their histology, which I made a long time ago, it turned out that practically every cell is different in both forms. Nematodes are cell-constant forms (Goldschmidt, Martini) and therefore many organs are built by a definite number of cells. For some organs it was found that this cell number was different in the two species. In other organs the intimate cell structure differed. At the time when I was engaged in this work (1902–10) I was able to assign with certainty a slide of any organ, or even an individual cell, to one of the two species

I do not know of any comparable work (though histological differences of a less extreme order between species are well known to experimental embryologists). Though parasitism and cell-constancy set the case of *Ascaris* somewhat apart, I wonder if a closer scrutiny would not reveal comparable features in ordinary specific differentiation.

As the studies of the authors mentioned above penetrated much deeper into the details of the situation than any purely taxonomic study could do, there can be no doubt that specific differences of a completely different type from that of subspecific variation actually exist. The question is whether or not these have nevertheless arisen by the same type of microevolution as is found within the species, which means that the existing gaps have to be filled by missing links.

Let us first see how the followers of the latter viewpoint account for facts of the type just discussed. There is one interpretation which closely resembles the type of phylogenetic speculation current in early Darwinian times. If two different forms were to be analyzed from the standpoint of evolution, a common ancestor was invented from which both forms had slowly diverged. In the same way species which are completely separated are supposed to be very old and therefore highly divergent. But as no intermediate forms exist, some kind of isolation must have taken place while the specific diversity was being established. But the two diverse species live side by side. It is therefore assumed (e.g., Rensch) that originally the specific diversification occurred at the extreme ends of a geographic series. The two now different forms secondarily migrated again and came to live side by side in the same area. I am afraid that this is hardly a plausible way to bridge an actually existing gap between two species. We found before that in most cases subspecific variation adapts the species to environmental clines. Wherever this is actually the case, a return of the subspecies at one extreme end toward the starting point could only be accomplished by retracing the steps of preadaptational mutation to its original condition; i.e., return to the original type. This situation is best illustrated in the cases already discussed in which different species have parallel subspecific clines within the same area, thus showing that all this variation is below the specific level. I may add that Mordvilko (1937), though himself a taxonomist and evolutionist of neo-Darwinian leanings, also realized this situation to a certain extent. He was confronted with the difficulty—usually shelved quietly—that species are supposed to originate by the accumulation of mutations, but that the species do not continue to produce further species in spite of the continuation of mutation and the production "of a multitude of local peculiarities, which greatly obscures the main path of species formation, because geographic forms can be discovered most easily." He adds that the geographic radiation can be only an unimportant way of species formation and, further: "In no case can this [the geographic radiation] explain the simultaneous occurrence of different species of the same animal or plant genus in one and the same region, species adapted to definite ecological features."

The situation would of course be different if nonadaptational traits were taken as the point of departure. A retracing of the steps would in this case be independent of eventual adaptational changes. But simple analysis makes it clear that in such a case a positive result would be completely independent of subspecific variation within a rassenkreis, as emphasized before. To repeat the argument: The ability of a subgroup in a population to live as a constant unit side by side with the rest of the population without interbreeding is exclusively controlled by the occurrence of genetic isolation. This might as well occur at the end-points of a geographic series or at a single point anywhere. If it can occur at all by slow accumulation of micromutations, it is bound to happen in either of two ways: either by isolation of a subgroup, with prevention of interbreeding and subsequent divergence of evolution up to the point of real difference, the specific gap, which would thereafter permit the forms to live side by side if isolation ceased; or by physiological isolation, which sets in suddenly and afterwards permits a diversification. In both cases the ultimate specific differences would be completely independent of the genetic status of the population at the moment isolation set in. A form without any subspecific variation, a single subspecific type anywhere in the area of distribution, or finally an extreme subspecies at the end of a series, would all be equally apt to allow individuals in their midst to start toward specific differentiation provided that isolation of one or the other type occurred. There is no reason whatsoever to suppose that such a diversification has to start with already differentiated subspecific strains. Given the existence of mutation, isolation alone is needed. The rank of subspecies as incipient species could be established only if it were shown that specific differences are nothing but an accumulation of subspecific ones produced in the same direction as the trend of subspecific diversification runs, or at least on the same general level. If this cannot be shown, subspecific differences, adaptational or not, remain permutations within the species, and the specific gap is produced by another type of evolution, which is not necessarily to be conceived of as microevolution by accumulation of micromutations. I am glad to find that Dice and Blossom (1937), who have studied, both taxonomically and experimentally, the species and rassenkreise of *Peromyscus*, have joined me in this conclusion. They write: "While a new species might differentiate from an isolated race through the production of an infertility with its relatives, there would seem no more likelihood of a new species originating from an isolated race than from any isolated division of a species. . . . Most geographic races, therefore, probably do not represent incipient species but are only responses to a local type of environment." (Dice nevertheless assumes the same type of microevolution for species as for subspecies; this point will be discussed later.)

I mentioned above that Rensch tried to dispose of my argument derived from the case of *Lymantria* by pointing out that I was dealing with very "old species." I should like to register my skepticism toward an explanation of the bridgeless gap by recourse to the rather gratuitous assumption of old and new species. Does not this argumentation actually run the following way: Why the bridgeless gap? Because the species are old ones. Why are they old ones? Because of the bridgeless gap between them. I may invoke as a witness a leading taxonomist in what probably is the best-known group of animals, the birds, and one who exposes this phylogenetic argument thoroughly while trying to defend it. Stresemann (1936) finds that ornithologists have become more and more convinced "that there is really something like natural species, surrounded by sharp structural boundaries" (see above, Kleinschmidt). But as most of these taxonomists are Darwinists, they face the difficulty of bringing together evolution of species and their constancy (see above, the psychological element). The way out of the dilemma for Stresemann is that most present-day species are much older units than had formerly been assumed to be the case and that "it is never the individual variation occurring *within* a population which gives rise to speciation. On the contrary, species multiply only on the basis of geographic variation and this is a very slow process." This statement by one of the leading and most progressive taxonomists is certainly remarkable. First it confirms with the authority of an experienced taxonomist the results obtained by the authors mentioned in this chapter: specific differences are on a different level from subspecific ones. It then shows the embarrassment of the neo-Darwinist in the face of incontestable facts and his refuge in phylogenetic assumptions which cannot be tested. Visible geographic variation, which is on the same microevolutionary level as any genetical variation within a population, is recognized as insufficient for an understanding of species formation; but as geographical variation has to be the basis of species formation, the only comfort available is recourse to an inaccessible, slow phylogenetic process. Unbiased by the wish to express the facts in neo-Darwinian terms, I prefer to conclude as follows: Species formation is based upon a different type of evolutionary procedure than that of subspecific differentiation, which latter is the result of selected or nonselected accumulation of micromutations. It will be seen below whether and in what form such a process is at present conceivable.

B. The Border Cases

A potent argument in favor of the derivation of species by an accumulation of subspecific micromutations is derived from cases in which the taxonomist is in a dilemma as to whether to call a form a subspecies or a species. Some such cases have already been mentioned. Stresemann *(loc. cit.)* describes such situations in the following way: Two similar forms are usually called subspecies if they replace each other in space. If they live in the same area without interbreeding they are called species. This, he thinks, is an artificial system. Many subspecies would behave like species if they could be settled in the same area. This can be proven, he assumes, in a few instances, as in the case of *Parus major*, which we discussed on page 120. "There are other and even more puzzling cases. The English sparrow *Passer domesticus* and the Mediterranean willow sparrow *Passer hispaniolensis* live side by side in Spain, in Greece, in Asia Minor, and in Palestine, differing somewhat in ecology. There they never interbreed and everybody would treat them as species. But if we proceed to northern Africa we will meet there a mixed population practically composed of hybrids only. Here they behave exactly like two members of a species. What to do with them?" The last question I should like to answer in the following way: As a taxonomist, do with them what appears to be practical. But as an evolutionist, treat them as members of one species, which they most clearly are, though they do not interbreed in some localities, just as the Brahmin does not interbreed with a Pariah, his own near biological relative. But the subspecies Brahmin or Indian could breed successfully with as different a subspecies as an Eskimo, if he wanted to.

Stresemann discusses another instructive example. In New Guinea, some highland birds closely related to lowland forms are found, but separated by constant differences with no intermediates. He asks, "Shall these mountain forms be treated as subspecies of the lowland species or do they merit the rank of a full species? *That is just a matter of taste.*[12] In these cases, at least, lowland and highland forms do not differ any more from each other, or they differ even less than horizontal representatives often do." Stresemann, a little later, answers the question appropriately for the taxonomist: "No fast line can be drawn here. But one ought to refrain from red tape. *Whoever wants to hold to firm rules, should give up taxonomical work.*[12a] Nature is much too disorderly for such a man. He would better turn to collecting postage stamps." This clearly means that there is no definite taxonomic technique. But does it mean that from the standpoint of evolution subspecies continually merge into species? Certainly not. Where species can be distinguished with certainty (see the last chapter) they are different and separated by a gap, if not by an abyss. If, however, the distinction is a "matter of taste," evolutionary conclusions are also a matter of taste, that is, worthless. Therefore caution is advised in regard to conclusions based exclusively upon statements arrived at by means of a highly subjective technique.

The same situation is encountered in another group of facts. There are cases in which a continuous geographic cline of subspecies is found, but in which there is also an isolated form somewhere, frequently upon an island or in a distant region. We have already met with such cases in *Crepis* and *Papilio.* Many taxonomists would describe such isolated forms as species. In Osgood's monograph on *Peromyscus*, mentioned before, examples of this kind are found, and the isolated location even enters the dichotomic key as a major distinctive feature! But there are other taxonomists who include such isolated forms as subspecies in a rassenkreis, as Jordan consistently does for Lepidoptera (see 1905). The extreme of this type of variation is the existence of different closely related forms or whole rassenkreise,

12. Italics mine.
12a. Italics mine.

termed different species, each in a definite geographic area, which is completely separated from the next one. In such cases different so-called species replace each other geographically, just as subspecies do in a simple rassenkreis. Many examples may be found in the books of Rensch and Reinig. In all these cases it is again "a matter of taste" whether the forms separated, but replacing each other geographically, are called subspecies or species. Rensch, in his book *(loc. cit.)*, mentions numerous cases in which different taxonomists followed one or the other rule, or even in which the same taxonomist changed his mind in subsequent studies of the same material. Rensch therefore proposes to speak in such cases of a circle of species (artenkreis), as opposed to rassenkreis, a concept which is severely criticized by Reinig on the basis of the same material.

Again we are confronted with the same situation as before. The taxonomic assignment of the category does not concern the evolutionist except when it is used to demonstrate that the subspecies grades slowly into another species. There can be no doubt that completely different species may sometimes live side by side, and in other cases may inhabit different regions. To take an example which we used before, *Lymantria dispar* and *monacha* live side by side over the Palearctic region, and *L. mathura* replaces them in southeastern Asia. But in Japan and northern China they overlap and do not interbreed. However, they are actually totally different forms, which nobody could take for geographic races. On the other hand, there is no necessity that distant forms replacing each other geographically always be species in the sense of evolution. Many of the cases described show that the differences are of the same order as subspecific differences. There are even cases in which an individual could not be assigned to one or the other "species" when its origin was unknown (see above, Rensch). Many cases are on record in which a more recent analysis has reduced numerous such species into subspecies; an example is the case of *Papilio machaon,* examined before. Clearly, then, the only decisive test is success in interbreeding.

Fortunately, we are not completely left to "matters of taste" but know something about the genetic behavior of such border cases. There is, for example, the case of the two closely related *Peromyscus* species, *leucopus* and *gossypinus*. Each forms a distinct rassenkreis of many subspecies, but they replace each other geographically; i.e., they constitute what Rensch would call an artenkreis. *Leucopus* ranges from Nova Scotia to southern Mexico (Osgood, 1909); *gossypinus* is found in Texas, Florida, and southern Virginia. The ranges overlap slightly, but Osgood did not find evidence of interbreeding. Nevertheless, Dice and Blossom (1937) found that both species, as well as their subspecies, produced perfectly fertile hybrids when crossed. Though it might be correct, then, to speak of two species from the standpoint of the taxonomist, as regards evolution there is no reason to consider these forms as anything but geographic subspecies which, by the way, may actually have formed subsubspecific clines spreading from an eastern and a western glacial age refuge, according to the conception of Reinig. The same applies to two "species" of *Peromyscus* which are still more isolated from each other: the deer mouse *Peromyscus maniculatus* and the old field mouse *P. polionotus* (the latter in Florida, the former widespread; see p. 96). According to Dice and Blossom (1937), they produce perfectly fertile hybrids.

I do not doubt that a proper study would lead to the same interpretation of all artenkreise. Through the courtesy of Dr. A. H. Miller I had access to his monograph (now in press)[13] of the genus *Junco*. Here a typical artenkreis, in the sense of Rensch, is found. The genus ranges from the Arctic tree line in North America south to western Panama, and is comprised of twenty-one forms. All of these are geographically complementary. Some distinct subgroups that are fully isolated and strongly differentiated from one another occur within the genus. Thus artenkreise of lesser scope within the entire group may be conceived. This material is assigned to eight species, some of which may be com-

13. In the University of California Publications in Zoology.

bined into a secondary artenkreis and some of which have formed a more or less diversified rassenkreis. Specific traits of ecological adaptation are not known. As already emphasized, each distinguishable form has its separate habitat. But where the so-called species have a chance to meet, *they interbreed freely*. Thus we face the old situation. The taxonomist prefers a definite nomenclature according to his taste. The evolutionist, however, must look beyond the nomenclature and then he finds that the artenkreis is nothing but a glorified rassenkreis, a case of microevolution without meaning for the problem of the specific gap.

A botanical example of the same type may be taken from the work on *Crepis* by Babcock and Cave. This is included in table 7 (p. 99). These authors analyzed a series of Mediterranean subspecies of *Crepis foetida* ranging east into Persia, and in addition two species, *Thomsonii* from India and *eritreensis* from Eritrea. The latter two are considered as species on account of their isolation. But their morphological differences are rather small, as the table shows, and some characters are identical with those of some of the subspecies of *foetida*, while others are divergent. There is a small difference in fertility relations, but it is so small that its significance may be doubted. Actually the hybrids between the species are fertile. The chromosomes also are identical, and the differential traits show the same simple Mendelian behavior as the traits of the subspecies (see p. 100). Babcock insists on calling the two isolated types species, and therefore concludes that the specific difference has arisen by only a few mutational steps. But the actual facts do not permit us to consider these "species" from the standpoint of evolution as at all different from other subspecies, whatever taxonomic term may be advisable. The facts actually show, in my opinion, that these border cases belong to the problem of subspecific diversification and do not carry any message regarding species differentiation.

There is another report from Babcock's laboratory by Jenkins (1940) in which comparable data are found for isolated endemic island forms. Jenkins studied four closely related "species" of *Crepis*, three of which are endemic to Madeira and the Canary Islands. The fourth is a widespread species from Africa and Europe, with a subspecies endemic to Madeira and another one introduced into Madeira *(C. divaricata, Noronhaea, canariensis, vesicaria)*. All of these are well characterized by traits similar to those found in the previous example. All have the same chromosomal complement, are fertile *inter se*, and produce fertile hybrids. The differences, constant within a certain range of variation, are heritable and seem to be based upon multiple factors. In a general way, the facts are of the same order as described for the other examples. The important point is that here well-isolated insular forms are studied which have probably been isolated for a considerable geological time. The taxonomist describes them as species, as they are isolated endemisms. But the genetic analysis, as I interpret it, reveals that these long-isolated forms have not differentiated beyond the subspecific level. They do not interbreed, for lack of opportunity. But Jenkins described one introduction into Madeira which does interbreed with the endemic form. This example is useful because it shows in a concrete case that so-called insular species may be nothing but ordinary geographic subspecies, and that the expectation of more extreme divergence after isolation has not been fulfilled. We do not know why this is the case. But it may be assumed that the reason is to be found in the absence of that type of genetic change which is able to bridge the specific gap. It will soon be apparent what is meant by "that type of genetic change."

Material of the type reported now; i.e., a combination of a taxonomic analysis with a genetical one in case of an artenkreis, is not yet abundant. In the animal kingdom the only case known to me paralleling that of *Crepis* is found in tropical fresh-water fishes. Breider (1936) analyzed an artenkreis of different "species" of the poeciliid fish *Limia* living in the Great Antilles. Three of these "species," *vittata* from Cuba, *caudofasciata* from Jamaica, and *nigrofasciata* from Haiti, were used. Their differences are of the type found in the taxonomy of fishes: measurements of body parts, proportions, skeletal differences in vertebrae and fins, and color. In addition, there is a difference in regard to sex determination. One form is bisexual; the others have that labile type of hermaphroditism which is unfortunately called phenotypic sex determination by Kosswig and his students (following Hartmann's nomenclature). Crosses produce fertile hybrids, though a certain amount of sterility is observed. In addition, the embryonic sexual differentiation of the hybrid shows certain aberrant features, which, according to Breider, have to be considered signs of real specific difference. These hybrid features may be described in a general way as changes in the velocity of differentiation. (Note, however, that in *Lymantria dispar* the speed of development of the gonads is one of the typical traits of subspecific distinction [see Goldschmidt, 1933b]). Fourteen different distinctive traits were analyzed in hybridization experiments. Statistical analysis of the data shows that the differences are based upon what is assumed to be series of multiple factors. There are, in addition, genetic differences controlling the different types of sex determination. Breider states directly that "there is no difference in principle in the genetic behavior of these species hybrids from that of racial hybrids based on Mendelian polyhybridism." He thus concludes that these forms are, if not already species, at least incipient species, which have been produced by geographic isolation on the basis of the already existing genetic variation. The fallacy of this conclusion, which actually begs the question, has already been emphasized.

Finally, we may consider a last example of an artenkreis, though no genetic work on it is available. The peculiarity of this artenkreis, that of the gall wasp *Cynips* as studied by Kinsey *(loc. cit.)*, which we have had repeated occasion to mention before, is a twofold one. First, the artenkreis is comparable to such cases as were mentioned before in which the "species" were isolated on islands. *Cynips* is bound to its host, white oak, and in the southwestern United States this tree grows on peaks surrounded by desert, a condition which might be termed insular by way of host isolation (see also Kinsey, 1936, 1937). Second, this artenkreis (which is not Kinsey's interpretation) is a very huge and diversified one and permits, therefore, the distinction of many taxonomic subunits. This, again, makes this artenkreis very useful for demonstrating the pitfalls attending evolutionary conclusions from just this type of variation.

The genus *Cynips* contains an immense series of distinguishable forms ranging over America and Europe. Kinsey distinguishes six subgenera (which adherents of the artenkreis concept would call species, and which the lepidopterologists, as well as I, would call groups of subspecies of a single species). They occupy different geographic areas: *Cynips*, *Philonix*, *Atrusca*, *Acraspis*—different parts of central, eastern, and southeastern United States, down into Mexico; *Antron* and *Besbicus*—the Pacific Coast of the United States. Each of these "subgenera" contains numerous "species" which may be grouped for taxonomic purposes into "complexes." It is obvious that these "species" correspond to the subspecies in the artenkreis concept, and to subsubspecies, if the whole formenkreis is assumed to be a single rassenkreis. Actually the relations of the individual "species" in the chains to each other are exactly the same as the relations between the different subspecies (and subsubspecies) of, for example, *Lymantria dispar*. Kinsey describes a series of morphological and physiological individual characters, none of which alone characterizes a subspecies and which vary independently over the range; formation of chains in which the nearest members are most similar and of the remote members most different; presence of transitions between adjacent "species," free hybridization where not prevented by geographical or host isolation; genetic differences frequently of a low order, not more different than are the differences between alternating generations of the same form, or genetic differences of a multiple-factor type, as concluded from the variability of hybrid populations (no genetic work is yet done); adaptive value of some physiological characters, no apparent adaptive value of others. A special feature in *Cynips* (maybe only a result of more intimate knowledge) is that the chains of forms called "species" are not simple linear chains, but they branch out in different directions, thus leading to more than two extreme ends. But all these features of the individual series of "species" within a so-called subgenus are also found where two of the "subgenera" become adjacent: here also the differences are not greater than between two typical adjacent "species." Fig-

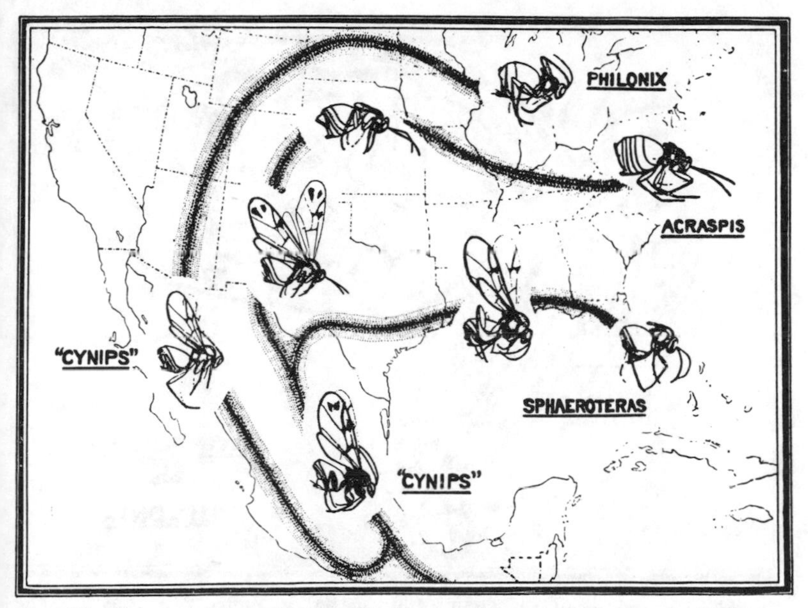

Fig. 27. Extreme types ("genera") in a continuous chain. See text. (From Kinsey.)

ure 27 represents the chains of geographic forms (called "species") the ends of which represent the distinct "subgenera" (called even genera in one of Kinsey's papers), namely, *Philonix*, *Acraspis*, *Sphaeroteras*, and *Cynips*. "But the limits of the genera are indefinable because of the continuity of the chain." Figure 28 shows a detailed elaboration of the same chains, indicating the localization of all eighty-six "species" involved.

Kinsey's conclusions from these facts are remarkable. He concludes that exactly the same characters differentiate

Mendelian races, species, and the "best-defined genera." (But he shows himself that one of the "best-defined" genera was characterized by wing length, which turned out to be a nonheritable seasonal modification!) In genetic terms the facts mean for Kinsey that mutations within a population, but not isolated, increase the variation within the species. If "the very same mutation" is isolated or selected, a new

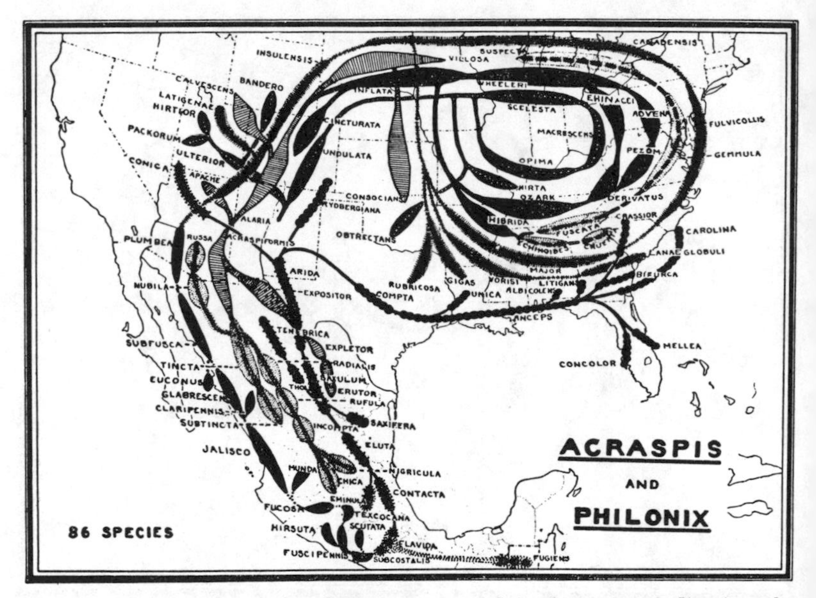

Fig. 28. The details of the chains represented in figure 27, indicating the location of the individual "species." The "genera" are marked with different shadings. Where two such are joined; e.g., *Acraspis* and *Philonix* in Mexico, the transition is not different in magnitude from that from any one "species" to another. (From Kinsey.)

species ensues differentiated by the same characters which before characterized the Mendelian races. "But, finally, if the specific differentiation involves major mutations which are continued through any series of species, we ordinarily consider that two genera have evolved." In all these cases the same characters are involved.

There is no need to go into further detail. All the facts discussed thus far, genetic and otherwise, demonstrate clearly the errors contained in these conclusions. Our interpretation of the facts, which is in harmony with the conclusions reached previously in this chapter, is that a very complicated case of microevolution within a species is involved, whatever nomenclature the specialist chooses to apply to the groups which he is able to distinguish. The case which probably contains the most elaborate data thus far presented for an individual chain of forms shows, in our opinion, how far microevolution within a species will go if circumstances are favorable. But the facts fail to give any information regarding the origin of actual species, not to mention higher categories. It ought to be added, however, that some of the data are difficult to understand. It seems clear that the so-called subgenera, mentioned above, are in part isolated from each other. But I could not make out whether some of them live also side by side without interbreeding.

We realize, then, that in a discussion of artenkreise from the standpoint of evolution we have to free ourselves first from the unconscious bonds imposed by a nomenclature which is apt to obscure the issue. We discussed before (see diagram, p. 166) the presence of different categories of distinguishable forms within a rassenkreis. Taxonomic usage permits the naming of subspecies only. But if a rassenkreis is well known, and especially if it spreads over a large area, more and more minute differences become known which permit a further subdivision. At this point there sets in what Stresemann termed a matter of taste, as far as the taxonomist is concerned. Let us describe the situation in symbols. The actual material is a large set of distinguishable forms, replacing each other geographically and isolated more or less, depending upon ecology and the geological features of the area. The forms are of different taxonomic value and can be grouped into more or less clearly discernible subgroups, which sometimes are delimited by natural barriers (e.g., islands) and therefore appear clear-cut, but sometimes are only chosen artificially in order to handle an otherwise unwieldy mass of material. Let us symbolize the entire formenkreis by the figure I, and the distinguishable lower units by 1, 2 . . ., then A, B . . ., then a, b. . . . The entire group then looks like the following diagram, if we assume a linear arrangement for simplicity's sake:

I	1	A	a, b, c	geographic arrangement
	major isolation ←———→			
	2	B	d, e	
		C	f	
		D	g	
	major isolation ←———→			
	3	E	h, i, k	
		F	l, m	
		G	n	
	4	H	o	
		I	p	
	5	J	q, r	
	6	K	s	
		L	t	

The taxonomists, then, might describe this group in different ways. The first student (e.g., of *Lymantria*) calls the group the "rassenkreis" of the species I. He describes A–L as subspecies, mentions that two major gaps (←——→)

permit one to distinguish three groups of subspecies (1, 2, 3–6) which he calls the Eastern, Western, etc., groups. Within the subspecies he distinguishes special, recognizable forms (a–t) which, however, are not named (the subsubspecies) and he may even recognize but not name further subsubsubspecific forms not contained in the diagram. Another taxonomist calls I a "genus geographicum" or "artenkreis" and feels justified in naming 1 and 2 as species, and 3–6 together a third species. In this case (e.g., *Crepis, Junco*) B, C, D would be the subspecies of the species 2, and a–g not named subsubspecies. But 3–6 would be called individually groups of subspecies, or as the case might be, subspecies, which would assign E–L either as subspecies to groups 3–6 or leave them unnamed subsubspecies of subspecies 3–6. In this or a comparable way an artenkreis would be obtained. Still another taxonomist would call a–t the "subspecies" (e.g., *Papilio*). They would be grouped into different geographic groups and subgroups, say, 1, East Asiatic; 2, American; 3–6, Palearctic groups. The result would be a huge rassenkreis. Finally, another one would call a–t "species," combine these into complexes A–L, these again into subgenera 1–6, these again either into the genera 1, 2, and 3–6, or into the one genus I (e.g., *Cynips*). Certainly the taxonomic "taste" would permit all these procedures, more or less dependent upon the type of material and its quality.

Now we come to the evolutionary significance. The first and third authors would claim that they deal with microevolution within a species. The second taxonomist claims that he has shown that subspecies gradate into species and are therefore incipient species. The last one, finally, insists that he has shown that the higher categories up to at least the genus are built up by slow gradation from the lower ones. My conclusion, however is: Whatever nomenclature the taxonomist deems necessary, the whole formenkreis is a single genetic unit (in agreement with, for example, Kleinschmidt), a case of microevolution within a single species, the subdivisions of which may interbreed and show ordinary

Mendelian differences. Nowhere have the limits of the species been transgressed, and these limits are separated from the limits of the next good species by the unbridged gap, which also includes sterility. Formenkreis, artenkreis, and rassenkreis, then, are quantitative variants of the same thing. They are distinguished from each other by some taxonomists for reasons of taxonomic technique. Their intrinsic difference, however, is, I am confident, nothing but the size of the area inhabited by the species, the more or less visible isolation of some parts of the area by geographic features and the amount of diversification—subdivision, subsubdivision, etc.—which the nature of the material permits one to recognize. Some taxonomists claim that the existence of artenkreise demonstrates the gradation of the subsubspecies via subspecies into species at the end-points of the subspecific range, and even of those into subgenera and genera. I think that such conclusions amount to self-deception produced by mistaking nomenclature for evidence of macroevolution. In an analysis of macroevolution the real gap—usually called specific, though the taxonomist might call it whatever his technique requires—is found where two species live or can live side by side without being able to interbreed or where they produce sterile hybrids if made to interbreed. This position, that of the classical Mendelian scholar Bateson, is also held strictly today by J. Clausen. At this point the "unbridged gap" of the authors (*loc. cit.*) is found to be related to the entire organization of the different forms; i.e., species in the sense of the evolutionists.

C. SPECIES FORMATION COMPARED TO SUBSPECIFIC DIFFERENTIATION

The conclusion at which we arrived from many different angles; namely, that the origin of species is not to be conceived of as occurring via geographic races or the members of a rassenkreis, does not, however, in itself exclude the possibility that the same type of genetical process leads to both types of diversification. This means that just as the subspecies are the product of accumulation of micromutations, presumably largely for the sake of adaptation to local conditions, a more extreme accumulation of such micromutations, independent of the subspecific differentiation and probably also independent of adaptation at the outset, will produce the specific differences. This viewpoint is certainly a direct consequence of the theory of mutation; i.e., of the conception that all hereditary differences must be expressible in terms of gene combination. Later we shall again challenge this neo-Darwinian conclusion, as we have already repeatedly done. In this paragraph we shall discuss some of the claims for its correctness, claims which are independent of the respective viewpoint regarding geographical races.

We have already quoted the conclusion which Dice (Dice and Blossom, 1937) has derived from a genetic analysis of *Peromyscus* species. He agrees that the hereditary traits distinguishing geographic races are not those which enter into the specific differences. But he expresses the opinion (held by most geneticists and taxonomists) that isolation followed by mutations producing sterility, and accumulation of random mutations, will lead to species formation. (Cautiously he repeatedly adds the word "theoretically.") Certainly nobody can deny the general theoretical possibility of such a neo-Darwinian process, and much of the recent work in mathematical evolution is intended to prove such a possibility on the basis of studies in population problems. We have referred to Wright's, Fisher's, and Haldane's calculations favoring this idea. I must confess that I am somewhat skeptical as to the significance of these ingenuous calculations which, as I emphasized before, are decisive only if the neo-Darwinian viewpoint of evolution by accumulation of micromutations is taken for granted. My skepticism is increased when I see that wherever facts of geographic variation have been studied in connection with geological history the result invariably is that the geographic races in question have been formed after isolation in the Tertiary or the Glacial Age (Anderson, Babcock, Breider, Goldschmidt, Kinsey, Rensch, etc.). Subspecific differentiation is apparently a slow process leading after a rather long time to nothing but small differences, permitting the form to spread into new areas. I wonder how long a species or a genus, not to speak of a family, would need for evolution if this were the usual method? I have my doubts, therefore, whether this method is probable and not merely a "theoretical possibility" (see quotation from Dice, p. 153).

Actual taxonomic or genetic facts which have been adduced to prove the neo-Darwinian point (apart from those which have already been discussed) are the following: Rensch (1929) emphasizes especially that the visible morphological differences by which species are distinguished are of the same type as those which distinguish subspecies, except that they are less variable and more numerous. This argument does not seem very convincing. The visible differences upon which the taxonomist must rely are limited. They will always tend to be measures, proportions, color, etc. If these are taken individually, they are of the same order. The beak of a bird within a single genus will be long or short or broad, etc., and if the beak varies within subspecies, this will also mean long or short, etc. But it is the combination of numerous such differences into a whole, separated by a gap from another comparable whole (in addition to sterility), which makes so many species so different that they have been characterized as different systems, isolated by bridgeless gaps. Rensch, who is predominantly an ornithologist, forgets the statements by such leading ornithologists as Kleinschmidt and Stresemann, who recognized the bridgeless gaps and tried to extricate themselves from their dilemma, one by recourse to theology, the other to phylogenetic speculations.

From the genetic standpoint it has been argued in favor of neo-Darwinism that crosses between two species, which may even differ in chromosome numbers and undoubtedly are good species, may exhibit a Mendelian segregation when part of the chromosomes are able to conjugate. J. Clausen (1931) has stressed this point in his *Viola* studies. But the word segregation may mean different things. It may mean segregation of Mendelian mutant loci. This is certainly to be expected in species crosses (provided that the chromosomal mechanism works) when the species differ in one of those Mendelian characters which are so frequently found as parallel mutations in different species. Numerous such cases are known in plants (and a few in animals, e.g., hawk moth) in which mutant characters show simple segregation in species crosses which contain such a mutant in one of the parents (for examples, see Renner, 1929). But these are accidental features independent of the actual species differences, features bound to occur when ordinary mutants are present and the chromosome in question is able to segregate normally. But there is also a segregation which is certainly not based upon simple gene differences; namely, the segregation of differences of whole chromosomes. The segregations in Oenothera, based upon interchromosomal linkage, may look like simple Mendelian segregation, and have been looked upon in that way, though actually whole complexes are segregating. The same may happen with trisomics. A segregation of differences in specific crosses is therefore not conclusive proof that the differences are due to gene differences; i.e., point mutations as in varietal crosses. I should think it rather dangerous to conclude from such facts that the actual specific differences are the same simple differences as in subspecific crosses. Already on the level of subspecies mono- or dihybrid differences are rare (in animals), and even colors segregate according to a multiple-factor scheme, as we have seen in a number of examples.

Leaving aside such cases in which mutants are involved, which exist as parallel mutants in different species, and also such crosses in which the so-called species is obviously a member of a rassenkreis, the results of species crosses in plants may be grouped in three major categories. We exclude from the discussion cases in which different chromosome numbers and all the complications of atypical chromosome behavior are involved. (Some of the latter cases have been used to demonstrate Mendelian segregation of species characters [see Lammerts' (1934) interesting work on *Nicotiana* species], but the facts are so complicated that an interpretation in terms of whole chromosomes or chromosome segments is also possible.) J. Clausen (1926, 1931) mentions Mendelian segregation of species characters in *Viola*, Honing (1923, 1928) in *Canna*, and Chittenden (1928) in *Primula*. A scrutiny of such data, however, seems to reveal a more complicated situation: part of the differential traits show in F_2 an unanalyzable behavior; others segregate more or less and can be conceived of as based upon a few Mendelian differences, though actual proof is missing; a few characters seem to exhibit a simple Mendelian segregation, but the ratios are not good and the results sometimes differ in individual crosses. Whatever the actual basis of the facts may be, a simple Mendelian behavior is only visible when expected, and a proof for simple genic differences between species characters is missing. It can hardly be expected, either, in view of the facts to be reported which are more frequently found in species crosses. These statements might be called hyperskeptical. That this is not the case may be illustrated by a quotation from a recent paper by Mangelsdorf and Reeves (1939), who try to prove the opposite point of view. They write: "Nevertheless there is not a single case . . . in which sufficient data have been accumulated to demonstrate that the differences between species or genera are governed by definite genes located in particular chromosomes." The authors then set out to furnish such proof for the "genera" *Zea* and *Euchlaena* by using marked chromosomes of *Zea* for crossing. They find what looks like linkage between the marker genes and the differentiating characters of maize and teosinte. But these characters appear concentrated upon a few chromosomes only. Nevertheless, the conclusion is drawn that this demonstrates gene differences between the genera, though the opposite conclusion appears to be more appropriate. But in a chapter which follows immediately after the one reported proof is furnished that teosinte (*Euchlaena*) is actually maize contaminated with a few translocations from *Tripsacum* chromosomes, and that the *Euchlaena* characters "are not due to single genes but to segments of chromatin which are usually inherited intact and to this extent behave as single genes." These results lead to a discussion of the next group of facts.

The second type which is encountered is the following. In F_2 an immense variability occurs, so that among even thousands of individuals there are hardly two alike. Not only are all kinds of conditions intermediate between the parents found, but also perfectly new types, among which many clearly pathological ones. Most modern authors who worked with these cases; e.g., Baur, Lotsy, Wichler, Honing, Winge, etc. (see Renner, 1929), tried to analyze such situations in Mendelian terms on the basis of their belief that species must differ from varieties in the possession of a larger number of genes. Usually they did not succeed and therefore assumed a polymeric segregation with many genes, which is difficult to analyze and therefore to prove or disprove. Thus, Baur needed an estimated number of more than one hundred gene differences to visualize the results in the F_2 of an *Antirrhinum* cross. But other authors, who attacked the problem without strong neo-Darwinian convictions, obtained other very remarkable results, which are usually left out of account in discussions of the subject. These authors succeeded in grouping the F_2 material into classes with a very significant result. Heribert-Nilsson (1918), in analyzing such a cross between species of willows, could arrange the F_2 individuals in classes according to a scheme involving three genes for breadth of the leaf. But each of these genes also controlled all other leaf characters, and in addition, size, color, periodicity—in short, all other distinguishing characters of the species. In discussing this material Renner remarks that he prefers to assume, instead of a few genes with such a diffuse action, three linkage groups in different chromosomes. Now this is a crucial point. A specific difference in whole chromosomes (or sections of such) which are so different that little if any crossing over occurs, will also lead to a kind of segregation. But this is not gene segregation and cannot be used to derive specific differences from accumulated micromutational differences.

Here a difference on another level is clearly indicated; namely, on the chromosomal level. I wonder if many of the rather vague polymeric differences of F_2 individuals described in species crosses would not lend themselves to analysis of a very different type, once the search for gene recombinations were abandoned. I am encouraged in this belief by quite a number of other facts recorded in work on species hybrids in plants and animals: appearance of segregation with very aberrant numbers; appearance of a majority of intermediates, or appearance of a majority of individuals resembling one of the parental types. To mention only a few cases: There is the much discussed case of speltoids in wheat and the specific difference between *spelta* and other species. Though all the details are not clear, it follows from the work of Nilsson-Ehle, Watkins, Huskins, Winge, and others (see review by Schiemann, 1932) that a difference on the chromosomal level is involved. Darlington (1939), when mentioning this case, remarks that "the characteristic groups of differences between species or races [? author] are often found to be closely linked or even inherited as a single unit." One such example is the following. Winge (1938) crossed the species *Tragopogon pratensis* and *porrifolius*, both having the same chromosome number. The F_1 was almost sterile, but some F_2 individuals were obtained. From these the pure species could again be recovered by selection. The same has happened after crossing *Verbena tenera* and *Aubletia*. Winge correctly concludes that here the species difference must depend on chromosome differences; but he adds, "i.e., on segregating gene differences in chromosomes." If the latter were true, the results would indicate that the gene differences are very few in number, which is of course utterly improbable. Chromosomal differences, however, would easily account for the facts if chromosomes segregate as a whole. We know from Navaschin's crosses (1927) of *Crepis setosa* × *capillaris* that the complete chromosome set of one species can be recovered in the cytoplasm of the other, and that the external effect is the recovery of the specific type. It might then be said that the recovery of whole chromosomes is identical with the recovery of all genes therein and that the facts are therefore in harmony with the theory of the gene. The cytological facts soon to be discussed will, however, show that such facts are in better agreement with chromosomal action which cannot be reduced to integrated action of individual genes. But this anticipates later discussions.

A third type of behavior in species crosses in plants which is frequently found is the following. The F_1 is intermediate and most characters in the F_2 appear very different from those in the F_1. East (1913, 1916) first analyzed such a situation in tobacco. His explanation was that the specific differences (for each individual character) were due to a series of polymeric genes. Of the innumerable recombinations possible in F_2 only a few are viable, for different reasons, and the majority of segregants therefore never become visible. The further development of this idea has led to a point where its own purposes begin to be defeated. The Swedish group of geneticists (see Rasmusson, 1933) claims that the number of genes involved in ordinary inheritance of quantitative characters is probably 100–200. As species are usually distinguished by numerous quantitative characters (see Zarapkin's count for *Carabus*, p. 67), a perfectly ridiculous number of small mutant steps becomes necessary in order to account for quantitative specific differences. Where such an assumption finally leads is clearly indicated in the latest analysis of this situation by Anderson (1939). He realizes clearly that in such a situation the numerous genes for one and for all quantitative characters must be closely linked. Why must we, then, subdivide the linkage group; i.e., the chromosome or chromosome segment, into innumerable polymeric genes, the existence of which cannot be proven in these cases, and which put an unbearable burden upon evolution by small steps? A study of the data, unprejudiced by the postulate of expressing the facts in terms of gene mutations, shows in this case, as in those mentioned before, that the facts point urgently to a chromosomal and not a genic difference between the species.

It is worth while to stop here for a moment and to recount the methods of specific differentiation by accumulation of mutations advocated by the different authors mentioned: (1) simple mutants if properly isolated (Kinsey); (2) a few Mendelizing mutants (Babcock, Chittenden); (3) a large number (order of magnitude: hundreds) of Mendelizing mutants (Baur); (4) a number of ordinary mutants plus an unanalyzed remainder, probably polymeric mutants (Clausen); (5) a large number of polymeric mutants (order of magnitude: hundreds) (Breider); (6) an immense number of polymeric mutants (order of magnitude: tens of thousands) (East, Anderson). Looking over this list we must realize the unsatisfactory situation into which neo-Darwinian preoccupation has led.

Returning again to the facts, it must be pointed out that what little material is available in animals points in the same direction. In animals most of the fertile species crosses made with good species and within a normal chromosome mechanism are found in Lepidoptera, especially in the hawk moth, where a number of real species hybrids which produced an F_2 have been described by Fischer (1924), Lenz (1926), Federley (1928).[14] (Cases which are usually listed as species hybrids but are obviously results of subspecific crosses have been discussed before for mammals and fishes. A similar example for birds is found in some of Phillips' duck crosses.) The description of the F_2 shows a very complicated array of types, which the authors vaguely describe as a sign of polymeric segregation. Looking at the detailed description, however, we find that the differences of the two species cannot be resolved into individual Mendelizing traits. There is, of course, segregation when the chromosomes conjugate (Federley), but the proof for the independent segregation of numerous Mendelian genes is lacking. There are actually indications that whole-character complexes segregate as a whole, an occurrence which Lenz, for example, registers but dismisses as "impossible." This

14. Crosses of species with different numbers of chromosomes or incompatible chromosomes are excluded from the present discussion.

is characteristic of neo-Darwinian prejudice which, in my opinion, blocks progress by erecting signs: "Verboten." (Here ought to be mentioned Cavazza's [1938] work on fertile mule mares which produce mules with a jackass but horses with a stallion! But the behavior of the chromosomes is not known.)

In other similar hybrids of hawk moths it has been claimed that genuine Mendelian behavior was found, for example, by Federley (1927) for the hawk moth *Chaerocampa elpenor* × *porcellus*, by Bytinski-Salz and Günther (1930) for *Celerio galii* × *euphorbiae*. The actual description of the backcrosses given by these authors, if accepted at their face value, shows a continuous series of transitional types which cannot be separated from each other, and also the appearance of new characters. Therefore the conclusion is that a multifactorial basis is present, a difference in many genes, as required by theory. This would then mean that specific segregation is not different from a subspecific one. In reviewing such material, P. Hertwig (1936) expressly states that it is not worth while to spend more time and money on such work, as no results which differ from ordinary varietal crosses are obtained, thus exploding the idea that there might be a different type of heredity for species. This sounds rather simple and final, but if we are not content merely to read the conclusions of the authors and if we look more closely at the material for which detailed reports are available, the picture is a rather different one, and actually resembles more closely Heribert-Nilsson's *Salix* case discussed above (see also the quotation from Darlington, 1939). Bytinski-Salz lists twenty-five distinguishing species characters for his moths. On the basis of these characters he divides his F_2 hybrids into nine classes ranging from one to the other species. Each class is characterized by a definite condition with regard to each distinguishing trait, and together they form a series. "As the individual traits within a class may vary, the individual is assigned to the group on the basis of all of the analyzed traits. These permit us nevertheless to assign individuals with a greater deviation of one or another trait to a definite group. Thus a class type of considerable uniformity appears with some fluctuation in a plus or minus direction." I am at a loss to understand how such a picture can be regarded as due to innumerable Mendelizing genes. It is obvious that this type of segregation is only possible if either one pair of genes with a few modifiers controls all the specific differences, or, just as in *Salix*, if sets of linked genes; i.e., whole chromosomes or major parts of them, are involved. As a matter of fact, Bytinski-Salz recounts a few instances in which a trait occasionally appears in a wrong class, or a few traits appear in what looks like a recombination. But this can hardly invalidate the foregoing conclusions, as within the small number of hybrids available each individual ought to show a different combination, but no arrangement in a few recognizable classes ought to be possible if Mendelian polyhybridism is actually involved.

Here is another example of the same type which has also been interpreted as an ordinary case of multifactorial Mendelian segregation. Lenz (1928) crossed the two moth species *Epicnaptera tremulifolia* and *ilicifolia* and obtained a fertile F_1. "A minority" of the F_2 individuals could not be distinguished from one or the other parent. The rest were intermediate with different grades. Also a few "recombinations not seen in F_1" were found. From this it was concluded —and the conclusion is accepted without criticism by other authors—that the two species differ only in a few genes. A similar example could be taken from Harrison's (1917) crosses of the moths *Poecilopsis pomonaria* and *isabellae*, or from one of Phillips' (1921) duck crosses. Harrison especially emphasizes a segregation of specific characters *en bloc*. Other botanical examples of the same phenomenon are found in Renner's review (*loc. cit.*).

Other fertile crosses which have been used as examples of Mendelian segregation have been made between domesticated and wild breeds of the same genus. Wolf and dog (Iljin 1934), wild and domesticated silk worm (Kawaguchi, 1934) the ostrich "species" (Duerden, 1920) hardly are reliable material for our problem, even if simple mutational characters of the domestic breeds are disregarded. These remarks also apply to a certain extent to crosses between species of mice. W. H. Gates (1926) and Green (1931–35) crossed the Asiatic *Mus bactrianus* to *Mus musculus*. Different mutational strains of the latter were used. Gates reported that even these mutants of the domesticated breed did not show a simple Mendelian inheritance: the characters of the Japanese waltzer (a domestic variety of *bactrianus*) tended to remain together in the F_2 instead of assorting freely (the results are based on very large numbers of individuals!). Not only did the Mendelian characters of the domestic breeds behave in this way, but also the quantitative characters which distinguish the species; i.e., measurements and structural differences of the skeleton gave the same result. Gates concluded that "all the characters of each parental species tend to associate together in inheritance." This agrees completely with the facts already reported for *Salix* and Lepidoptera. It is true that Green (1935) does not agree with Gates. He claims an ordinary random assortment of mutant characters ntroduced into the cross by the house mouse, at least as regards some of the mutant color genes. This, however, is not the decisive point, which is the behavior of the specific lifferential traits. Green himself finds a tendency for certain color and size characters of *musculus* to remain together, and interprets this as linkage between color genes and size genes. I prefer to use Gates's interpretation for this situaion. I suspect that many more data of the same type would be available if their analysis had not been made on the basis of the firm conviction that Mendelian behavior of individual raits has to be demonstrated. The question of mutant traits which are independent of the actual specific differences (e.g., oat-color mutants of house mice introduced into a species ross) will come up again later when the problem of soalled identical genes in different species is discussed.

Only one more point need be added in order to show that he facts regarding species hybrids may lead to interpretaions which are rather different from the "conventional" ones, s East (1935) puts it. East, who has studied innumerable

F_1 hybrids from a great many different species crosses in *Nicotiana*, describes their behavior in terms which he himself calls "rather heterodox," though he seems to agree with the orthodox opinion that accumulated micromutations make up the specific differences. He finds in these hybrids a "phenotypic reaction pattern" (a term used by Sinnott). If I correctly interpret its meaning, the term indicates that the hybrid reacts as a whole and not as a mosaic of individual genes. Gene changes producing the qualitative effects which are used in ordinary Mendelian work appear different from those which accumulate in specific differences, the latter being quantitative and difficult of demonstration. The former (qualitative, varietal) "are, by their nature, usually incapable of playing a part in natural evolution, though they may be very advantageous in building up knowledge of the hereditary mechanism." (This phrase sounds almost word for word like Johannsen's statement quoted on page 8.)

The pattern of the hybrids shows that each species genome controls "normal orderly," as opposed to "restricted," processes. (I interpret this as meaning the control not of small features at the periphery of organization, but of the general processes of orderly growth and development.) Therefore, East—obviously unwilling to take the step beyond the genes and their accumulated micromutations—concludes that the various genes of each genome produce slight changes in developmental patterns in different organs. He recognizes that the standard type of mutation is without significance for evolution; he realizes that species differences are differences of the whole developmental pattern and, as individual gene mutations can hardly be recognized on this level, he assumes that a multitude of micromutations must have accumulated to build up the pattern-controlling new genome. I think that these observations, which had been anticipated to a considerable extent by Goodspeed and Clausen (1916) (see below), fall in line with all the data discussed before. All the examples, then, demonstrate clearly that the facts, if closely scrutinized, are not at all what they appear to be in reviews and textbooks. I further think that the real meaning of the facts will become clear only after the decisive step has been taken of completely discarding the concept of accumulated small gene mutations as the material of macroevolution. The following chapters will show what is meant by this. But one point may already be emphasized here. The facts reported indicate differences between species which are on a chromosomal level and, maybe, frequently even on a genomic level. If species are formed by accumulation of gene mutations, they must possess numerous homologous genes which will Mendelize. Ordinary Mendelian segregation will follow with the usual complications of linkage and crossing over. If, however, whole chromosomes or groups of them segregate, it means, according to all our cytogenetic knowledge, that the homologous chromosomes do not have the same pattern of loci, that they are actually not homologous in detail. We shall later go into the details of this situation. Here we shall note only that specific differentiation has actually turned out to involve a chromosomal reorganization. Assuming, for argument's sake, that the chromosome is a string of genes, we are confronted with the following alternative: Either the mutant genes are alone responsible for the specific differences, and their different order in different species, which accounts for the special features reported, is a chance condition without any significance; or the intrachromosomal pattern is a feature which plays an active part in specific differentiation. In the latter case the reported facts are highly significant. We believe this to be true and shall soon discuss the reasons for our conviction.

7. Conclusions

We have repeatedly indicated in the course of our discussion the conclusions which we have to draw from our survey of the facts of microevolution, and have, I think, covered all important angles of the problem. Only a short summary is therefore needed before we turn to the problem of macroevolution.

A survey of the facts relating to microevolution; i.e.,

evolution within the species (or whatever two different, nearly related forms separated by "an unbridged gap" may be called severally), has led us to reaffirm the conclusions which we have drawn in former papers: Microevolution within the species proceeds by the accumulation of micromutations, in addition to occasional upshoots of local macromutations, or polymorphic recombinations of such. The lowest taxonomic unit used for practical purposes (that is, for the sake of unequivocal labeling), the subspecies or geographic race may in many cases be subdivided into subgroups distinguishable with different degrees of certainty, and resubdivided even as far as individual colonies. In addition, what are subspecies in one form may be on the same genetic level as subsubspecies in another, according to the amount of information available, the usefulness of the respective traits for taxonomic description, and the special type of subspecific spreading over smaller or larger areas. The differences between two subspecies are usually clinal, merging into each other, except when isolation produces sharper differences. But the clinal character may be obscured if subspecies located in separate centers form clines of subsubspecies radiating from these centers. The subspecific and lower differences are based upon a number of hereditary traits, most of which do not show the simpler types of Mendelian inheritance. The character of the individual subspecies is the result of a definite combination of these traits, each of which may vary independently within a rassenkreis of subspecies. Many, if not most, of these traits are directly or indirectly adaptational, and their intraspecific variation follows the corresponding variation of the different climatic or other conditions to which adaptation is made. These geographic races are frequently arranged in the form of continuous chains with a continuous linear type of variation of the individual characters. This type is found only when some of the conditions to which adaptation is vitally necessary have an arrangement of a gradient type. If this is not the case, or if nonadaptational traits are involved, a correspondingly irregular pattern of distribution and of traits may occur

The series of subspecies, or rassenkreis, is separated by a gap from the next one; while the characters of subspecies are of a gradient type, the species limit is characterized by a gap, an unbridged difference in many characters. This gap cannot be bridged by theoretically continuing the subspecific gradient or cline beyond its actually existing limits. The subspecies do not merge into the species either actually or ideally. Border cases which have been interpreted in a positive way can be brought into line with these conclusions. Nor can the gap be bridged by the assumption of slow accumulation of micromutations independent of subspecies formation. Microevolution by accumulation of micromutations—we may also say neo-Darwinian evolution—is a process which leads to diversification strictly within the species, usually, if not exclusively, for the sake of adaptation of the species to specific conditions within the area which it is able to occupy. This is the case for microevolution on the subspecific level of formation of geographical races or ecotypes. Below this level, microevolution has even less significance for evolution (local mutants, polymorphism, etc.). *Subspecies are actually, therefore, neither incipient species nor models for the origin of species. They are more or less diversified blind alleys within the species. The decisive step in evolution, the first step toward macroevolution, the step from one species to another, requires another evolutionary method than that of sheer accumulation of micromutations.*

IV. MACROEVOLUTION

At the lower level of macroevolution, evolution of species, genera, and even families, there is still available some information based upon collaboration of genetics and taxonomy. Above this point, however, experimental genetics is ruled out as a source of information, except for that part of genetics which is called physiological genetics, as we shall see later. Conclusions will have to be based upon generalizations derived from general genetics together with such insight as can be derived from embryology, comparative anatomy, paleontology. Such information is sometimes looked at askance by experimental geneticists as being speculative. But this is a narrow-minded viewpoint. Progress in biology is derived from coöperation of observation, experiment, and constructive thinking, and none of these can claim primacy. A good observation may lead to results which a meaningless experiment cannot achieve, and a good idea or analysis may accomplish with one stride what a thousand experiments cannot do. This truism, obvious as it is in the history of all sciences, is frequently forgotten in this era of overestimation of new techniques, which are tools of progress only when in the hands of constructive thinkers. We must therefore take whatever material is available in any field and try to use it to its full extent, subject to critical evaluation.

At the decisive point of incipient macroevolution; i.e., at the point of emergence of the separate species, a body of important facts is found which relates to the chromosomes as bearers of hereditary traits. The meaning of the facts in question is largely dependent upon our viewpoints regarding the architecture of the hereditary material. It is to these facts that we have to turn first in our search for a method of macroevolution, after we have discarded evolution by accumulation of micromutations.

1. CHROMOSOMES AND GENES

A. General

A CONSIDERABLE part of recent cytogenetic literature is concerned with the chromosomal differences between different species (and higher categories) and the behavior of the chromosomes in crosses. It has been frequently said that evolution is identical with the evolution of the chromosome set. The majority of the pertinent facts, important as they are in many respects, do not however, furnish any information regarding our chief problem; namely, whether there are methods of evolution other than accumulation of micromutations. There is no doubt that visible chromosomal differences are sometimes among the distinguishing characters of higher categories. This is especially true among plants. The facts are widely known and are represented in all textbooks of cytology and genetics, and are treated in detail in the books by Darlington (1937, 1939). Different species may have the same or a different chromosome number; their chromosomes may be distinguished by structural details (satellites, point-of-spindle fiber insertion); their chromosomes may show signs of being derived from each other by subdivision or, in other cases, by fusion of individual chromosomes; they may be present in heteroploid and polyploid series. But, on the other hand, considerable evolutionary differences may not affect the visible chromosome structures at all. The most extreme examples in animals known to me from my own experience, as well as from the literature, are found among the Lepidoptera and the mammals. In Lepidoptera the majority of all forms studied have chromosome numbers of 29–31 (haploid), all chromosomes of nearly the same size, shape, and arrangement in metaphase. This is the case in Rhopalocera (butterflies) as well as in Heterocera (moths). Federley (1938) records 58 per cent of Finnish Rhopalocera as belonging to this group; Beliajeff (1930) records twenty-one out of thirty-eight species from different families as having thirty-one chromosomes (haploid). I could place side by side two microscopes with meiotic figures taken from two quite divergent families of Lepidoptera, and the chromosome sets could hardly be told apart. On the other hand, one finds individual species with a small number of chromosomes within a family in which a large number is the rule, the size of the chromosomes indicating a fusion. Or one finds two closely related species with very different chromosome numbers. In other cases a group of chromosomes is united into one large unit in one of two related species (*Lymantria monacha* and *dispar*, according to Seiler and Haniel), or in one of two races of the same species (*Phragmatobia fuliginosa*, according to Seiler), or in one of the two sexes of one species (Seiler, Federley), or only temporarily as a result of fusion during gametogenesis (Seiler, Goldschmidt, Kawaguchi). In mammals the situation is similar, $n = 24$ being a number characterizing innumerable species from mouse to man. For details and literature see Vandel, 1938.

It is obvious, then, that the visible differences in regard to chromosomes are just one morphological character by which different species may or may not be recognized. It is also clear that these visible differences are not necessary features of evolution, as they may be completely absent. As a matter of fact, any experienced cytologist may extend this statement to many general cytological features. It is true that reptile and bird sex cells may generally be recognized by the arrangement of their chromosomes, and amphibian sex cells, by the configuration of chromosomes in mitosis; that the chromosome groups of Lepidoptera, Diptera, Orthoptera, and Crustacea have certain features of morphology and arrangement which set them apart from each other even when the same numbers are represented; that the maturation division spindles in the eggs of flat worms are quite characteristic of the group. On the other hand, similarities occur which bridge the largest gaps: the four chromosomes of a Protozoan (*Monocystis*, a gregarine) look exactly like the eight long and slender chromosomes of some higher plants and animals. It is further true that these differences, where they exist, may furnish important clues as to the phylogenetic relationships within a group (for instance, the work of Navashin and Babcock in *Crepis*). Furthermore, the data on chromosome behavior in crosses involving different chromosome sets are very important for the analysis of chromosomal pairing, etc., and some of these facts will soon be applied to our analysis. But in a general way the facts of cytogenetics, with the exception of certain phenomena restricted to plants, do not furnish information on the decisive question of whether macroevolution uses the same methods as microevolution.

There are a number of special facts, however, which are highly significant and tend to show, in our opinion, a method of macroevolution different from that of microevolution. These we shall select from the large body of cytogenetic information.

B. Chromosomal Races and Species

We have already mentioned cases in which small quantitative differences in the chromosomes of geographic races have been found (*Lymantria, Crepis*). There are a few other cases of a nearly related type, showing another kind of difference in races which otherwise may or may not be distinguishable (grasshoppers, moths, mole crickets; papers by Helwig, Seiler, Voinov, de Winiwarter). In this group of facts considerable importance is attached to those cases in which no visible racial divergence is found within the species, but in which subspecific differentiation occurs solely on the basis of chromosome structure. Of these I shall mention only de Winiwarter's (1937) cytological analysis of races of *Gryllotalpa grylloides* (the mole cricket). In three European races, not morphologically separable, the typical haploid chromosome numbers are 12, 15, and 14–17, in addition to differences in the sex chromosomes. It is assumed that the higher numbers are derived by breakage of larger chromosomes into smaller ones. Such cases indicate the existence of a type of genetic diversification within a species which is not at all linked with so-called gene mutation. The best information on this topic has been obtained for Diptera. An elaborate case was analyzed by Dobzhansky (1935, 1937) in *Drosophila pseudoobscura* in the United States. Here, despite a phenotypic uniformity of the species over its range, considerable variation in shape of the Y-chromosome is found. Seven types can be distinguished as V-shaped or J-shaped and the like, and each occupies approximately a definite area. One is found only in Southern California; another, only around Puget Sound, while others occupy larger areas.

More important is the work of the same author in collaboration with Sturtevant (Sturtevant and Dobzhansky, 1936; Dobzhansky and Sturtevant, 1938) on another case of geographically varying chromosome behavior in *Drosophila pseudoobscura,* again independent of visible racial differentiation. The well-known salivary-gland method of chromosome study in Diptera, introduced into genetics by Painter, made possible the comparison of the intimate structure of chromosomes: the point-by-point attraction between homologous chromosomes is made visible by this method, and therefore differences at any point can be detected by the lack of normal attraction. In this way it was found that one of the chromosome abnormalities found in laboratory work, the inversion of part of a chromosome, is a frequent feature in wild populations. I must emphasize at this point that in these cases there is no indication that anything has changed but the serial order of the constituent parts of the chromosome, usually described as the serial linear order of the genes within the chromosome. In other words, nothing but the chromosomal pattern has been changed. A large number of such different patterns have been discovered and assigned severally to strains from definite localities. All possible cases have been encountered: rearrangements present in some individuals in many localities; others which are always found in a percentage of individuals in a given locality; and still others which characterize a single locality.

Another very remarkable feature is that these inversions tend to be additive. Among the seventeen different types found there is a series of what are called overlapping in-

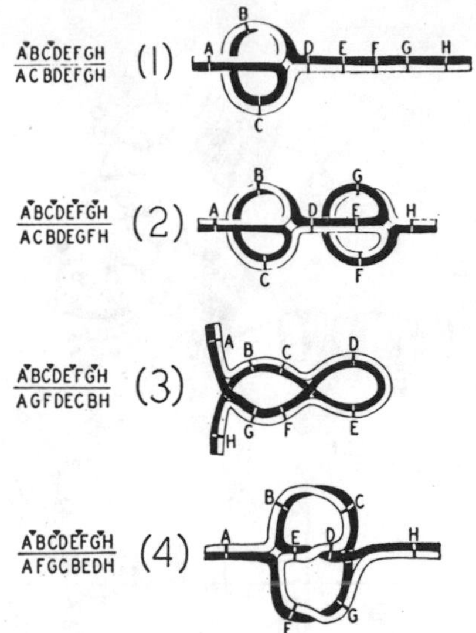

FIG. 29. Diagrammatic representation of the configuration of salivary-gland chromosomes in Diptera containing heterozygous inversions. The letters indicate the linear order of the loci along the chromosome, and the loop formation makes union of identical loci possible. 1, single inversion; 2, two independent inversions; 3, two included inversions; 4, two overlapping inversions. Left: arrangement of loci in the black and white chromosomes, respectively; triangles: points of breakage. (From Dobzhansky–Socolov. Courtesy *Journal of Heredity*.)

versions. This means that a second inversion has occurred which has one break inside the first one and the second outside. For example:

Start:	A B C D E F G H I
1. Inv.:	A—EDCB—FGHI
2. Inv.:	A—E—HGF—BCD—I

These are cytologically recognizable, and the breaking points can be established. (Figures 29 and 30 illustrate the situation. Figure 29 shows the type of conjugation of two homologous salivary chromosomes if one of them, the black one, contains an inversion; 2–4 represent the more complicated types of inversions [see legend]. Figure 30 represents the actual microscopic appearance of such inversions and the configuration of an overlapping inversion.) In carefully comparing the individual cases, it has been found that a kind of phylogenetic order can be established. Definite sequences are found which on geometrical grounds cannot be formed by a single event and which are only possible in two consecutive steps, a circumstance which then establishes a

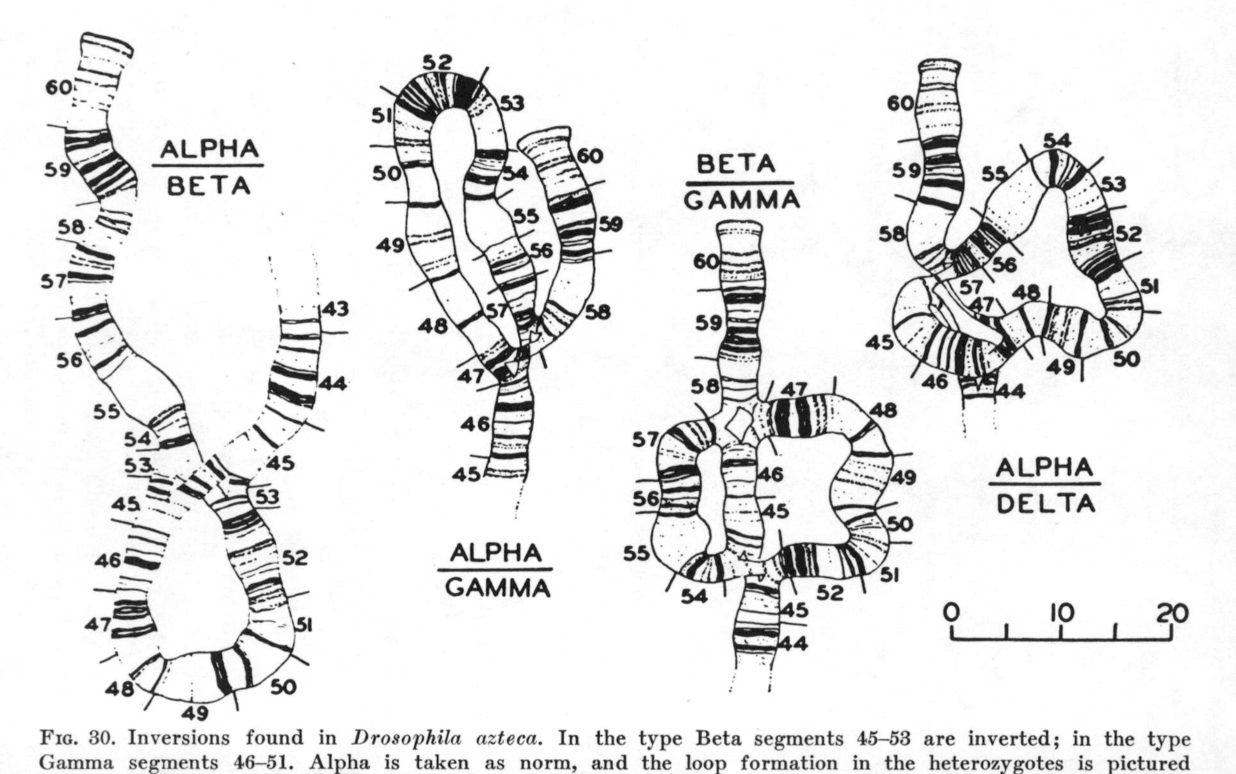

FIG. 30. Inversions found in *Drosophila azteca.* In the type Beta segments 45–53 are inverted; in the type Gamma segments 46–51. Alpha is taken as norm, and the loop formation in the heterozygotes is pictured (also for a fourth type, Delta). Conjugation of the Beta–Gamma chromosomes gives the figure-8 characteristic for overlapping inversions. (From Dobzhansky–Socolov. Courtesy *Journal of Heredity.*)

definite phylogenetic order in the origin of complicated intrachromosomal patterns. The all-important point in this ingenuous analysis is that within the species the internal chromosomal pattern may slowly change in a series of steps without any visible effect on the phenotype and without any accumulation of so-called gene mutations, small or large!

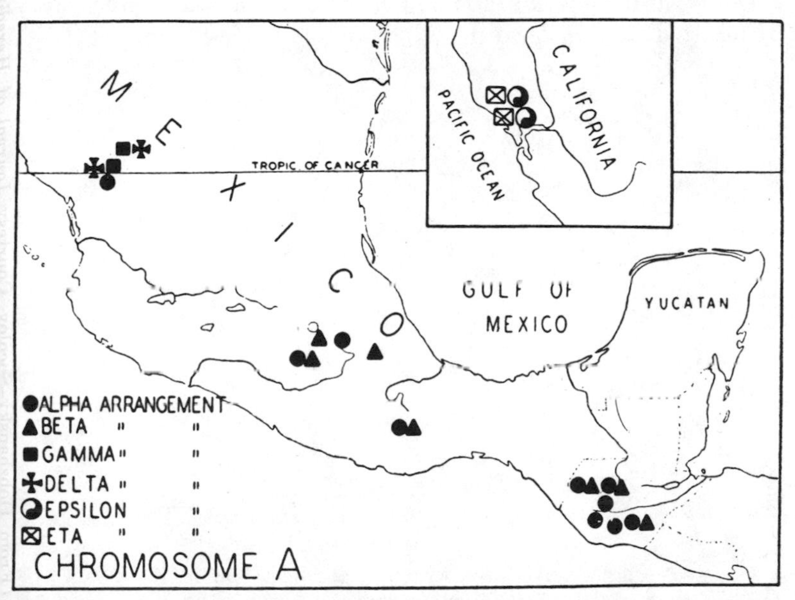

FIG. 31. Geographic distribution of chromosomal races of *Drosophila azteca* in California, Mexico, Guatemala, in regard to the inversions pictured in figure 30 (and two more). (From Dobzhansky–Socolov. Courtesy *Journal of Heredity*.)

Dobzhansky and Socolov (1939) have recently analyzed a second case of exactly the same type in the species *Drosophila azteca* (from which figs. 29, 30 are taken). Here again the "chromosome races" are distinguished only by inversions and not by any other chromatin rearrangements. Again there is a certain, not very regular relation to geographical distribution. The distribution in California, Mexico, and Guatemala of the inversion types which were distinguished can be seen in figure 31. But it must be kept in mind that these "racial differences" are limited mostly to quantitative variations in the proportions of the types composing a population (see above, p. 24, on the same condition for visible characters).

It might be assumed that we are dealing here with very exceptional conditions. But this is not the case. Dubinin, Socolov, and Tiniakov (1937) have made an extensive study of many *Drosophila* and *Chironomus* species from very dif-

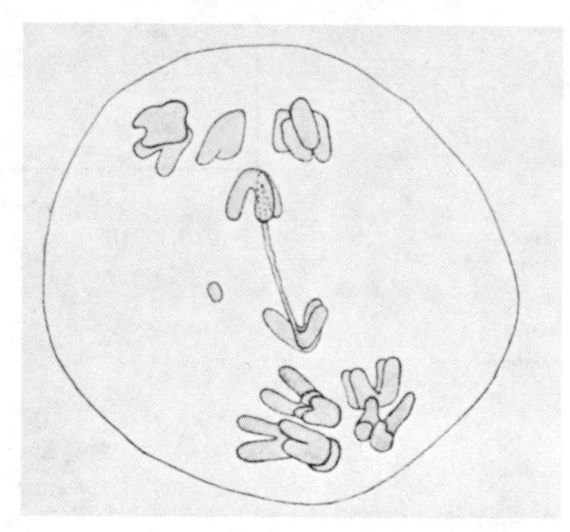

FIG. 32. Anaphase with bridge formation in the meiosis of rye, heterozygous for an inversion. (From Muentzing.)

ferent localities. They found a very large percentage of inversions, and in one species a wealth of different chromosome rearrangements. Single individuals could be found carrying four inversions, two intrachromosomal rearrangements and two complex aberrations.

Though an exact analysis of these features is thus far possible only in Diptera, by use of the salivary-gland method, there can be no doubt that similar features are widespread. In plants the presence of inversions is expressed in characteristic aspects of the meiotic chromosomes, which form bridges in anaphase (fig. 32). (For details and literature, see Darlington, 1939). Thus, groups otherwise indistinguishable but characterized by the presence of inversions, are frequently found within a species. Let us mention only two recent papers: Geitler (1938) found about forty-five different inversions in different chromosomes of *Paris quadrifolia* from many different geographical habitats. He has reason to assume that these inversions are phylogenetically old. In this case there is, however, a suspicion that the species is actually a structural hybrid (tetraploid on the basis of hybridization), and the case therefore belongs properly to the following paragraph, in which we shall discuss chromosomes of species. This difficulty, however, does not exist in the case of *Campanula persicifolia* (Darlington and Gairdner, 1937), which shows a close parallelism to the case of *Drosophila pseudoobscura*. Here different local "races" exist which, aside from the presence of trisomics, tetraploids, and segmental interchange (see below), exhibit different inversions.

It is very probable that the same phenomenon will be found more frequently in animals, if attention is given to it. Darlington (1936) mentions inversions in the grasshopper *Stauroderus*, and I have frequently noticed the presence of typical bridges in meiotic anaphases of insects, pictures which used to be discarded simply as artifacts. A modern investigation of a whole rassenkreis of animals, including the hybrids between the races and using the methods introduced by Darlington, is badly needed. But the few examples mentioned are sufficient to show that we are dealing not with an exceptional behavior but with a very definite feature of great evolutionary significance.

There is another case which closely resembles those just reported, though another type of chromosomal repatterning is involved, that of translocations between nonhomologous chromosomes (segmental interchange). This phenomenon also can easily be checked cytologically, even in the absence of genetic information, and without the salivary-gland technique which is restricted to Diptera. The presence of ring-and-chain formation of meiotic chromosomes in plants indicates translocations, according to Belling, who showed that rings and chains are produced as a consequence of the attractions between parts of homologous chromosomes which have changed place after segmental interchange (see textbooks of cytology and genetics). In the Jimson weed *Datura stramonium*, in which these cytological observations were first made, the phenomenon to be described also occurs (Blakeslee, Bergner, and Avery, 1937). This weed spreads over considerable parts of the world without clearly forming distinct geographic races, though simple Mendelian differences in flower color occur. Chromosomal rearrangements due to segmental interchange are found everywhere if tests are made for ring formation of definite chromosomes, the norm of which is arbitrarily set by a tester race. Depending on the chromosomes which have interchanged parts, these new pattern types are called prime type 1, 2, etc. It turned out that, just as in *Drosophila pseudoobscura* and *azteca*, populations in a given region may be of one type only, or preponderantly so, or of two different types (1 and 2, 2 and 3), or, finally of three or four different types without any clearly visible regularity. Again we are faced with the formation of new chromosomal patterns without concomitant accumulation of mutant genes or visible effects, and this within a species. Attention must be drawn to the difference between the two cases; i.e., preponderantly inversions in Diptera, translocations in *Datura*. Again we have to emphasize that we are not discussing a somewhat freakish behavior of one species, but an obviously widespread phenomenon. I need refer only to the above-mentioned work of Darlington and Gairdner (1937) on *Campanula*, where exactly the same situation exists as in *Datura*. (The presence of inversions has been previously discussed.)

The next step in this line of facts is the existence of different species, distinguishable mainly by chromosomal patterns. The most important information is derived from the work of Dobzhansky and his collaborators (see Dobzhansky, 1935; Dobzhansky and Tan, 1936; and other papers), work which, in my opinion, has revealed the most illuminating facts thus far available for bridging the gap between microevolution and macroevolution. *Drosophila miranda* is a new species closely related to *D. pseudoobscura*, with which it may be crossed. They are so similar in appearance that "taxonomists would hesitate to separate them on the basis of morphological differences alone." But they present two different reaction systems, marked by clear-cut physiological differences, in addition to a difference in sex determination. The morphological differences involve size, number of teeth on the sex combs, and numerous small quantitative features which have been described in detail in Dobzhansky (1937). The physiological differences are speed of development, time of maturity, reaction to temperature. The chief remaining difference is found in the totally different X-chromosomes. The hybrids between the two species are sterile, and the males are thoroughly abnormal.

The salivary-gland method permits comparison of the architecture and pattern of the chromosomes of the two species by observation of normal pairing or abnormal behavior in the hybrid. The result is most amazing, in view of the fact that the chromosome groups of the females in both species look identical (i.e., in the ordinary metaphase plate, usualy studied in cytogenetic work). In the salivary glands the chromosomes either fail to pair or pair in very complex configurations. The details indicate that homologous loci are either situated at different points along the same chromosome, as the result of inversions, or they are found on different chromosomes, as the result of translocations. Further, certain chromosome sections are obviously present in only one or the other species. It is assumed that the latter situation is the result of such complete rebuilding of some chromosome sections through repeated inversions and translocations that homologous loci cannot attract each other any more. Detailed study of these differences, which are shown in figures 33 and 34 (see legends), proves that the actual differences must be the consequence of a series of consecutive inversions and small translocations, resulting in a completely rebuilt chromosomal pattern. Dobzhansky and Tan have estimated that altogether about one hundred breaks were necessary to complete the picture. Very similar facts have been disclosed for another species, *athabasca* (Bauer

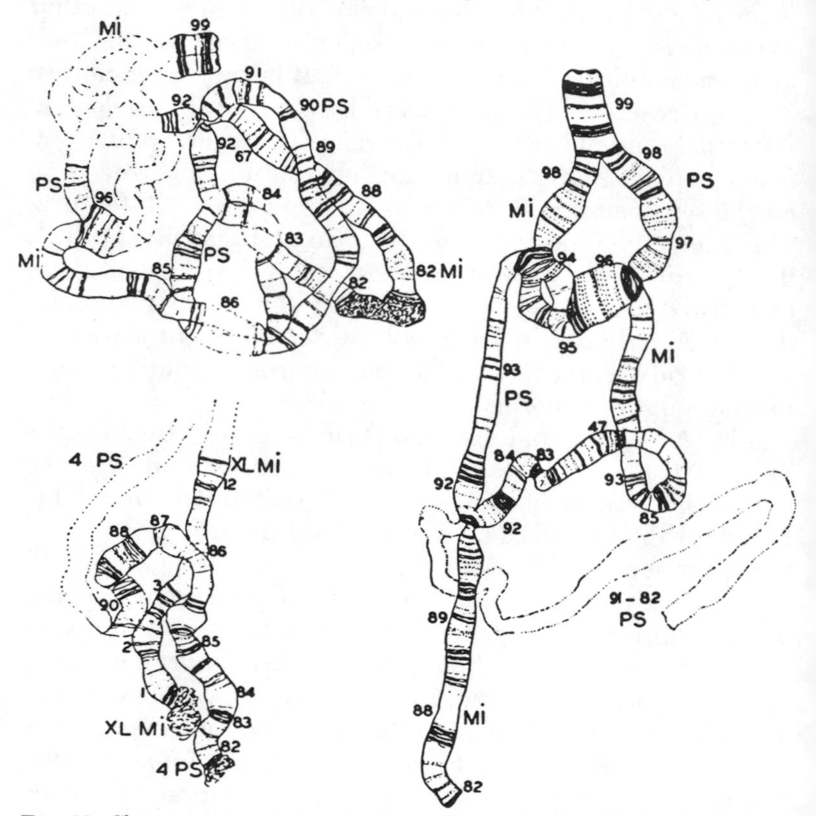

Fig. 33. Chromosome pairing in the salivary-gland cells in the hybrid *Drosophila pseudoobscura* × *miranda*. Upper left, fourth chromosomes; lower left, the fourth chromosome of *pseudoobscura* partly paired with the left limb of the X-chromosome of *miranda;* right, the fourth chromosomes. PS and Mi, the respective chromosomes derived from the two parent species; XL, left arm of X-chromosomes. The numbers mark the serial loci according to a standard map. (After Dobzhansky and Tan. Reprinted from Dobzhansky, *Genetics and the Origin of Species,* by permission of Columbia University Press.)

and Dobzhansky, 1938). The available information actually shows that the differences between the good species *Drosophila melanogaster*, *pseudoobscura*, *miranda*, *azteca*, which either do not cross at all or produce completely sterile hybrids, are extreme in regard to the intimate chromosomal pattern. No chromosomal sections have identical patterns, not to speak of different chromosome sizes and numbers (e.g., *melanogaster*, 4 pairs; *azteca*, 5 pairs). Therefore, a complete repatterning of the chromosomes by transloca-

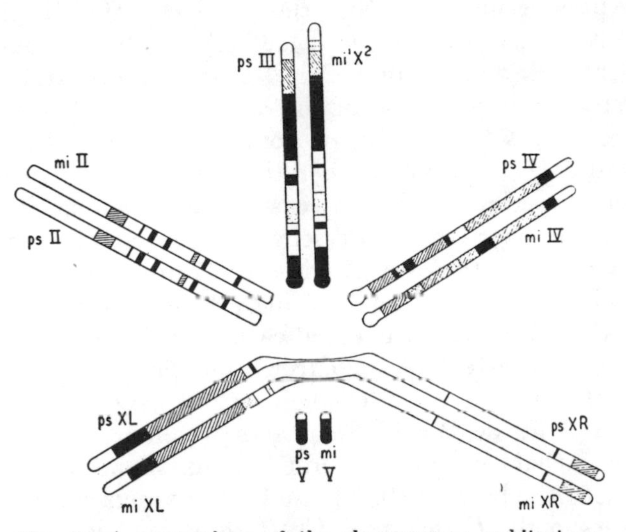

FIG. 34. A comparison of the chromosome architecture of *Drosophila pseudoobscura* and *miranda.* Sections which conjugate normally and are supposed to have the same serial pattern are white; inverted sections cross-hatched; translocations, stippled; sections of which the homologues are not detectable in the other species, black. ps IV, fourth chromosome of *pseudoobscura;* mi XR, right arm of X-chromosome of *miranda,* etc. (Reprinted from Dobzhansky, *Genetics and the Origin of Species,* by permission of Columbia University Press.)

tions, inversions, etc. must have taken place in the phylogeny of these species. "Chromosomal differences between species of the same genus prove to be far more extensive than one might have dared to suppose" (Dobzhansky and Socolov).

Though not many species crosses in Diptera have been analyzed by the same method, comparable results have been found whenever tested. We shall mention later another case in *Drosophila.* Metz (1938) has analyzed species crosses in *Sciara* and comparable results were obtained. There is, however, a difference. In *Sciara* crosses the pattern differences of the chromosomes involve mostly only a single or a few bands, and frequently a structural difference in what seems to be the same band appears. (On a smaller scale these differences occur also within the species, just as is the case with inversions in *Drosophila.*) There can be no doubt that further work will bring to light many more variations of the repatterning process in the chromosomes of species.

Before we set out to draw conclusions from these facts, we must ask ourselves whether these are unique features of Dipteran species, or if it is reasonable to assume that here a typical process has been discovered. A strictly comparable analysis is thus far only feasible in Diptera because of the salivary-gland method, but a number of facts are known, especially in plants, which indicate that rearrangements in the intimate pattern of the chromosomes accompany specific differentiation. A considerable part of recent cytogenetic work in plants deals with such facts, which are too numerous to be reported here. (For detailed accounts and literature see Darlington's book [1937] and the annual reports by Oehlkers in the *Ergebnisse der Botanik.*) In a general way it may be said that inversions may be recognized in the meiotic divisions of plant species hybrids by bridge formation within separating tetrads (see above, p. 192). Reciprocal translocations—segmental interchanges—are recognized by the formation of rings or chains uniting two or more tetrads. These types are actually found very frequently. This is especially true of segmental interchange, which seems to be more frequent than inversion in plants. We have already mentioned the facts regarding segmental interchange in otherwise indistinguishable races of *Datura.* In species crosses of *Datura* (Bergner and Blakeslee; Bergner, Satina, and Blakeslee, 1932–35) the same situation is found to a still larger extent. How far this goes and whether the number of translocations parallels the diversities of the species and whether definite chromosomal patterns may be referred to definite phenotypic effects are, however, difficult to ascertain, as ring formation is due only to exchanges at the free ends of the chromosomes, the rest of the chromosome remaining completely or almost unanalyzed. But there can be no doubt that all over the plant kingdom specific differentiation is most frequently associated with such pattern changes within the chromosome set as are recognizable with the present technique. Many other facts might be mentioned which lead less directly to the assumption that comparable chromosomal pattern differences are distinguishing features of species. I shall mention only the behavior of the chromosomes in hybrids of Lepidoptera in cases in which the complication of different numbers of chromosomes is absent. We know, especially from the work of Federley (1913), that from one to all chromosomes, as the case may be, may fail to conjugate. As we know that conjugation is the result of point-to-point attraction, the most probable inference is that in these cases also the specific differences involve intrachromosomal pattern changes in one or more chromosomes. (Further cases, literature, and details in Darlington's book, 1937.)

C. Interpretation

We come now to the decisive question of how these facts are to be interpreted. We started this chapter with the conviction, gained from an unbiased analysis of all pertinent facts, that microevolution by means of micromutation leads only to diversification within the species, and that the large step from species to species is neither demonstrated nor conceivable on the basis of accumulated micromutations. We have long been seeking a different type of evolutionary process and have now found one; namely, the change within the pattern of the chromosomes. Though present technique restricts detailed analysis to species of Diptera, and allows only a certain amount of insight in the case of inversions and of translocations involving chromosome ends in plants, the widespread occurrence of similar features when plant species are compared and the indications of identical facts in animals point to the presence of a major and general principle. The question now arises as to how a change of the serial pattern within the chromosomes can be conceived of as having evolutionary significance. Only a few years ago such an idea would have had to be considered utterly senseless. The chromosome was the carrier of a string of genes, independent, atomistic units of at least molecular order, each of which controlled certain developmental processes, though the final result was brought about by an integration of these processes. A genetic change could, therefore, be conceived of only as a change in one or more individual genes, a mutation. Evolution in terms of chromosomes, therefore, could mean only accumulation of gene mutations, loss of existent genes, or creation of new genes. This conception had actually become an article of faith, a credo which remained unshaken, when occasionally some critical mind tried to follow the idea to its ultimate conclusion, as did my predecessor in this lectureship, the great skeptic Bateson (1914). His shocking conclusions were accepted as a kind of grim joke, though he probably was very serious about carrying the genetic theory of his time to its inevitable consequences. The origin of new genes has since been discussed with extreme reluctance, and not until recently has any one tried (to my mind the product of embarrassment) to derive new genes from duplicated old ones which undergo various mutational changes until they are completely different (see below for criticism).

However this may be, the classical theory of the gene and its mutations did not leave room for any other method of evolution. Certainly a pattern change within the serial structure of a chromosome, unaccompanied by gene mutation or loss, could have no effect whatsoever upon the hereditary type and therefore could have no significance for evolution. But now pattern changes are facts of such widespread and, as it seems, typical occurrence that we must take a definite stand regarding their significance. Three possibilities present themselves: (1) The pattern changes may be a kind of freakish occurrence without genetical or evolutionary significance, leading the investigator astray simply because they are conspicuous; (2) The pattern changes are in themselves insignificant, but act as a visible sign for the appearance or disappearance of genes, together with numerous mutational changes; (3) The pattern changes are in themselves effective in changing the genotype without any change of individual genes. If the first of the three possibilities is true, we may dismiss the whole subject and consider the time of all investigators just quoted as wasted. The second possibility would not change the situation either, beyond pointing to a curious and hardly comprehensible phenomenon. The third possibility, however, finds no place in classical genetics. The dilemma is increased—provided the pattern phenomenon is at all relevant—when we realize that point-mutations have never been known to change the point-to-point attractions between the homologous chromosomes in the heterozygote. According to the elementary facts of genetical investigation, any number of mutations might accumulate within the chromosomes, theoretically affecting every single gene, without changing the normal attraction and the normal pattern. Therefore, as I said, only a few years ago, the facts regarding pattern change could not have been regarded as significant.

Meanwhile, however, the genetical situation has begun to undergo a complete change. Without going into the highly technical details (see the discussion in Goldschmidt, 1937, 1938, 1940 in press), here are the essential points. More and more facts are accumulating which show that the intimate serial pattern of the chromosome is important for the action of the hereditary material. Chromosome breaks which lead to new serial arrangements of the parts of the chromosome; namely, deficiencies, inversions, duplications, and translocations (see fig. 35) may produce definite genetic effects, which are not different from the typical effects of mutations. Such effects have been called "position effects," a term implying that the genes have some kind of action upon each other and that, therefore, it makes a difference whether they are located side by side or separated. Not all rearrangements have visible position effects, but invisible physiological effects will probably always be found if proper investigation is made (see Goldschmidt, Gardner, and Kodani, 1939). Though many geneticists still cling to an explanation of these pattern effects in terms of gene neighborhoods, it is becoming more and more evident that the effect has nothing to do with the theoretical units, the genes,

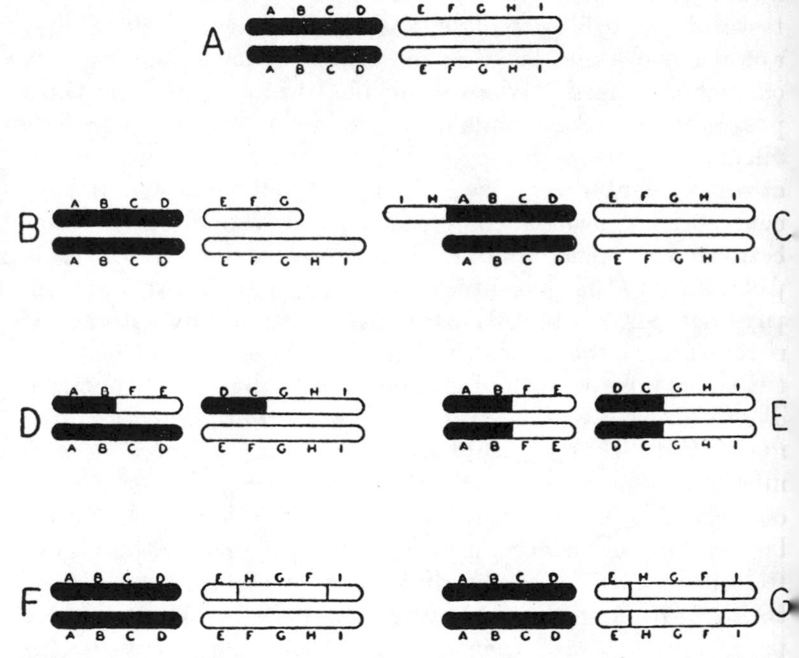

Fig. 35. Diagrammatic representation of simple types of rearrangement of the serial pattern of the chromosome. A, normal; B, deficiency; C, duplication; D, heterozygous translocation; S, ditto homozygous; F, heterozygous inversion; G, ditto homozygous. (Reprinted from Dobzhansky, *Genetics and the Origin of Species,* by permission of Columbia University Press.)

but is an independent effect of the whole chromosome or of more or less small sections of it. The normal serial pattern has a definite genetic effect which is changed with a change of the pattern, and eventual units such as genes have nothing whatsoever to do with the effect. If we were to illustrate this conception with a simile, we might use one of the following models. A violin string as a whole may produce the tone A; if the string is stopped at a certain point, the tone becomes C. The constitution of the string has not been changed, but only its vibrating length; i.e., pattern. Or the letters e, o, r, s read "rose" in one order, and "sore" in another. Or the male sex hormone has well-known physiological actions, but some of its stereoisomeres; i.e., different patterns of the same atoms and radicals, have no such effects at all. I shall not discuss here the viewpoint, which in my opinion is daily becoming more probable, that actually no particulate genes exist, but that all mutations are based on very small pattern changes (see Goldschmidt, *loc. cit.*). For our present discussion the decisive point is that pattern changes as such may have definite visible effects of the same type as the standard mutations, and that, whatever the explanation of the pattern (position) effects, these effects can play a role in evolution. A repatterning of a chromosome may have exactly the same effect as an accumulation of mutations. And even more, a complete repatterning might produce a new chemical system which as such; i.e., as a unit, has a definite and completely divergent action upon development, an action which can be conceived of as surpassing the combined actions of numerous individual changes by establishing a completely new chemical system. Model: two different pictures produced with the same set of mosaic blocks, the new picture "emerging" only when all blocks are in their proper place. It is certainly most remarkable that the new developments in genetics lead to the same conclusions which are derived as postulates from an unbiased analysis of the evolutionary facts. This encourages me to believe that the dead end reached by neo-Darwinian theory based upon the conceptions of classical genetics can now be passed successfully.

Let us now return to the interpretation of the facts regarding chromosomal patterns in relation to specific differentiation. It does not seem that many geneticists realized the difficulties of the situation. Dobzhansky, I think, came nearest to this realization. In 1935 he wrote: "All these data indicate that the process of race and species formation may be resolved into at least two components: the genic differentiation and the differentiation in the chromosome structure. These two components seem virtually independent, but the phenomena of position effects may lead us to a recognition of an ultimate connection between the two." A paragraph in his book (1937b) also bears witness to an uncomfortable feeling regarding the meaning of those facts: "To what extent the differences between such species as *Drosophila pseudoobscura* and *D. miranda* are due to position effects is also a matter of speculation; the greatly different gene arrangements in these species may be responsible for many alterations in the morphological and physiological properties of their carriers. In any event, position effects show that gene mutations and chromosomal changes are not necessarily as fundamentally distinct phenomena as they at first appear." It is obvious that the author realized, more or less, that a decision had to be made between the mutationist, neo-Darwinian view of the accumulation of gene mutations and a repatterning of the chromosomes independent of the assumption of genes as units. But he has not yet been willing to cut the Gordian knot.

Blakeslee, as mentioned before, furnished the largest body of evidence for a single plant genus, *Datura*, both for races where the segmental interchanges do not have a somatic effect (?) and for many species in which new complicated chromosomal patterns accompany the specific differentiation. The dilemma certainly has also been felt by him, but, as far as I can find, he has not committed himself. On one occasion he wrote (Bergner and Blakeslee, 1932): "Segmental interchange, therefore, may be assumed to have taken place in the differentiation of these species. Just what role, however, this process has played in speciation is not yet clear." And on another occasion (Blakeslee, Bergner, and Avery, 1937): "Since we believe the problem of evolution of species can best be stated in terms of the evolution of their chromosomes. . . ." These remarks are rather noncommittal, evading the fundamental issue: chromosomal pattern versus additive gene mutation. This issue, however, must clearly be faced.

The facts which we have discussed thus far have led step by step to the conclusion that we must look for a genetic agency able to accomplish separation of species without slow accumulation of gene mutations. In the changes of chromosome pattern by rearrangement of the parts of the chromosome such a new genetic method has been revealed. We must, therefore, find out whether a method of macroevolution has been discovered here which does not need the slow accumulation of minute steps by gene mutation. For a long time I have been convinced that macroevolution must proceed by a different genetic method, and I have occasionally pointed to this necessity (see Goldschmidt, 1933, and the next chapter for similar views of other authors). The discoveries just discussed, together with the facts of position effect, and the elaboration of a conception of gene mutation differing from the classic one, have led me to believe that a pattern change in the chromosomes, completely independent of gene mutations, nay, even of the concept of the gene, will furnish this new method of macroevolution.[1]

This conception, which will be elaborated now, may be stated in a few words somewhat as follows (the statement in its general form is independent of whatever conception we have of the nature of the gene and of gene mutation, though its importance is enhanced by the recent developments in genetical theory which have been sketched above): So-called gene mutation and recombination within an interbreeding population may lead to a kaleidoscopic diversification within the species, which may find expression in the production of subspecific categories, if selection, adaptation, isolation, migration, etc., work to separate some of the recombination

1. This conclusion, together with some of its consequences, was elaborated in a general address delivered at the International Zoological Congress in Lisbon, 1935. It was presented also, in more or less different form, in a number of lectures in Holland and England, at the New York Academy of Sciences, and at a number of university lectures in eastern United States universities in 1936. None of these papers have been published, because I feared that my position would be misunderstood if not properly documented.

groups, as described in preceding chapters. But all this happens within an identical general genetical pattern which may also be called a single reaction system (see below). The change from species to species is not a change involving more and more additional atomistic changes, but a complete change of the primary pattern or reaction system into a new one, which afterwards may again produce intraspecific variation by micromutation. One might call this different type of genetic change a *systemic mutation,* though this does not have to occur in one step, as we have seen.

A systemic mutation (or a series of such), then, consists of a change of intrachromosomal pattern. This is what is actually found taxonomically (the bridgeless gap) and cytologically. Whatever genes or gene mutations might be, they do not enter this picture at all. Only the arrangement of the serial chemical constituents of the chromosomes into a new, spatially different order; i.e., a new chromosomal pattern, is involved. This new pattern seems to emerge slowly in a series of consecutive steps, as the work just reported indicates. These steps may be without any visible effect until the repatterning of the chromosome (repatterning without any change of the material constituents) leads to a new stable pattern, that is, a new chemical system. This may have attained a threshold of action beyond which the physiological reaction system of development, controlled by the new genetic pattern, is so basically changed that a new phenotype emerges, the new species, separated from the old one by a bridgeless gap and an incompatible intrachromosomal pattern. "Emergent evolution" but without mysticism! I emphasize again that this viewpoint, cogent as it is and, in my opinion, necessary to an understanding of evolution, is to be understood only after the fetters of the atomistic gene theory have been thrown off, a step which is unavoidable but which requires a certain elasticity of mind.

Once this viewpoint is accepted, it must be realized that the problem of selection assumes a different aspect. As Dobzhansky and Socolov *(loc. cit.)* have emphasized, a repatterning of the chromosomes—our systemic mutation—necessarily leads at first to nonviable groups (homozygous translocations, deficiencies, etc.). The new pattern, therefore, cannot survive in the population except in the absence of selection pressure against the heterozygote and under proper conditions of interbreeding. But this applies only to some of the initial steps corresponding to the simple pattern changes by so-called chromosome mutation. The fact that, for example, in *Drosophila miranda* a chromosomal pattern perfectly different from that in *pseudoobscura* is viable in homozygous condition proves that at some point in the repatterning process the condition of a new system, viable in homozygous state, must have been accomplished (of course, provided that one pattern is evolved from another one, which can hardly be doubted). It is not known at which point this decisive condition is reached, and it can be neither proven nor disproven that it may be accomplished in a single complete shakeup of the chromosomal pattern. However this may be, as soon as the condition of homozygous compatibility is accomplished, selection acts only upon the new system as a whole, with exclusion of nonviable classes or swamping by interbreeding. The process of selection thus becomes reduced to the simple alternatives: immediate acceptance or rejection. But even if a slow repatterning takes place and the first steps are already subject to selection, they may lead to positive selection in the heterozygote, before the level of the homozygous viable system is reached. There are a number of facts known regarding position effects which indicate that pattern changes are frequently associated with physiological actions upon development (see Goldschmidt and collaborators, 1939). Invisible, purely physiological, selective features may therefore assist in the increase of pattern changes. If such changes are the decisive features in building up new reaction systems, the rate of species formation will obviously be a much quicker one than that required by the theory of slow selection of innumerable small mutants, even with a relatively slow process of repatterning.

The facts regarding the evolution of species in the relatively short time since the Tertiary certainly are in good agreement with such a conclusion. The most conspicuous example of this type of quick evolution, and one which has been mentioned repeatedly, is found in the fauna of the Pacific volcanic islands of rather recent geological origin. The closest relatives of such faunas, found on the nearest continent, have undergone very little diversification, but a kind of explosive evolution has taken place upon the islands. The decisive differential feature is certainly lack of selection pressure and the presence of numerous niches ready for occupation by properly fitted forms, as emphasized by many writers. But the results are hardly conceivable if accumulation of micromutations be required, whereas a relatively small number of systemic mutations, occurring in small populations without selection pressure, would suffice. A good material with which to work out this point would be the Hawaiian drepanids (see recent discussion by Mordvilko, 1937), or the geospizids of the Galapagos Islands. In both, the immense variation on a macroevolutionary level and the many adaptational features in regard to feeding habits, etc., are conspicuous. I cannot help feeling that this explosive evolution under specific circumstances is a special feature which is very difficult to describe in ordinary taxonomic terms. I would not be surprised if, in such cases, even so-called genera turned out to be interfertile, and segregation in terms of chromosomes or chromosome complexes were found. But no such data are available. Such cases, then, may serve as a model for quick macroevolution which parallels on a higher level the microevolution within a species. We shall mention in the next chapter Guppy's work on Pacific floras which led him to the assumption of large steps in species formation.

The difficulties presented by this group of facts to the ordinary neo-Darwinian conceptions were, it seems, also realized by J. Huxley (1938), though he otherwise defends the neo-Darwinian standpoint of slow selection of micromutations. He points especially to the ground finches (Geospizidae) of the Galapagos Islands, and their irregular behavior in regard to variation. He mentions that Swarth, the latest taxonomist to study the group, "after classifying them into five different genera with over thirty species and subspecies, adds that it would be almost as logical to put them all in one genus and species!" Huxley assumes the same selective conditions as referred to above for an explanation, and thinks that the ground finches show a very peculiar form of evolution, diversification of type, without proper species formation. (This would be a parallel of human diversification, I think.) He proposes the term reticulate evolution (as opposed to branching) for this type. I agree with the general appraisal of the case from the standpoint of taxonomy and evolution, but prefer an explanation free of neo-Darwinian bias, such as the one presented above.

A conception of macroevolution by systemic mutations, a concept properly applicable also to the higher categories up to phyla, requires a number of basic assumptions which must be shown to be justified, before conclusions can be drawn from the new concept. These assumptions are: (*a*) Macroevolution cannot be conceived of on the basis of accumulation of micromutations. The material bearing on this point has already been discussed for the lower end of the taxonomic categories and will be studied again. (*b*) Macroevolution is accompanied by repatterning of the chromosomes. The factual material has been presented. (*c*) An intrachromosomal pattern change may exert a considerable phenotypic effect independent of genic changes. In the quotations from Dobzhansky, we have already hinted at the pertinent facts (position effect), and additional information has been given. (*d*) Such a thing exists as a complete change of the reaction system based upon a genetic change different from an accumulation of micromutants. We called such a change systemic mutation. (*e*) It is possible to produce immense phenotypic changes of a macroevolutional order by relatively small systemic mutations not involving the creation of anything new within the germ plasm. (*f*) The classical atomistic theory of the gene is not indispensable, for genetics as well as evolution. It is this theory which blocks progress in evolutionary thought, as was demonstrated before when we mentioned the impasse reached by Dobzhansky (see quotation, p. 204). We have already foreshadowed the twilight of the gene. (*g*) Models are available which make it possible to visualize the systemic effect of pure pattern changes in the germ plasm. The following chapters will discuss these points, insofar as they have not already been analyzed.

2. MACROEVOLUTION AND MICROMUTATION

At many different points of our analysis we have emphasized that macroevolution cannot be understood on the basis of the neo-Darwinian principle of accumulation of micromutations. This is true for the first step of macroevolution, and still truer when the higher categories up to phyla are concerned. We have derived this viewpoint from a close scrutiny of numerous facts including those of genetical analysis. The latter point is of importance because the current neo-Darwinian conception is almost exclusively based upon the present genetic theory. All the work on the action of selection in populations is based upon the axiom that evolution proceeds in the neo-Darwinian way, and that the theory of the gene and its mutations furnishes the necessary basis for this assumption. We have tried to show that this basis is slowly but certainly slipping from under our feet, as far as the genetical facts and the facts of macrotaxonomy are concerned. Thus we have been forced to assume large evolutionary steps, and we are now engaged in proving that modern genetics is furnishing the necessary facts for an understanding of this process.

Though the analytical procedure and the interpretation of the several facts are claimed to be novel, it must now be added that the necessity of large steps in evolution, involving the whole system of the organism, has been realized by many evolutionary writers who have not been in possession of most of the facts presented here. I shall not delve into history in order to trace the idea back to Geoffrey St. Hilaire, etc., as the idea has become of major importance only since the development of genetics has permitted a discussion on a factual basis. But whereas the genetical facts are of importance only in connection with the taxonomic work, it is important to show that modern taxonomists with great field experience have also reached the same conclusions, without thinking of the possibility of a genetical interpretation. We have already mentioned Willis' work on *Age and Area* (1922), and we now add that Guppy (1906) had already anticipated many of Willis' ideas as a consequence of his own study of the flora of Pacific islands. The conclusions relating to our present problem (aside from the derivation from facts of distribution) are well expressed by Willis (1923), who says that the change from one species to another must be in one or, at most, a few large steps, changing many or all characters of the plant at once. Knowing that the geneticists will not agree, he makes the situation clear in the following words:

"The current attitude of the Mendelians towards questions of evolution is one of an aggressive agnosticism. Since investigations upon Mendelian lines have not as yet been able to throw as much light upon the problem as had been at one time expected, they seem to think that no other line of attack upon the question will be any more likely to find a way that may possibly lead to something in the nature of a solution of the problem at some future date. They seem inclined to think that because they have not themselves seen a 'large' mutation, such a thing cannot be possible. But such a mutation need only be an event of the most extraordinary rarity to provide the world with all the species that it has ever contained. As I have pointed out (*Age and Area*, p. 212), one large and viable mutation upon any area of a few square yards of the surface of the earth, and once in perhaps fifty years, would probably suffice.[2] The chance of seeing such a mutation occur is practically nil, whilst if the result were subsequently found it would probably be called a relic. Darwin's theory of Natural Selection has never had

2. Dr. Guppy has suggested that it is by no means unlikely that the many species once seen and never afterwards discoverable may often be such mutations. The case of *Christisonia albida*, described in *Age and Area*, p. 151, is almost certainly a case of a nonviable mutation, and it may be noted that Hooker, who was not a "splitter," accepted it as a Linnean species.

any proof except from a priori considerations, yet has been universally accepted, and has led to great advances in biology; and until the Mendelians show us how to control mutation (a thing that will evidently be some day possible), the proposition now put forward will presumably go without actual demonstration by verified fact. What I contend is that the facts brought up here and elsewhere go to show that neither of the extreme suppositions—Special Creation and Natural Selection—contains all the truth, and that therefore this, or similar, compromise between them is rendered necessary by the present condition of our knowledge.

"The small mutations that are all that the Mendelian school will allow are obviously in the highest degree unlikely to give rise to mutual intersterility, such as so commonly characterizes specific difference, and if they were to be accumulated it is difficult to see where the sterility would come in, for each would seem as likely to be fertile with its successor as with its predecessor—A with B, B with C, C with D, and so on. But let a big step, say, from A to M, such as dropping of endosperm, be taken, and one would feel inclined to expect mutual intersterility as a matter of course.

"If so large a difference as having, or not having, endosperm, rudimentation of endosperm, few or indefinite stamens, etc., etc., can occur, as it does occur, over and over again between genera which are obviously closely allied, we are evidently simply making difficulties for ourselves by supposing such differences to be gradually acquired. It must never be forgotten that gradual acquisition is an assumption of the theory of Natural Selection. Whether the differences were infinitesimal (or due to the universally occurring fluctuating variation), or whether they were more of the nature of sports, they were never supposed by Darwin and his followers to be anything but small, and evolution of new species was by their accumulation, whilst the larger groups were due to further accumulation and to destruction of the intermediate forms. Now the work which has been done to establish the theory of *Age and Area* goes to show that destruction of intermediates can no longer be invoked

There has been vast destruction of individuals, and probably of species which were only represented by a few individuals, but not of intermediates, unless these species which were destroyed were of intermediate type; and in that case it is difficult to see how they could give rise to the later and more successful forms. Even in the earliest known geological horizons that contain the group there can be recognized many families of flowering plants that exist today, and that cover a very large part of the systematic range at present existing. They are as well and as widely separated as those now existing, and into families that no longer exist, and if these gaps were due to destruction, then Natural Selection must have operated with great rapidity and decision in the earlier ages of the flowering plants. If the earliest known flowering plants already show such differences as that between Monocotyledons and Dicotyledons, then evolution upon the Darwinian plan must have been going on previously (in flowering plants) for an enormously longer period than has since elapsed, or selection and destruction must have been much more rapid."

Willis then mentions many relevant examples from plant organization showing that characters of all kinds, however important they may be in classification, may be acquired over and over again by single genera, and, therefore, that they can be easily acquired without needing an immense period for their acquisition. I think that the zoologist must agree with this conclusion (which is independent of the age-and-area hypothesis, though derived from it). The underlying facts in both kingdoms are exactly the same as those which have always raised difficulties for the Darwinian conception of evolution. These difficulties we are trying now to remove by the new conceptions which are being presented here.

We just mentioned that Guppy had preceded Willis in his views. Guppy's views were derived from his expert knowledge of the flora of the Pacific islands. We might, therefore, return once more to this fascinating subject. Present-day opinion does not favor the older theory postulating vanished land bridges between these islands. It is, moreover, assumed that all islands appeared independently in their present place and that they were settled by plants and animals arriving from different directions over vast stretches of water. (For details and discussion see Setchell, 1935.) The endemic fauna and flora of these islands are therefore not very old and must have gone through their evolutionary process only in recent geological time. Where endemic species are not very different from their continental relatives or those found on the probable path of migration, any theory of species formation fits the situation. This is, however, not the case when, in some instances, a kind of outbreak of evolution has occurred. Let us take as an example the much discussed honeysuckers of Hawaii (see A. Gulick's review, 1932a). (The comparable case of the ground finches from the Galapagos Islands has been mentioned before.) Probably derived from a tropical American immigrant, the indigenous family of Drepanididae consists of eighteen genera and forty species. What has happened is expressed by Gulick in the following words:

"When the ancestral pair of drepanids first arrived, and found this land rich with beetles and various other insects, and gay all the year around with lobelias and other honey-bearing flowers, yet with no rival or enemy in sight, it had before it an evolutionary opportunity such as can scarcely have been duplicated in the whole history of avian life. The standardizing effect of close natural selection was removed. Food was very abundant but probably differed in one way or another from what those birds had formerly been used to, so that they may even from the start have begun to turn aside from their previous habits of feeding. And when pressure of population finally began, there came to be a direct premium on physical variability, and on versatility of feeding. In addition they were subjected to considerable geographical isolation as between the several islands. This must have led over into a condition comparable to that found today among the Geospizidae of the Galapagos—several new-formed genera, all of them very variable and rather versatile feeders, but each with a norm for beak, claws, etc., corresponding fairly well to its major tendencies in food and habits. These several genera must have spread, each independently, over all the islands that furnished the appropriate foods, and each developed its own geographically restricted races and species. But in Hawaii all this happened ages ago, and since that time constant competition and selection and specialization have worked the extreme results which the accompanying diagram visualizes [see fig. 36].

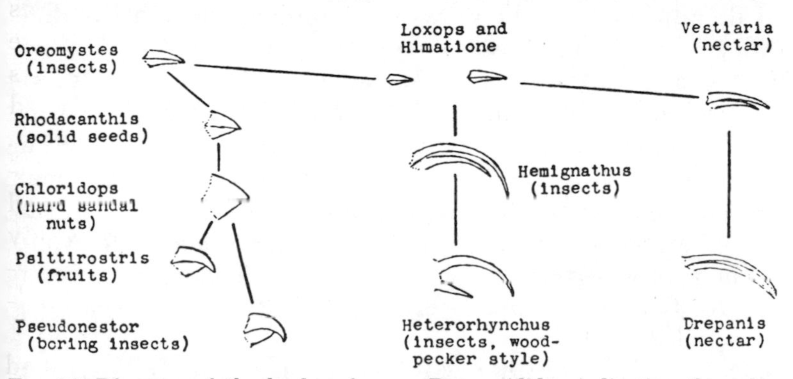

FIG. 36. Diagram of the beaks of some Drepanididae indicating the relation of beak form to food. Most of these "genera" are represented by distinctive "species" on each island. (From Gulick.)

The ancestrally homogeneous family now has its 18 genera ranging from stocky, seed-eating birds with bills like crossbeaks and parrots, through finch-like birds that glean smaller seeds, to little creatures with long, thin, flexible bills for gathering honey and insects out of tubular lobelias, and even birds with a short, stout lower mandible and a very long, slender upper one, the first usable for pecking away loose bark, and the other for probing out the grubs of the native boring beetles. Adaptation has become completely and narrowly specialized for feeding upon the nectar, seeds, and insects native to their Hawaii."

It is well worth while to inquire whether such an evolution can possibly have taken place on neo-Darwinian lines. How the woodpecker or stonecrusher type, which involves, of course, the whole anatomy, may have evolved from a honey-sucker type by a series of micromutations controlled by selection is simply unimaginable, and one can understand why so many nongeneticists stick to the Lamarckian explanation. But with all our knowledge of mutation, selection, population pressure, we cannot understand, either, why all these different lines of evolution should have blossomed out simultaneously, even if a neo-Darwinian interpretation of the resulting type were feasible. One may try to work out the story in detail, but one will always be confronted with the difficulty of understanding the beginning and all the intermediate stages of the conditions, which lead up to such diversifications of adaptation. As the geneticist is unable to accept the Lamarckian viewpoint, there is only a single solution left: the origin by large steps, our systemic mutation, which leads at once so far toward the new type that selection can immediately be efficacious, and which permits a large evolutionary process to take place in a time as short as, or even shorter than, is ordinarily required for the production of a subspecies.

Only one more author will be quoted, again a taxonomist. (I like to underscore "Taxonomist," since geneticists frequently assume that taxonomic knowledge leads necessarily to the neo-Darwinian conception.) Robson (1928), after a very detailed presentation of the facts pertaining to the species problem, finds difficulties with the neo-Darwinian standpoint. He says:

"Our survey has been confined to the processes that are currently believed to give rise to the differentiation of species and lesser groups. We have confined our attention to the slight divergences, principally in structure, that distinguish allied species and races, and we have neglected those larger aspects of morphology that are concerned with the history of organs and their progressive improvement or deterioration. Taking the narrower view and confining our attention to interspecific differentia, we may perhaps be inclined to believe that all such differences may be satisfactorily explained by one or the other of the theories just considered, or by them all collectively. On this assumption we must believe that the history of an organ such as the vertebrate eye, or of a group of animals such as the Cephalopoda or the Reptilia, consists of a number of small steps, some of which might be treated by the systematist as 'specific.' We ought, before we conclude our inquiry, to ask if modification by the environment, Natural Selection or chance mutation can bring about the origin of complicated organs or those sustained and progressive evolutionary episodes which we see in the history of any of the great groups of animals. In short, if one of these principles, or all of them collectively, can produce the sort of differentiation which we have been studying, is there any particular difficulty in believing that, e.g., a complex organ is simply the end-term of a series of such change, each stage being more complex than the preceding one?

"We have, however, reason to doubt whether any one of these principles by itself and without very important modification constitutes a satisfactory explanation of the whole of the phenomena of evolution. It has even been suggested that they do not do so collectively, and that the evolution of complex organs on the one hand, and of certain evolutionary trends on the other, seem to require a special explanation. If that is the case, if such phenomena testify to the activity of a totally different evolutionary principle, may not the latter, held as it is by some to be apparent in the production of complex organs and orthogenetic trends, also intervene in the divergence of species. I do not say that need for this additional principle is proved or even required. There is enough evidence available to suggest that the current hypotheses, together with the effects of isolation, may be indeed valid explanations within the limits of our experience. But we cannot disguise the fact that they involve many difficulties and labour under many disabilities, so that we are bound to consider what alternatives have been proposed."

Though this author does not offer any solution for the difficulties, it is obvious that only two solutions are conceivable. The first is the mystic one (emergent evolution, *évolu-*

tion créatrice, etc.), the second, the one presented here. We shall have to discuss later on how modern paleontologists have come to a similar conclusion.

3. THE REACTION SYSTEM

We have repeatedly used this term, which is also found in one of the quotations from Dobzhansky, to express the following viewpoint: According to the theory of the gene, each individual gene exercises a definite influence upon the development of morphological and physiological characters. These influences are mutually interdependent and the end result is the specific type. Although a definite visible trait will be controlled by many, if not all, genes, nevertheless certain genes will be predominantly concerned with definite and localized actions, so that the action of the individual gene is, after all, the decisive feature. A mosaic of individual actions is thus assumed to produce the final whole. It is sometimes denied that modern geneticists still think of the gene in such a way. The best proof that the theory of the gene necessitates this atomistic assumption is found in the cases of deficiencies opposite a so-called hypomorphic gene. In the absence of the allele, such a "wild-type gene" produces the type of mutant character ascribed to the same gene when mutated! For example, a single wild-type gene at the vestigial locus in *Drosophila* causes a kind of vestigial effect if the allele is absent due to a deficiency covering this locus.

The idea of the reaction system in the sense in which this term will be used is opposed to the idea of integrated genic action. It means that the germ plasm as a whole; i.e., predominantly the chromosome complex, controls the general features of development which lead to a definite type, the species in question. This idea dispenses completely with the individual gene and its individual action, with the attending difficulty of integrating mosaic action into a unified whole. It considers only a single unit action of the whole germ plasm, with more or less independent action of the individual chromosomes. Whether the intricate pattern of this germ plasm is a pattern of genes, or whether there are no genes at all, is another problem; the point here is that the germ plasm as a whole controls a definite reaction system, which, then, is not a mosaic of separate effects but a single developmental system controlled as a whole by one agency. (The important problem of how such a conception tallies with the facts regarding definite actions of sections of the chromosomes is discussed in Goldschmidt, 1940 [in press]. We shall not go into these technical details here.) It is certainly difficult for many geneticists to think in such terms, as most of them are so completely wrapped up in the axiomatic belief in the atomistic gene theory that they are unable to think in other terms. But embryologists, physiologists, and probably taxonomists will, I trust, not find any difficulty in accepting such a conception, and may even welcome it.

The term reaction system was introduced into genetics by Goodspeed and Clausen (1916), who realized at that early date, right at the height of the new triumph of the theory of the gene, that something more than the additive action of individual genes must be involved in genetic determination. It is highly significant that they derived their new concept from experiences with species hybrids. As a matter of fact, these authors did not take the decisive step away from the gene mosaic conception, but they tried to expand it by adding the new idea of reaction system. "For if this conception of genic interaction be valid then it should not be possible, in certain cases at least, to shift and recombine the elements, from which systems have been built up in the haphazard way that some advocates of Mendelism have attempted to do. *If, for example, it is possible to obtain hybrids involving not a contrast between factors within a single system, but a contrast of systems all along the line, then it is obvious that we must consider the phenomenon on a higher plane, we must lift our point of consideration as it were from the units of the system to the systems as units in themselves.*"[3] These conclusions were derived from a set of facts which we might have mentioned above when discussing species hybrids and emphasizing their peculiarities, espe-

3. Italics mine.

cially in the segregation of whole complexes of characters (see p. 180). In crosses of the Nicotiana species *sylvestris* and *tabacum* the F_1 shows exclusive expression of all *tabacum* characters, irrespective of the presence of recessive or dominant mutational characters in the parents. This applies to all details of structure and habit, so that there is actually a dominance of the entire reaction system of *tabacum* over *sylvestris*. This applies also to all varieties (subspecies) of *tabacum*, including individual mutants. The F_1 is partly fertile and may be backcrossed to *sylvestris*. Among highly abnormal offspring a considerable proportion of pure-breeding *sylvestris* is produced, which was interpreted as a segregation for a whole reaction system. Meanwhile it has been found out that the situation is more complicated. Clausen (1927) has come to the conclusion that *N. tabacum* is an allotetraploid hybrid, one of the genomes being derived from the species *sylvestris*, the other from *tomentosa*. By continuous backcrossing to *sylvestris* the chromosomes derived from *sylvestris* can be tested, because they form tetrads with the *sylvestris* chromosomes. They have been found to be completely different genetically. The idea of the reaction system thus becomes less generalized and actually applies to the whole architecture of individual chromosomes, in the same sense as we developed it before. We have already referred to similar cases which cannot be explained by linkage or by survival of certain combinations only, in view of the innumerable gene differences which would have to be postulated between the two species. Among other examples, we have mentioned the perfectly parallel case in animals, described by Harrison (1917) for crosses of geometrid moths.

At the time when these views were developed there was no reason to express them in any other terms but as an extension of the theory of genic collaboration. A whole chromosome or a whole set of chromosomes meant nothing but a number of genes, and a concerted action of a genome meant a type of interaction of individual genes in a definite combination. When, at the same time, Renner (1917) established the theory of the complexes in Oenothera which was later understood in terms of segmental interchange between chromosomes, phenomena of the type under discussion were reduced to specific abnormalities in the chromosome behavior (though this does not apply to the F_1 results from which the concept of reaction systems was derived). But with the present knowledge of the importance of chromosome pattern, independent of genes, all facts pointing toward the existence of reaction systems assume a new significance, as was just mentioned in relation to the *Nicotiana* case, though this original case has since been found to be based on specific conditions of hybridism (allotetraploidy).

A. Reaction Systems and Genes

Goodspeed and Clausen's conception was first derived from the fact that a whole genome appeared dominant over that of another species, irrespective of the dominance relations of individual genes in varietal crosses. Such a condition does not seem to occur frequently (apart from triploidy, which is involved in *Nicotiana*), though it is observed rather often that individual mutational differences show a changed dominance relation in species crosses. Such facts, however, can be completely explained within the theory of the gene, either by assuming modifiers for dominance (Fisher) or by postulating different grades of wild-type action beyond the minimum threshold (see discussion in Goldschmidt, 1938). But it must be emphasized that explanations of the facts in terms of individual gene differences have been construed under the assumption that gene differences are the only differences known and worth discussing. Therefore explanations had to be found in such terms. This, however, does not mean that the same facts might not be understood, and even better understood, if analyzed from the standpoint of reaction systems. Let us take as an example Harland's (1933, 1934) work on cotton, though the question might be raised in this connection as to whether the species which have been crossed are actually different species or only subspecies. This, however, is not very important for the present problem, as we have seen that chromosomal pattern differences may develop within subspecies independently of mutational differences. Harland crossed *Gossypium hirsutum* × *barbadense*, species which are said to be different in every discernible character. Within each of the species a number of characters have mutated in a parallel direction, as is so frequently the case. For example, a large spot at the base of the petals may be present or absent in both, or red versus green leaf may be found in both species. In *barbadense* these differences are single gene differences, though intermediate grades occur on the basis of multiple alleles. The same applies also to *hirsutum*, but the multiple alleles are not found here! *Barbadense* spotted × *hirsutum* spotless produces an F_2 with all gradations between the two extremes, differences too minute to be classified. Reversal of dominance of spot over spotless was also observed in later generations. Transference of the *barbadense* spot to a *hirsutum* background by repeated backcrosses gave a weak spot. The conclusion was that there is a basic gene for spotting, not identical but allelomorphic in both species. The *hirsutum* gene for spotting is able to produce its effect upon the *hirsutum* "background," but the same background almost nullifies the action of the *barbadense* gene. In addition, there must be a group of intensifying modifiers present in the *barbadense* genotype but absent in *hirsutum*, which may have a different set of modifiers. This would then mean that homologous characters in the two species became built up in evolution in a different genetical manner. (In old Darwinian language this would be called convergence.) A test for nine different characters gave the same results, from which it was concluded that most, if not all, genes are different (though allelomorphic) in the two species and that each is accompanied by a different modifier complex. Harland therefore reached the conclusion that the development of new modifier complexes is the decisive step in evolution, that "the modifiers really constitute the species."

This is a rather complicated explanation of the facts, derived in order to explain the distinctive features of the case in terms of gene mutations. It seems as if a restatement of the facts in terms of reaction systems would be appropriate. If the normal type (spot, etc.) is produced by different reaction systems; i.e., chromosomal patterns, in both species, and if chromosomes in the hybrid conjugate and segregate, as is the case, though with little or no crossing over, the system of each chromosome will segregate as a unit and therefore will act like a pair of alleles. If the chromosomes segregate independently, a complex chromosome arrangement involving all homologous chromosomes, different in pattern in both species, is found in the F_2. This mixture of incongruous patterns appears as a multiple-factor segregation, though whole chromosomes, and not gene differences, are involved. This view can be tested by an analysis of all species differences in the F_2, an analysis which has not been reported. I expect that the results will be expressed not in terms of innumerable gene segregations, but rather in terms of more or less intricate reaction systems involving whole character complexes, as reported above for Lepidoptera and *Salix*.

This last example leads to a discussion of a method frequently employed for the comparison of the genetic makeup of two species which do not produce fertile hybrids but which have developed phenotypically identical mutants, just as in the cotton example. Three methods for comparison are available:

(*a*) If both species produce mutants and may be crossed, the identity of similar-looking mutants may be tested. If a cross between two mutants results in the mutant type, whereas mutant × wild type shows normal dominance relations, it can be said that the mutant types are allelomorphic. In addition, these mutants can be localized on the chromosome map and their loci can be compared. The best-known example of this type is found in Sturtevant's (1920, 1921) crosses of *Drosophila melanogaster* × *simulans*. The species differences are very small; the chromosomes are very much alike, but the hybrid is sterile. In both species a large number of similar-looking mutants are known, and the majority of these behave as alleles in the crosses. In addition, they are found to be located in the same chromosomes and in the same order. The obvious conclusion is that both species have numerous genes in common. But it is not obvious that any conclusion as to the actual genetic differences which must exist between the species can be drawn. This is a vital point. In this case, however, the salivary glands give decisive information (Pätau, 1935; Kerkis, 1936). There is a normal attraction, point for point, throughout most of the length of the chromosomes, showing the identical nature of the minute pattern. But, in addition, a large inversion and a number of very small differences in detail are found. The latest study of the same case by Horton (1939) actually reveals ten clear chromosomal rearrangements; namely, five short inversions, one large one, and four very minute changes at the ends of the chromosomes. In addition, there are fourteen regions in which the chromosomes do not pair ordinarily in synapsis, indicating minute rearrangements. There is, then, a considerable pattern difference, which might be made alone responsible for the specific differences. Otherwise one would have to assume that the genic differences between the species were all confined to genes which did not happen to mutate, an assumption which sounds rather improbable.

(*b*) The second method applies to species which cannot be crossed but in which comparable mutations are found. The linkage maps or linkage relations between some mutants showing a parallel effect may be compared. Haldane (1927) has reviewed a number of instances in mammals where similar-looking mutations had also similar linkage relations, and Green (1938) has recently complemented the list. In *Drosophila* a more extended comparison has been made for *melanogaster* and *pseudoobscura* (Sturtevant and Tan, 1937). If similar-looking mutants are assumed to be based upon the same genes (which, however, is not the fact, as is easily seen in the standard *Drosophila* mutants where frequently a number of different mutants are indistinguishable phenotypically), parts of the chromosomes or whole chromosomes in the two species seem to contain the same genes. But in this case—two indubitable species—the arrangement of these mutant loci is a very different one in both species, so that the authors conclude that many inversions must have taken place in the derivation of one type of intrachromosomal pattern from the other. The salivary-gland chromosomes of the two species compared by Dobzhansky and Tan (1936) are entirely different, and no parts can be considered completely alike. This shows that analysis by way of comparing mutants does not give reliable information regarding the species differences, and, I might say, even leads to considerable misinformation. It stresses what are probably most unimportant similarities and overlooks the vital point, the pattern difference. (This, however, is not Dobzhansky's opinion, nor that of Harland (1935), who has reviewed a number of comparable instances.)

(*c*) The third method applies to fertile hybrids where mutants in one form may be crossed to the other wild type. If they show simple Mendelian segregation, an identity of the genes is assumed. We mentioned this case before and many similar examples have been reported since the first work of Correns on *Mirabilis* crosses. The furthest conclusion which may be drawn from this material concerns the existence of one or a few homologous loci, but in most cases the facts do not permit any other conclusions beyond those on the functioning of the mechanism of chromosome pairing and disjunction. I cannot see that this helps our understanding of species differences, aside from the question of whether species are actually involved.

B. Reaction Systems Versus Genic Balance

In plants, and sometimes also in animals, closely related species differ by a single chromosome or by a few of them. In some cases (Orthoptera, McClung's school and others) it may be shown that one of the chromosomes of a set (or even all the chromosomes) seems to have been broken into two (which would require the acquisition of a new spindle fiber). In other cases (Lepidoptera; Seiler and others) it seems as if two or more small chromosomes had united into a larger one. There are also all imaginable conditions of fragmentation, ranging from a single fragmentation in one chromosome to fragmentation of all chromosomes. The latter produces what appears to be polyploidy, but is something different. No definite rules can be laid down, but

Figs. 37, 38. Normal diploid *Nicotiana tabacum* compared with 10 trisomics and one more complicated chromosome type (Tertiary). Names from left to right: fig. 37, diploid, recurved, enlarged, puckered, bent, stubby; fif. 38, narrow pointed, late, compact, inflated, sticky (= Tertiary of recurved and compact). (From Goodspeed–Avery.) [Courtesy of the late Dr. P. Avery.]

Vandel (1938) points out that in a great many instances the increase in chromosome number goes hand in hand with specialization. But the opposite relation also occurs

Fig. 38

in *Cyclops*. This pseudopolyploidy, as we may call it, certainly has some relation to sexual propagation, as it favors parthenogenesis (see above concerning *Artemia* and Vandel's discussion in 1936. A full account of all the facts and a review of chromosome numbers in many groups of animals have been given by Vandel, 1938). If the cytological explanation (fragmentation) is correct, it means in terms of evolution that either the chromosomal change is nothing but a haphazard feature accompanying species differentiation—in other words, one of the differential specific characters, without being the cause of specific differentiation—or the chromosomal change is one of the causes of specific difference. In the latter case we are facing a change in chromosomal pattern without a change of the so-called genes, as the cytological facts do not point to any structural changes except the union or breaking of chromosomes. This then means that in these cases the genetic effect of the same chromosomal material changes as the material is divided into a different number of constant units. We might call this a kind of position effect. I do not know of any experimental facts in favor of such an explanation; experimental fragmentation of chromosomes (by X-rays) does not produce stable conditions which can be properly compared with the case in question. We must, therefore, content ourselves with stating the problem without drawing conclusions from known facts. But we may point out, at least, that the facts of chromosome fragmentation have to be kept in mind, as their eventual interpretation in terms of position effect will agree with all the other deductions to be made here from similar facts.

Another type of change of chromosome number is found when a single chromosome has been duplicated so that the 2n chromosome set is changed into 2n+1; i.e., so-called trisomics. It is known that this actually occurs in many instances in plants, and the facts of the case have a very definite bearing upon the problem of reaction systems and chromosomal patterns versus accumulation of atomistic gene mutations. The existence of one chromosome in triplicate instead of duplicate condition was early found (Gates) to be the cause of a special form of Oenothera, one of the so-called mutants. Since then, numerous cases have been analyzed, including some in which trisomics for every single one of the chromosomes of a set have been found and distinguished. Blakeslee's pioneer work on the trisomics of *Datura* is generally known and reported in all textbooks: each of these trisomics involving a different chromosome bears a definite name relating to its visible features and can be distinguished at sight. A similar situation has been studied by Goodspeed and Avery (in press) in tobacco. Through the courtesy of these investigators I have had an opportunity to see the material in the field and to observe its striking features. The accompanying table (table 9) and figures 37 and 38, which I owe to Dr. P. Avery, present some of these features. Each of the different trisomics (eleven out of twelve possible ones have been realized) is characterized by a special habit which is expressed in all parts of the plant and which makes the different trisomics appear very different to the eye. If the genetic makeup of these plants were not known, one would expect that their differences were due to a single gene with manifold (pleiotropic) effect. Trisomics were actually among de Vries' mutants, and were recognized by him as a single departure from the original form. Though he was mistaken as to the type of mutation involved (chromosomal versus point mutation), he was more perspicacious than many of his followers in realizing that the trisomic mutational change was a change not in one trait but in a complete growth pattern. As far as can be ascertained from descriptions, other trisomics in plants behave like our example. There is an additional feature in the tobacco case: Secondary trisomics with more than one chromosome in triplicate are new types, which, however, permit us to see the combined influence of both primary types (Goodspeed and Avery, *loc. cit.*). In animals comparable cases have not been analyzed, due mainly to the lethality of hyperploids. But a few exceptions exist, as, for instance, the trisomics of the fourth chromosome in *Drosophila* (triplo- IV). These have small smooth eyes, narrow, more pointed wings, darker body color, suppressed trident—in short, a complex deviation from normal.

TABLE 9

SOME MORPHOLOGICAL CHARACTERISTICS OF PRIMARY TRISOMICS OF *NICOTIANA SYLVESTRIS*

From Goodspeed and Avery (in press)

Primary trisomics	*General character differences from diploid*	*Conspicuous diagnostic characters*
Recurved	Slender throughout; leaves, flowers, calices, capsules longer and narrower.	Recurved corolla limb. Limb blade without extra fullness.
Enlarged	Length of parts increased; flowers, calices, capsules, pedicels, internodes, much longer.	Midrib elongates more rapidly than leaf blade, causing the blade to fold back at the midrib.
Narrow	Narrow leaves, flowers, calices, and capsules.	Many basal laterals produce a bushy habit. Leaf blade without extra fullness except for slight undulations on upper margin.
Pointed	Tips of leaves and of corolla and stigma lobes conspicuously acuminate.	Excess fullness of leaf tissue gathered along midrib; leaves very erect, close-set, numerous.
Late	Very slow growing; 10 to 18 months later in blossoming, becoming biennial; rosette leaves very close-set and numerous with bases tightly interlaced and internodes almost eliminated.	Leaf tissue becomes necrotic between veins.
Puckered	Lateral branches few and short; branches of inflorescence short with few flowers.	Extra fullness of leaf tissue accumulates in "puckers" between veins; small necrotic areas develop at apex and along margins of mature leaves.
Bent	Large, full leaves and flowers; internodes, flowers, calices, and capsules somewhat longer.	Margin of leaf bends at intersection of veins where tissue accumulates to form a "blister."
Stubby	All apices blunt; leaf and limb of flower orbicular; main shoot, branches, leaves, flowers, calices, capsules, pedicels reduced in length and number.	Extra fullness of blade accumulates along veins.

TABLE 9 (*Continued*)

Compact	Short internodes, leaves, branches of inflorescence, flowers, calices, and capsules.	Leaf tissue full between sunken veins and midrib; full, waved margin.
Inflated	Infundibulum of corolla tube and calyx much inflated; lobes of corolla limb and capsule also broad.	
Lax	Leaves, flowers, and anthers droop.	Apices of leaves and corolla lobes acuminate.

The current explanation of these cases is based upon the so-called "theory of genic balance." I had found (1911–15, quotations in 1934) that sex is determined by a quantitative relation or balance between male and female determiners in the sex cells, thus showing that the phenotype, in this case primary and secondary sex characters, is determined by a balance between two competing and quantitative genetic conditions, which may be shifted in one direction or the other. Soon afterwards it was realized that this balance works through the control of properly adjusted reactions, and this permitted us to apply the same principle to all genic actions (1920). The balance theory thus assumed the form of a balance between closely adjusted developmental reactions steered by the genes. As it was assumed that the speed of these reactions was controlled by the quantity of the genes, it followed that a disturbance of the balance was produced either by a change in the quantity of a single gene or by summation of quantities of different genes of identical action. This theory was subsequently formulated in a somewhat different way by Bridges (1922), who looked at the balance more from the standpoint of the genes than of gene-controlled reactions. He thought that in the formation of each hereditary trait all genes were involved. Some may pull the processes leading to a definite character in one direction, others in the opposite direction, and the end products will depend upon the number of genes in one or the other group; i.e., on their balance. For each actual trait this balance is properly set, and a change in it will lead to a different somatic effect. Whichever of the two formulations of the balance idea one prefers (a choice which is of no importance in the present discussion), the decisive point remains that a definite quantity of genic material controlling one type of developmental process has normally to be present, in order to insure a balanced, well-adjusted interrelation of all the morphogenetic reactions which occur in development.

This theory was also used to explain the effects of the presence of one chromosome in triplicate. Goodspeed and Clausen (1916) were the first to use such an explanation for trisomics, thus anticipating the later development of the balance theory, especially the formulation by Bridges. The idea is that the addition of one more chromosome into the genome introduces an extra set of genes, and if these genes are preponderably of a type pulling in one direction (Bridges plus and minus modifiers) the whole type of the individual will be shifted in this direction. This explanation seems reasonable at first sight, but it cannot stand closer inspection. One of the basic facts of genetics is that mutant genes controlling definite characters are usually spread over all chromosomes. Well-known examples are the minute bristle mutations and the eye-color, eye-shape, and eye-texture mutants in *Drosophila*, flower colors, chlorophyll types in many plants, etc. There is no fact known which suggests that genes influencing one trait are all or preponderantly linked in one chromosome. But as there are twelve pairs of chromosomes present in tobacco, one of which is in triplicate, and as it is assumed that each chromosome carries numerous plus as well as minus modifiers, the appearance of a distinct and specific type in each trisomic requires, on the basis of the balance theory, the accumulation of modifiers of one type in each individual chromosome and of different types in the different chromosomes. In other words, twelve groups of modifiers would be required, each with different action and each present preponderantly in one of the chromosomes. This is in contradiction to all our genetic knowledge.

If, then, genic balance does not account for the facts, we must turn—and this is the reason for discussing these facts here—to an effect of the chromosome as a whole, an effect of the whole chromosome with its specific serial pattern upon the whole reaction system of the plant. No genic balance is needed, and, as for that, no genes either, to understand the reported facts, as soon as we have decided to assume that a chromosome of a definite structural pattern is acting as a unit. I know, of course, as I have already emphasized repeatedly, that it is difficult for many geneticists, brought up to think exclusively in terms of integrated gene action, to cut themselves loose from this preconceived idea. But here are the facts, and the whole discussion contained in this chapter, as well as in former chapters, leads to the same conclusion, which is all-decisive for an understanding of evolution.

Another remarkable problem ought to be pointed out in this connection, though it seems at first sight to be rather remote from the problem of species formation. When I first tried to formulate ideas upon evolution based upon the then new insight into the action of the gene (Goldschmidt, 1917, 1920); namely, its action by controlling velocities of reactions concerned with differentiation, I started the analysis with an evaluation of genetic sex determination. It was pointed out that sexual differences within a species may be of such a nature that, if found distributed among different organisms, they would provide a basis for classification into different species, families, or even higher categories. These differences frequently touch upon practically every single character of the organism, morphological and physiological. Two forms found in nature, which showed morphological differences of such degree as that existing between male and female insects in genital armature or, in other cases, antennae, wings, segmentation, would never be considered as belonging to the same species, or even genus (not to speak of such differences as are found in *Bonellia,* Cirripedia, etc.). In the sexual differences we have, then, two completely different reaction systems in which the sum total of all the differences is determined by a single genetic differential. We shall return later to this interesting set of facts from the point of view of development. The genetics of sex determination ought, therefore, to furnish information on how a completely different reaction system may be evolved on the basis of existing and known genetical agencies. Insofar as a problem of chromosomal number is involved, to wit, the mechanism of one versus two X-chromosomes, the agency in question is of the same order as that involved in heteroploidy.

This is not the place to go into the details of the problems of sex determination. Only those points will be mentioned which lead to a conclusion comparable to that derived before. There is no doubt that sex is genetically determined by a balance between female and male determiners, one being located within the X-chromosomes, the other outside of them (Goldschmidt, 1911–20, see 1934). There is no doubt, either, that the X-chromosome mechanism regulates this balance by opposing one or two quantities of the determiners within the sex chromosomes to the determining condition outside of the sex chromosomes, which remains constant in both sexes, and that this balance works by regulating the amount of activity of sex-determining stuffs (Goldschmidt, *Loc. cit.*). The decisive point which concerns us now is the genetic meaning of male (*Lymantria* type) or female (*Drosophila* type) determiners within the X-chromosomes. I used to describe these determiners as single genes (or closely linked groups of genes) because in the *Lymantria* experiments only a single action and a 1 : 1 segregation without any evidence of crossing over were observed. Bridges (1921–22), in his studies on triploid *Drosophila* intersexes, not only assumed (without a single fact in favor of the assumption) that these determiners exist in the form of a large number of sex modifiers, but in addition assumed that these modifiers pushing sex in one or the other direction are present in all chromosomes. The X-chromosomes, under this assumption, would differ from the autosomes only in containing a higher number of one or the other type. Dobzhansky and Schultz (1934) later tried to prove this conception experimentally by comparing the degrees of triploid intersexuality obtained after adding duplicated pieces of fragmented X-chromosomes. I have shown (Goldschmidt, 1935a) that the facts presented by these authors fail to substantiate their claim. On the other hand, adherents of Bridges, like Winge (1937), have tried to show that my experimental material can be interpreted in line with their position. Hämmerling (1938), however, found it easy to demonstrate that Winge's interpretation does not work, if applied to the details of the case. There are, as far as I know, only a very few facts available which may be interpreted as genetic proof for a localized sex gene. These are the experiments of Möwus (1933) with the flagellate *Chlamydomonas*, where crossing over between the locus for sex determination and another locus is claimed, and the experiments in fishes by Kosswig, some of which we can interpret in a similar way by assuming the occurrence of crossovers between individual sex determiners (Goldschmidt, 1937a; see, however, Kosswig's reply, 1939). Recently two groups of facts have become known which furnish some positive information. Completing the work of Patterson and collaborators (see 1938), Bedichek (1939) has covered every single locus of the *Drosophila* X-chromosome by duplications added to triploid intersexes. The result is that no individual locus which switches sex from one to the other alternative has been found. But simultaneously Knapp (1939) found something very different in the liverwort *Sphaerocarpus*. He broke the X-chromosome into different fragments by X-raying and found that some of these splinters of different length changed the sex in a completely clear-cut way without any intersexual conditions. This shows that the smallest of the sex-determining fragments contains all the female determiners, and that the rest of the chromosome does not contain any indispensable sex determiners. These new facts may mean that in *Drosophila* many sex determiners are sprinkled over the whole X-chromosome, and that in *Sphaerocarpus* they are confined to a small section or a point, as I had assumed for *Lymantria*. But there is also another interpretation, which is the reason why these facts have been mentioned in the present context.

Looking over the facts just reported and comparing them with the facts discussed before regarding trisomics and species differences, I am now inclined to assume that both Bridges and I were wrong in the discussion of one gene versus many. It is, I think, neither a single differentiating gene at one locus of the X-chromosome which decides the sexual alternative, nor is it an array of male and female "modifiers" sprinkled over the X-chromosome and the autosomes (a rather poor mechanism, by the way, to be evolved all over the living world in order to secure a simple alternative!). The sexual alternative seems, moreover, to be decided by a pattern effect of the whole sex chromosome or, as the case may be, of a more or less small portion of it. It is not this or that gene or array of genes which is acting to produce the extreme morphogenetic differences of the sexes, but rather the typical serial pattern within the X-chromosome, or definite parts thereof. The chromosome as a whole is the agent, controlling whole reaction systems (as opposed to individual traits). The features which are assumed by many geneticists to prevent a scattering of individual sex genes by crossing over (one X in heterogametic sex, or inert Y) actually prevent major changes of the pattern within the chromosome as a whole. Once more I must emphasize that such a conception offers mental difficulties to those steeped in the classic theory of the gene. To them an X-chromosome cannot mean anything but a collection of genes influencing many different somatic traits, interspersed with genes controlling sexual differentiation. An action of the chromosome is, therefore, nothing but the additive action of the individual genes (see Bridges' plus and minus modifiers). Once this conception has been understood as only one possibility, the other possibility (the action of a whole pattern arrangement without individual action of the serial ingredients) becomes intelligible. This is not the place to discuss the merits of this viewpoint of the genetic basis of sex determination. It has been brought out only to show that facts are accumulating from all sides in favor of major morphogenetic changes being based more on architectural features of the chromosomes as a whole than upon accumulation of small integrating differences.

A few words ought to be added in this connection regarding the phenomenon of polyploidy, which, in plants, seems to play a considerable role in some types of species differentiation. I do not intend to enter into a discussion of the general role of polyploidy in the evolution of plants. The pertinent facts may be found in Darlington's books (1937, 1939). There is no doubt that the different types of polyploidy (autopolyploidy, allopolyploidy) have something to do with species differentiation in many plant genera. In animals true polyploidy by doubling of the chromosome set is either not found or is of very limited significance (see Vandel, 1938). But in plants also it cannot lead very far in macroevolution, as the possibilities of such differentiation are quickly exhausted. In our present discussion, however, the facts have a special significance; namely, in regard to the question as to whether the chromosome or the chromosome set has a definite action which cannot be expressed in terms of integrating genes and which therefore comes in as a factor in evolution outside of the neo-Darwinian conception of the accumulation of gene mutations. We are not speaking of allotetraploidy, the combination of two different specific or generic genomes in one individual by hybridization followed by tetraploidy. We are referring only to autotetraploidy, doubling of the chromosome number within a species, resulting only in a change of the number of chromosomes and their constituent parts without a change of the gene-mutation type or a juxtaposition of different genomes. It is generally known that the mutation to a tetraploid condition is accompanied frequently by giant size (gigas forms in numerous plants). It is further known that this is sometimes the result of an increase in cell size proportional to the chromosome number. This certainly is due to an action of the whole genome as a unit, which thus reveals a nongenic action upon cellular growth. It is further known that frequently this relation does not exist at all. But more important is the fact that the doubling of the chromosome number, in nature or in experiment, without any other known change, may also result in morphological and physiological effects of the same type as otherwise attributed to gene mutations. All the facts pertaining to this question in plants have been assembled and tabulated by Muentzing (1936–39) (see also Melchers, 1939). The general points relevant to our discussion are: Experimental and natural tetraploids are slower growing and exhibit a tendency to become biennial or perennial. In a general way the rate of cell division is decreased and specific physiological properties of the cell—for example, the osmotic system—are changed. The whole metabolism in influenced, and starch and vitamin content increased. Even the basic chemical constitution may be changed. A most remarkable fact has recently been added to this discussion by Sinnott and Blakeslee (1938). They find in experimentally produced autotetraploids of cucurbit fruits the size difference so frequently observed in tetraploids. But, in addition, in every case the tetraploid fruit is distinctly shorter and wider than its corresponding diploid type. This is the result of changes in the shape of the early fruit primordium and in the relative growth rates of length and width during development. Sinnott had previously demonstrated that shape in these fruits is determined by "genic" action independent of the determination of size. Here the action is obviously one of the whole genome.

It is most remarkable that the natural polyploid races of plants which are autotetraploids frequently show a definite geographical or ecological habitat, which means that the physiological traits, just mentioned, adapt the species to a definite environment. Many facts of this type are tabulated by Melchers (1939). This predilection for certain habitats is significant and demonstrates the evolutionary importance of the phenomenon. Muentzing, in his analysis, comes to the conclusion that in numerous cases the evolution of perennial and specially adapted species is due exclusively to a doubling of the genome. In a particular case the diploid and tetraploid races *Dactylis Aschersoniana* and *glomerata* were compared morphologically in all details, with the result that the differences were found to be caused only by the difference in chromosome number and not by specific genes present in one and absent in the other "species." We do not intend to discuss further details. The points which we want to emphasize are that considerable diversification may be brought about by a quantitative change in the chromosomes without any participation of gene mutations or changes in so-called genic balance, and that the facts relating to polyploidy, therefore, contribute to the conception of the determination of definite reaction systems by chromosome-pattern action.

Tetraploidy is not frequent in animals, but in the few cases studied (*Artemia*, Artom, Gross; *Solenobia*, Seiler) the whole reaction system of the individual is changed, in addition to morphological changes which do not conform to the simple formula: more chromosomes, larger cells. The work of Gross especially falls completely in line with that in plants. However, changes in chromosome number in animals result more frequently from fragmentation than from doubling, and it seems as if fragmentation is somehow connected with specialization in evolution. A complete review of the facts is found in Vandel (1938), and a few cases have been reported above (p. 187). Though no definite rules are clearly visible, in a general way it may be said that in animals also the chromosome number alone, excluding genic changes, has a definite morphological and physiological effect which is independent of eventual changes in cell size. Here is a large field awaiting an experimental attack.

We do not intend to go beyond the discussion of the points which seem important for the conception of what we called systemic mutations. We shall especially refrain, as already indicated, from a discussion of the polyploid series in plants, which are certainly connected with specific diversification, and also from a discussion of allotetraploidy or structural hybridism, which also lead in plants to species formation. Since it seems that comparable features are absent, or at least unimportant, in animals, the process of polyploidy cannot be regarded as a general evolutionary principle. The facts can be found in Darlington's books. I might point out only that the facts concerning tetraploidy which have been mentioned seem to indicate that the effects of polyploidy may also be understood as due to concerted action of chromosomes and whole genomes, independently of the action of individual genes. Winge (1923) has tried to formulate an explanation on the basis of multiplication of genes (polymery) resulting from tetraploidy. I cannot see how such a conception can work if all genes are involved.

In concluding our present discussion I wish to point out that the new conception of systemic mutation might pave the way for an understanding of some facts which have always defied an explanation in neo-Darwinian terms. Foremost among these are complicated mutual adaptations, like adaptations of flowers to insects, or the favorite hobbyhorse of teleologists, the "fremddienliche Zweckmässigkeit" (Becher) as found in plant galls with a preformed exit. Here also belong many of the facts generally included under the term mimicry; e.g., the resemblance of orchids to bees, flies, or bumblebees, the general similarity to ants or termites of some of their commensals, the leaf-butterflies and leaflike Orthoptera, and many similar examples. There is a possibility that one such case can be considered as suggesting that the entire adaptational complex has been produced in a single step, a systemic mutation. I mean the much discussed cases of mimetic *Papilios*, as, for example, *Papilio dardanus*. Here a number of different female types exist within the same species, whereas only one male type is found. These females, hatching from the same egg batch, are as different in wing pattern, shape, size, and color as other species or even genera. In addition, they mimic poisonous butterflies of very different families with amazing exactness, though different wing-pattern elements may produce the similarity in appearance in the two cases. In addition, the different geographic races of *Papilio* mimic different models in their respective habitats. In this case something of the genetic situation is now known: The different patterns behave in heredity as if they were controlled as a whole by one or two ordinary Mendelian differences, which, however, act only in the female body. From innumerable crossing experiments in Lepidoptera we know that pattern differences of such an order, if analyzable at all, ought to turn out to be the result of collaboration of many mutant genes. Ford (1936) has tried to solve the difficulty by assuming, following Fisher's ideas, that the simple Mendelian difference is in fact the result of a long selection of invisible modifiers to which the one gene difference is finally added only to act as a differential. I wonder whether the proper solution is that the simple Mendelian difference is in fact a difference of a whole chromosome, the architecture of which has been changed by a systemic mutation, thus demonstrating *within* a species in a case of special adaptation the happenings otherwise found *between* species. A cytological test of this explanation ought to be possible where the material is available. I point to this case not in order to offer a hypothesis but to draw attention to the constructive possibilities of the new conception in instances where the accumulation of micromutations, which can hardly be of selective value before the whole pattern is practically finished, does not work, as is proven by the never-ending discussions on the origin of mimicry and the proposed radical way out of the difficulties; i.e., complete negation of mimicry, to which many authors have resorted.

4. PATTERN EFFECT AND SYSTEMIC MUTATION

Having shown that numerous facts tend to demonstrate that the basic feature of macroevolution is a change of intrachromosomal pattern as opposed to slow accumulation of gene mutations, we have to answer the question as to whether such pattern changes are known to produce the required effects. We have already quoted (p. 204) Dobzhansky, who, on realizing the importance of intrachromosomal pattern changes, has found himself rather embarrassed by his own discoveries. He argues: "There can be little doubt that chromosomal changes are one of the mainsprings of evolution. . . . It may be paradoxical that inversions and translocations are so important as evolutionary agents. Indeed, they change only the gene order, but not the quantity or the quality of the genes. . . ." But then there is the classical theory of the gene, which he thinks to be beyond doubt and as well established as the theory of molecules and atoms. Evolution therefore can proceed only by gene mutation and the origin of new genes. How to solve this dilemma? In the book from which we quoted it is quietly shelved. But in an earlier paper (1935) the dilemma is solved, by denying its existence (see above, p. 204). He assumes that there are two different elements of species formation, genic differentiation and differentiation in the chromosome structure. Both procedures are independent, but the phenomenon of position effect "may lead us to a recognition of an ultimate connection between the two."

I think that the time has come to face this situation without permitting ourselves to be prevented from further analysis by a dogmatic belief in the inflexibility of the classical theory of the gene (as emphasized over and over again in the previous analysis). The appeal to position effect leads, I believe, to a dodging of the issue. This issue is: Does the assumption of the accumulation of small gene mutations explain the facts of macroevolution? If not, are other facts available? If the other facts demonstrate effects without appeal to gene mutations, are genes and gene mutations needed at all to understand evolution? My answer to this last question is: No. I am firmly convinced that, except in microevolution, the facts already available today force us to drop completely from evolutionary thought the idea of the so-called gene mutation, whatever it turns out to be physically or chemically. We have discussed these facts, and we have concluded that the really decisive change of the genetic material involved in macroevolution is the change of chromosomal pattern. The problem thus left is to find out not how such a pattern change may be fitted into the classical theory of genes and slow accumulation of their mutations, but whether this type of genetic change is able to produce the effects required. If this leads to a change in our attitude toward the classic conceptions, we shall have to adjust our thoughts to the unavoidable new conceptions. We have already pointed out (see p. 201) the critical situation in which genetical theory finds itself, if the classic theory of the gene is accepted as an unalterable fact. We have seen that the direction of the present development is toward an abandonment of the particulate gene in favor of a conception which emphasizes the serial pattern of the chromosome and its parts. We have seen that an unbiased analysis of numerous facts relating to evolution falls completely in line with the purely genetic facts.

Now that the evolutionary facts have been discussed, we return once more to the genetical situation. We had to use repeatedly the conception of position effect, a term which is meant to describe a phenomenon which actually does not fit into the classical conception of the gene in terms of this theory; namely, in terms of changed distances between genes. Rearrangements in intrachromosomal pattern, like inversions and translocations, occur in nature and can be induced experimentally by some of the agencies; e.g., X-rays, which are assumed to produce so-called gene mutations and according to the same quantitative laws. Some of these pattern changes produce visible effects which are called position effects because it is assumed that the change of the normal position of the individual gene, bringing it into a different neighborhood, is responsible for the effect. These phenotypic effects (in animals) are of exactly the same type as those of so-called gene mutations: dominant and recessive effects upon the same organs and of the same visible type as those produced by mutants; simple and pleiotropic effects; multiple-allelic effects fitting into a series of mutant multiple alleles; lethal effects; modifier effects. (Details and literature in Goldschmidt, 1938, 1940 in press). There is, then, no doubt that an accumulation of chromosomal pattern changes as they are found actually may lead to considerable phenotypic effects without the necessity of a single gene mutation. Here again we must beware lest our discussion founder on the rocks of prejudice. We have repeatedly emphasized that the concept of pattern effect is intelligible only after the particulate concept of the gene string is no longer considered to be the only possibility. The parallel prejudice regarding evolution; namely, that it must be understood in terms of accumulated gene mutations, has in turn also blocked the path of progress for the theory of the gene. Muller, who discovered many of the decisive facts upon which our ideas regarding the pattern effects are based, also considered the possibility of changing the particulate gene conception into one of intrachromosomal pattern (see 1937, 1938). But it is evident that one of the main reasons why he hesitated is the conviction that evolution cannot be understood without genes.

Leaving aside for the moment the question of genes and gene mutations as such, it appears that the pattern effect —if we use this term instead of the term position effect which implies adherence to the classical gene theory— saves us from some of the insurmountable difficulties attending the concept of evolution by means of gene mutations. A complicated change of intrachromosomal pattern may occu instantaneously or in a few consecutive steps, and, if it lead at all to a stable condition, may at once produce the nev reaction system, the species. Thus all the difficulties i the way of a slow, step-by-step selective accumulation o innumerable mutants vanish, difficulties which appear whe we get away from generalities and try to apply the neo Darwinian explanation to concrete cases. Selection will wor upon a whole balanced system which may be rejected o accepted, but not upon one minute change of minor sig nificance after another. The insurmountable difficulty of the production of new genes is avoided. (The idea, proposed first, I think, by Bridges [on the basis of gratuitous interpretation of repeated parts of the pattern as duplicated genes] that new genes are formed by duplication of old ones, with consequent diversification by mutation, cannot be taken seriously. One of the main tenets of the theory of the gene is that mutation changes a gene into an allele. How could mutation in a duplicated gene produce anything else, but for a new localization? One might call the new locus a new gene, but what could it accomplish in evolution except a repetition of the action of the original gene and its mutants?) If only the serial pattern as a whole is decisive, an unlimited number of patterns is available without a single qualitative chemical change in the chromosomal material, not to speak of a further unlimited number after qualitative changes (model: addition of a new amino acid into the pattern of a protein molecule). Macroevolution may proceed by large and rather sudden steps which accomplish at once what small accumulations cannot perfect in eons, and this on the specific as well as on any higher level. The systemic pattern mutation—as opposed to gene mutation—appears to be the major genetic process leading to macroevolution; i.e., evolution beyond the blind alleys of microevolution.

We have repeatedly said that the repatterning of the chromosome produces at once a new genetic system with all its consequences. This "at once" is an important point. From the work on intraspecific chromosome changes (see p. 189) we know that inversions and rearrangements may occur without having any noticeable effect, even when they are accumulated, as in the case which was quoted from Dubinin and collaborators. From these facts alternative conclusions may be drawn: These pattern changes may be an accident, without any significance except for creating new conditions of genetic isolation by chromosomal incompatibility, as the adherents of neo-Darwinism assume. A population problem and not a strictly genetical problem is involved here.

Dubinin and collaborators seem to favor this view. On the other hand, the pattern changes may be such that in the beginning they do not lead to a new pattern above the threshold of pattern action (systemic mutation). Only when by chance a pattern, viable in homozygous condition and above the threshold, has been reached; i.e., such as the patterns actually found when comparing species, does the new system of reaction suddenly emerge, though prepared by subliminal steps.

At this point of our discussion I would like to draw attention to a fact which seems to be of great importance for the problem of chromosomes and genes. The neo-Darwinian theory requires numerous mutating genes besides the formation of new genes. The basic fact regarding genes is their serial location in the chromosomes. Since the ingenuous analysis of Roux (1883), we know that the form of the chromosomes and their longitudinal division make sense only if they provide a mechanism for an exact division of determiners arranged serially in the chromosome. The classical theory of the gene from Weismann to Morgan assumes the same. From this it follows that the number of genes in a chromosome is roughly proportional to the length of the chromosome when completely stretched. The chromosome maps and the structure of the salivary chromosomes of *Drosophila* bear out this conclusion. Now the hardly organized protozoan *Monocystis* has chromosomes of the same order of magnitude as those of the highest animals and plants, not to speak of the large and numerous chromosomes of Radiolaria, which may not be strictly comparable to other chromosomes. If Roux's analysis is correct—and how can it be otherwise—*Monocystis* must have approximately the same number of genes as some higher animals and plants. Let us not deceive ourselves and try to evade the question by an appeal to inert material (see modern textbooks of cytology), for which the cogent argumentation of Roux is just as binding as for the rest of the chromosome. The difficulty vanishes if the pattern replaces the genes. In this case different chemical effects may be produced without a change in the number of loci making up the serial pattern if in the lower organism the constituents of the chromosomal pattern are chemically simple, and if in the higher organism they are complex (model: simple amino acids in a chain molecule versus highly substituted ones).

Most of the foregoing conclusions, however, may be drawn without touching upon the problem of the gene and its mutations, as repatterning happens on the chromosomal level and not on the level of the gene locus. It is possible that within the major pattern changes, the individual loci, called genes, may undergo local changes, called mutations, which may play their local role in microevolution, as analyzed above. So-called gene mutations are known to occur within inversions and translocations, and in a similar way, within the different chromosomal patterns of two species (say, *Drosophila pseudoobscura* and *simulans*, or two *Oenotheras*). But I do not think that we can stop here. The analysis of the genetic basis of evolution has actually led us to a point which is identical with the point which has been reached from another avenue of approach, as pointed out before. In a few short reports (to be followed soon, I hope, by a comprehensive one) I have come to the conclusion that all the recent developments of genetics tend to show that the classical theory of the gene as an actually existing unit, lying in the chromosome like a bead in a string of beads, is no longer tenable; that the linear order of the loci in a chromosome is an internal pattern of integrating elements which does not necessarily involve the existence of separate units of a molecular order; that the mutational change at a definite locus, which alone informs us of the existence of this locus, does not prove that a particle is located at this point, a break, for example, being also able to account for the changed action. In spite of all opposition which such a viewpoint is bound to encounter, I am fully convinced that geneticists will have to accept it eventually. The coincidence of the derivation of such conclusions from both a genetical and an evolutionary point of attack supports me in my strong conviction that I am thinking along the right lines. We have already tried to visualize the effects of repatterning (a fact which has to be understood in terms of its effect) by models which help in forming a mental image. We may now use one of these models as a simile for the systemic pattern mutation. Let us compare the chromosome with its serial order to a long printed sentence made up of hundreds of letters of which only twenty-five different ones exist. In reading the sentence a misprint of one letter here and there will not change the sense of the sentence; even the misprint of a whole word (*rose* for *sore*) will hardly impress the reader. But the compositor might arrange the same set of type into a completely different sentence with a completely new meaning, and this in a great many different ways, depending upon the number of permutating letters and the complexity of the language (the latter acting as "selection"). To elevate such a model to the level of a biological theory we have, of course, to restate it in chemical terms. I do not think that an actual chemical model can yet be found. But we might indicate the type of such a model which fulfills at least some, though not all, of the requirements. It is not meant as a hypothesis of chemical chromosomal structure, but only as a chemical model for visualizing the actual meaning of a repatterning process.

Let us compare the chromosome to a very long chain molecule of a protein. The linear pattern of the chromosome is then the typical pattern of the different amino-acid residues. Let us assume that this chain molecule acts as an autocatalytic proteinase (an assumption required for any model of the germ plasm). As it is known that each protein (and therefore probably each proteinase) is characterized by the length of the chain, the type of amino acid residues, and the specific order or pattern or rhythm of the repetition of these residues along the chain, innumerable types of protein may be obtained by permutation of these three variables, without any change within the individual residues, the loci of the chain; still more may be obtained if different polypeptids are united end to end into a superchain. The mechanics of the possible changes from one type of protein to another by a pattern change involving the three variables may be described in terms equivalent to the words breakage, inversion, translocation, deletion, rearrangement. A series of steps will probably be needed to transform one stable pattern into another, though the details can hardly be understood yet. As soon as this transformation is completed, a new protein, proteinase, chemical system has been achieved. It is possible and conceivable that within one such long chain small local pattern changes (stereoisomerisms) occur which do not change in a general way the catalytic activities of the whole though they impair it. I do not know of chemical examples involving proteins, but such a thing is known for sex hormones, where, according to Ružicka, different stereoisomeres—pattern changes—sometimes produce very different effects; namely, action as sex hormone or no such effect. A similar condition, applied to small parts of a chain molecule, would be a perfect model for mutations, if mutations were actually identical with position effects, as we claim. But larger and complete repatterning effects, producing a new chemical system though using nothing but the same residues, would be the model for those complete pattern changes within the chromosome, or systemic mutations, which account, as we believe, for the two major steps of macroevolution. Whether this model is good or bad, possible or impossible, *the fact remains that an unbiased analysis of a huge body of pertinent facts shows that macroevolution is linked to chromosomal repatterning and that the latter is a method of producing new organic reaction systems, a method which overcomes the great difficulties which the actual facts raise for the neo-Darwinian conception as applied to macroevolution.*

In the whole preceding discussion the chromosomes alone were discussed in the light of their evolutionary significance. There is no doubt that the action of the chromosomes is primarily an action upon the cytoplasm, and therefore we might ask what role ought to be attributed to cytoplasm in evolution. Unfortunately, this question cannot be answered. There are many cases known in which cytoplasmic differences can already be found at the level of microevolution. We mentioned before cytoplasmic differences between geographical races of *Lymantria*. Other cases will be found reviewed in Goldschmidt (1938). But there is no indication that whatever differences exist are of essential value to evolution. (We speak, of course, only of cytoplasmic differences which are not under chromosomal control. Most of them probably are, as the inheritance of serological features proves.) Actually, recently experimentation has shown that tissues of different orders (Amphibia and fishes, Oppenheimer, 1939) may be combined into a whole, which would hardly be possible if the cytoplasmic constitution were so very different. For the present, therefore, the evolution of cytoplasm—not under chromosomal control—may be neglected.

5. EVOLUTION AND THE POTENTIALITIES OF DEVELOPMENT

We emphasized before that direct genetic information stops almost at the point where macroevolution begins, though a considerable body of evidence is still available right on the borderline. But where the higher categories begin, and especially where huge differences of the entire architectural plan are involved, direct genetical information ceases to exist, though indirect information may be found, as we shall see. But this does not mean that no exact method for further analysis is left. Exact analysis is not confined to experiments in hybridization, as some geneticists want us to believe, but may be based upon any body of reliable facts. Such a body of facts was used in my essays of 1920, when I tried to link them with definite genetical conceptions. These facts were mainly taken from the realm of development, in the widest sense of this word.

Evolution means the transition of one rather stable organic system into a different but still stable one. The genetic basis of this process, the change from one stable genetic constitution to another, is one side of the problem. No evolution is possible without a primary change within the germ plasm; i.e., predominantly within the chromosomes, to a new stable architecture. But there is also another side to the problem. The germ plasm controls the type of the species by controlling the developmental processes of the individual. Whatever may be our conception of the germ plasm, mosaic of genes or chromosomal pattern, the specificity of the germ plasm is its ability to run the system of reactions which make up the individual development, according to a regular schedule which repeats itself, *ceteris paribus*, with the purposiveness and orderliness of an automaton. Evolution, therefore, means the production of a changed process of development, controlled by the changed germ plasm, as well as the production of a new pattern of germ plasm. A change within the germ plasm, therefore, is of evolutionary significance only if the subsequent different processes of development are again properly integrated to produce a balanced whole, the new form. It is, therefore, of decisive importance for the understanding of evolution to take into consideration the potentialities of the developmental system for a more or less radical change. In other words, the *action* of the germ plasm, the genes, or what you will, in controlling orderly development has to be taken into account when we try to link genetical changes with the resulting evolution. Continuing the line of argument derived in the foregoing chapters, we must find out further whether the developmental system is capable of being changed suddenly so that a new type may emerge without slow accumulation of small steps, but as a consequence of what we called a systemic mutation.

Such an analysis may be carried out in complete independence from the detailed conceptions which we developed concerning the architecture of the germ plasm and its changes. It does not make any difference whether a single macroevolutionary step is caused by a major change within the chromosomal pattern, a systemic mutation, or by a special kind of gene mutation with generalized effect, if such is imaginable. The decisive point is the single change which affects the entire reaction system of the developing organism simultaneously, as opposed to a slow accumulation of small additive changes. As a matter of fact, when I first tried to derive ideas concerning macroevolution on the basis of specific genetic changes (Goldschmidt, 1920), I did so within the classical theory of the gene by making use of the concept of gene quantities and their relation to reaction velocities. But all the facts reported above which push the systemic mutations into the foreground point to the necessity of regarding these as the effective agents of macroevolution. In the following discussions we mean, therefore, systemic mutations when we speak of genetical changes, though we admit the possibility that the same facts may be discussed, at least theoretically, in terms of single large gene mutations.

A. The Norm of Reactivity and Its Range

In early Mendelian days Woltereck introduced the term "norm of reactivity" (Reactionsnorm) to describe one of the basic conceptions of genetics. The genotype cannot be described simply in terms of the phenotype, since the description must contain the whole range of reactivity of the phenotype under different external or internal conditions. A genetic condition controlling, for example, large size, is in fact a condition which produces large size, provided that a series of environmental conditions is present, like nourishment, temperature, light, normal production of hormones. The genotype is, therefore, the inherited norm of reactivity to the ensemble of conditions which may influence the phenotypic expression. This concept of norm of reactivity, under natural as well as under experimental conditions, is founded on a huge set of facts which are of basic importance for the discussion of our present problem, the potentialities of development.

a. Examples

We do not need to discuss the innumerable modifications produced by the environment which furnish the material for the statistical treatment of nonhereditary variation. But within this group of facts we meet one rather general feature which parallels features of evolution. Species and varieties

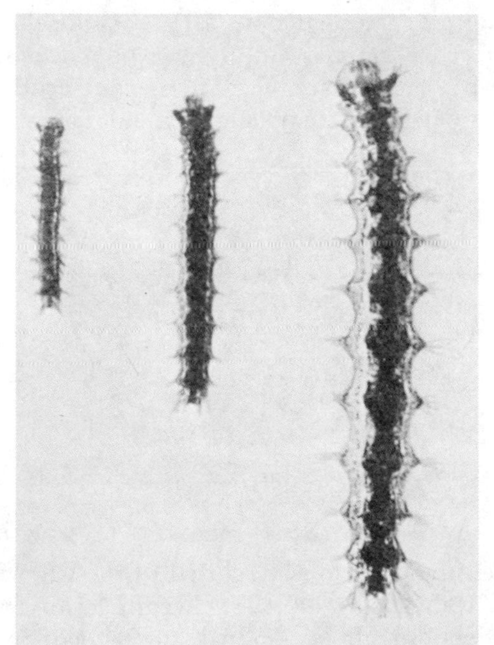

FIG. 39. Three sister caterpillars of *Lymantria dispar,* of equal age (stage), raised in normal, optimal and pessimal conditions. (From Goldschmidt.)

differ in many cases in typical size and proportions. But frequently, though not always, the range of modificability of one species under experimental conditions transcends the maximum of the ranges of all species combined and studied under natural conditions. For example, a stunted large species may be not only smaller than a luxuriant small one, but also smaller than the smallest individual recorded for the small species. To use our old paradigm, figure 39 represents sister caterpillars of a pure race of *Lymantria dispar*, one bred under optimal, the other under pessimal conditions. The size difference goes far beyond the limits of hereditary size differences observed in different races bred under average conditions. A long chapter could be written on such facts, all illustrating the same point. (For further examples see Cuénot, 1911; Goldschmidt, 1911, 1929a). Only one more example may be mentioned as relevant to the present discussion. The hermit crab *(Pagurus)* has a highly modified, asymmetrical abdomen with many

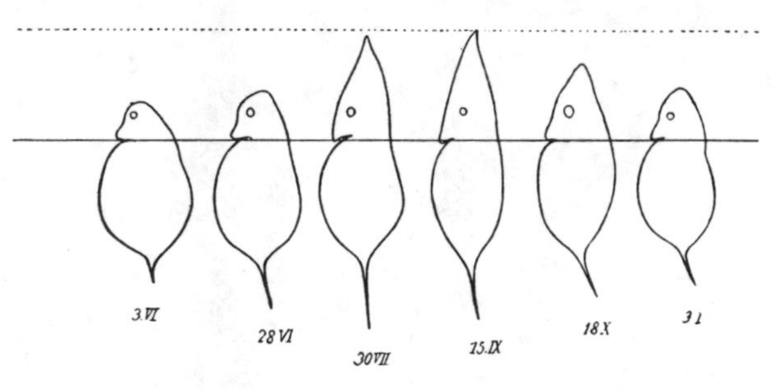

Fig. 40. Cyclomorphosis of *Hyalodaphnia*. (From Woltereck.)

abnormal features, obviously related to the life within shells. Przibram (1907) showed that hermit crabs which were forced to live outside a mollusk shell transformed their abdomen into a hard, more symmetrical structure, closely resembling that of free-living pagurids like the coconut-thief, *Birgus latro* (details may be found in Harms, 1934). Much more significant, however, are cases in which the norm of reactivity is typically alternative or polymorphic under different external conditions. A well-known example in plants is the case of *Limnophila*, which has broad leaves if grown in air, and finely laciniated leaves if grown in water. Here the same genotype is able to produce under alternative conditions two phenotypes which differ in order of magnitude to an extent typical of macroevolutionary differences. Many analogous cases in animals are known. We may consider, for example, all the cases of so-called cyclomorphoses in fresh-water animals like daphnids and Rotatoria. Here the generations which follow each other in the seasonal cycle are morphologically completely differ-

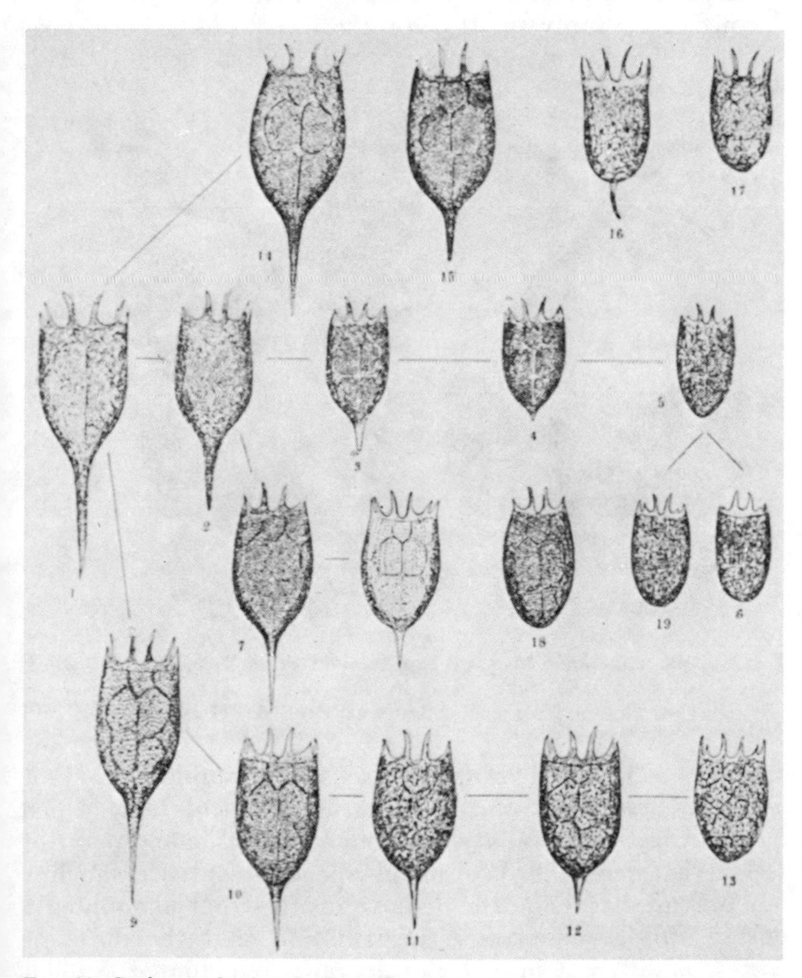

FIG. 41. Cyclomorphosis of *Keratella* (Anuraea) (Rotatoria), first form giving rise to four different series of variation in time. (From Lauterborn.)

ent, and the order of magnitude of the differences is again sometimes on the specific level (see figs. 40, 41). An example of the alternative type is the seasonal dimorphism of many Lepidoptera (*Vanessa, Papilio,* and others). Here the spring and summer forms, or, in the tropics, the forms of the dry and the wet season, may exhibit such different patterns, colors, and shapes that the order of magnitude of the difference is certainly on the specific if not the generic level

FIG. 42. Seasonal dimorphism of tropical butterflies. 4, 5, *Prioneris Watsoni;* 4, dry season; 5, wet season; both from underside. 6, 7, *Precis octavia natalensis;* 6, wet season; 7, dry season. (From Hesse-Doflein.)

(see fig. 42). Another much discussed example is that of castes in bees, ants, and termites, a group of facts which has been repeatedly analyzed from an evolutionary point of view ever since the Weismann-Spencer controversy. For our present discussion the decisive point is that undoubtedly only a single genotype is present and that the different castes, as different in order of magnitude as higher systematic categories usually appear to be, fall within the norm of reactivity of this genotype. Though the details are not yet completely known, there can be no doubt that the development of the same individual can be pushed in the direction of one caste or another by external influences. (We shall return later to this point.)

A rather remarkable type of the norm of reactivity is the one in which the retracing of very old phylogenetic steps lies within the range of the reactivity under known or unknown external conditions. We shall return to this point later and mention only a few cases here. Experimenting on the blind newt *Proteus*, Kammerer (1912) obtained large open eyes in individuals raised in yellow light.[4] The mountain newt *Salamandra atra* is viviparous. The embryos develop peculiar gills which are used for intrauterine absorption of food. Under definite experimental conditions the larvae develop in water and form the usual water-gills (von Chauvin, 1877; Kammerer, 1919). This is certainly an immense range of reactivity, retracing phylogenetic differentiation. A final example: The earwig *Anisolabis annulipes* is a completely wingless form. In the same group of animals an African genus exists, *Psalis*, which is winged. Pantel (1917) found a specimen of *Anisolabis* with complete wings of exactly the structure found in *Psalis*. In this case it is not known whether the condition was hereditary or not, though the latter is more probable.

These examples should suffice to show that under definite external conditions (including the internal environment) development may be easily changed, typically in some cases and exceptionally in others, in such a way that the order of magnitude of the shift is on the level of a higher categorical difference. (We mentioned before Kinsey's case, where it turned out that forms of *Cynips*, considered to be distinct genera, were seasonal variants of the same species. Many seasonal forms in butterflies were for a long time regarded as distinct species or genera.)

4. As many of Kammerer's claims are under suspicion, I may say that I have seen the specimens. Of course, I could not swear that the good eyes had not been transplanted into the specimens.

b. Experimental analysis

The important point now for our discussion is to ascertain those features of development which make it possible for the range of variation within the same genotype to be large enough to include the extent of morphological and physiological features found to distinguish higher systematic categories. Though this is predominantly a problem of development, it will obviously convey important information regarding the possibilities of evolution. Most of the material to be discussed will be taken from the animal kingdom. This does not mean that in plants the situation is a different one in principle. But it is different in detail, owing to the fact that from the standpoint of development plants are open systems, and animals closed systems. One of the consequences of this fact is that experimental embryology in plants has not yet furnished the same degree of insight as has been achieved by Entwicklungsmechanik in animals, though recent developments in our knowledge of plant hormones bid fair to lead to important discoveries in the near future which may clarify the differences on a rather simple basis. For this reason I shall mention here only one experimental contribution to the problem of norm of reactivity in plants, which demonstrates the huge range of variability within the same genotype, of an order of magnitude found genetically in macroevolution, though it does not lead to an understanding in terms of development.

The numbers of the organs of flowers; i.e., petals, sepals, anthers, etc., are more or less constant and characterize the species and higher categories of plants. In experiments performed with *Sedum* and *Sempervivum*, Klebs (1907) could shift these numbers considerably, far beyond the range of natural variation. The experimental conditions were, broadly speaking, different types of nutrition, including light. The typical number of anthers is ten, with a normal variation down to five anthers in about 20 per cent of the individual flowers. This range of variation could be altered experimentally. The result was variation from three to sixteen anthers, the modal class being different in different cases. The normal and practically constant number of five petals could be shifted to from two to fourteen.

In the animal kingdom the facts of the type discussed in this chapter which lead to insight into the potentialities of evolution are mostly, though not exclusively, found in cases of alternative norm of reaction. This means those cases in which normally, in nature, the same genotype is hidden behind alternative phenotypes with differences of a huge order of magnitude. Some of the cases have been given as examples above, and their meaning must now be discussed. Let us consider first the case of seasonal polymorphism, found in the form of cyclomorphosis in fresh-water organisms, as described above. The experimental work of Woltereck (1919) has demonstrated that the same range of variation may be produced by controlled feeding, including also the feeding of the parental generation. However, the very interesting details, which have led to a considerable amount of theorizing, do not permit as yet an actual insight into the developmental processes which are changed by the environmental action.

But there are other cases in which such an insight is available to a certain extent. Let us take up first the case of seasonal dimorphism in butterflies. The classical example, known since the pioneer work of Dorfmeister (1864) and of Weismann (1875), is the case of *Araschnia levana-prorsa* (see fig. 43). The wing patterns of the two generations differ more strikingly than do in many instances the patterns of two good species of Lepidoptera. The early experimenters could already show that the action of a series of temperatures upon certain stages of development resulted in the production of a corresponding series of intergrading conditions between the two standard patterns, as seen in figure 43. If such a series based upon hereditary differences were found in nature, it would be hailed as an example of species formation by gradual accumulation of small steps. If, however, the two extreme seasonal types *levana* and *prorsa* were found in nature as two distinct hereditary forms, they would certainly be called good species, and if a cross between the two produced in the F_2 the gradated series just mentioned, the case would be proclaimed a fine example of purely multiple Mendelian differences between two species. These two unreal *if's* show at once why we are discussing this material; the norm of reactivity within this genotype produces exactly the same degree of visible effects as is produced in other cases by a series of evolutionary steps based upon genetic changes; i.e., mutations and their accumulation. In other words, within a constant genotype the po-

FIG. 43. *Araschnia* (*Vanessa*) *Levana* (left upper) and *prorsa* (right lower) connected by experimentally produced intermediates. (From Goldschmidt.)

tentialities of individual development may include a range of variation of the same phenotypic order of magnitude which otherwise characterizes large evolutionary steps based upon changes in the genotype. The norm of reaction thus shows what paths are available for changes in the genotype (mutations in the broadest sense) without upsetting normal developmental processes, changes concerned with the same developmental processes in both cases, modification as well as mutation.

In this case some knowledge is available as to how the external agencies may act upon development, an action which, according to our interpretation, is identical with the type of action of genetic changes in other cases. The facts, however, have turned out to be much more complicated than might be assumed on the basis of the short description given above. *Araschnia levana* and *prorsa* had been considered as two good species until their real nature was recognized. This finding made the case of great interest for Darwinistic speculations of a phylogenetic nature and simultaneously stimulated what might be called the first major contribution to experimental evolution. Dorfmeister (1864), and especially Weismann (1875), performed the important experiments, later followed by Merrifield (1912). The basic facts discovered by these early authors are the following: *A. levana-prorsa* has two generations, a spring and a summer generation. The spring generation is the form *levana* with a yellowish brown-and-black pattern. The summer form *prorsa* is almost entirely black, with a white band across both wings; the underside of the wing is also different. These two generations alternate in the following way. *Prorsa* produces offspring which develop up to the pupal stage, when the diapause sets in for the hibernating pupa. In the spring, *levana* hatches and its offspring develop without diapause into *prorsa*. The authors just mentioned took it for granted that it is the influence of winter cold upon the pupae which produces the *levana* form, and the influence of summer heat which produces the *prorsa* form. They were led to this conclusion by their experiments with temperatures. By action of cold upon the pupae of the summer generation which were to produce *prorsa*, *levana* was produced; by action of different low temperatures a series of intermediates could be produced. The summer form was obtained by action of heat upon the hibernating pupae. The situation is, however, more complicated, as is shown by experiments performed in my laboratory by Süffert (1924). Actually two different features come into play; namely, the alternative of diapause or no diapause and the action of temperature. Hibernating pupae (with diapause) always produce *levana*, even if kept all the time in warm temperature. *Levana* is, then, absolutely linked with the occurrence of a diapause, and in the same way, *prorsa*, with the absence of a diapause. If in nature or under experimental conditions the diapause is occasionally prevented, *prorsa* invariably hatches; and if after the diapause, producing *levana*, a second diapause is induced experimentally, *levana* is again the result. This induction can be obtained by the action of cold upon young caterpillars or grown caterpillars before pupation. The temperature-effective period (critical period) for the control of the diapause occurs before pupation. The result is always the *levana* form. Süffert could also induce partial diapause by less extreme action before pupation (*before pupation* is important, as we shall see at once; after pupation the pupae were kept in normal temperature). In this case transitional stages between *prorsa* and *levana* were produced of a grade roughly in proportion to the length of delayed development. Thus it is clear that the physiological processes which cause the diapause produce a condition which simultaneously controls the wing pattern, and the same applies *mutatis mutandis* to the conditions preventing diapause.

But there is a second phenomenon which produces similar phenotypic effects but on a different physiological basis. The same old experimenters who were mentioned before had shown that it is generally possible to change the wing pattern of Lepidoptera by applying different temperatures to pupae of a definite age. (A complete review of the facts may be found in Biedermann, 1912.) Thus the temperature-effective period for pattern changes (the critical period) was discovered. The same method can now be applied to the pupae of the summer generation of *levana-prorsa*, destined to be *prorsa*. By action of cold the *levana* form may be produced without diapause, and by graduating the action, the transitions to *prorsa*. (The specimens photographed in figure 43 were produced by this method.) For this action the sensitive period was found to be a pupal age of twenty-four hours, which agrees with the results obtained in other Lepidoptera. It is obvious—and borne out by all the later work on sensitive period and wing pattern in Lepidoptera (reviewed in Goldschmidt, 1938)—that in this case the action of temperature influences directly definite developmental processes occurring during the period in question and connected with wing-pattern formation. In the other set of facts, however, the same result was obtained by enforcing (or changing) a period of rest in development occurring before the time of the sensitive period involved in the present case. The problem is to find an explanation which covers both sets of facts.

I have derived (1920, 1927) an explanation of the situation on the basis of a generalized insight into the action of the genetic material in controlling development. This action can be described in terms of balanced reaction velocities. Developmental, differentiating processes proceed at a definite speed, and the different processes are properly synchronized. The genetic material controls the velocities of production and the time of action, of the determining stuffs which control differentiation. The proper timing of these processes is the decisive feature in the general control of development. This idea, which has since been elaborated and has been found to agree with numerous experimental facts (see complete review in Goldschmidt, 1938), was applied to the case under discussion. The elaboration started from the fact that it is possible to shift the wing pattern of all Lepidoptera by applying extreme temperatures at a definite time after pupation, the sensitive period, though the shift is not as extreme as in the *levana-prorsa* case. (A huge body of detail is found in Biedermann, *loc. cit.*) As a similar, though not quite identical, action can be produced by extreme heat and cold, suffocation, narcosis (Standfuss, Fischer, von Linden, etc., and recent work by the school of Kühn), it was concluded that in these experiments a decisive process of wing-pattern formation occurring during the sensitive period was changed in its velocity (slowed down relative to other developmental processes), different processes of this sort having different temperature coefficients or susceptibility to inhibition, as the case may be.

After the sensitive period the processes in question would otherwise have led to a final determination of the typical pattern. (Details and further development of the problem are to be found in Goldschmidt, 1938.) This idea could be immediately applied to that part of the experiments with seasonal dimorphism in which the direct effect of temperature twenty-four hours after pupation was shown. The difference between the two phenomena is a twofold one. First, the temperature effect in ordinary cases is less extreme than in the *levana-prorsa* case. In the latter case a different norm of reaction is present, which permits a pattern shift of a higher order of magnitude. We do not know what the underlying mechanism is: either a larger range of differences in the integrating reactions still capable of cooperation, or different threshold conditions. The second special feature of the *levana* case is that another hereditary norm of reaction, which determines the alternative of diapause or no diapause according to the prepupal temperature, indirectly also determines the wing pattern. As the same patterns are produced by direct temperature action after pupation and indirect action via control of diapause, we might well conclude that the latter process also results in some way in the same slowing down of the patterning process. We do not need to discuss these special features in detail. For our discussion it is sufficient to know that a change of pattern (in some cases also wing shape and size) of an order of magnitude found in macroevolution can be produced by a slight shift of definite developmental processes, enforced at a time when these processes take the decisive step in embryonic determination.

In order to visualize the importance of such facts for our evolutionary discussions, let us enumerate the different groups of relevant facts. (1) External agencies which affect the speed of developmental processes concerned with the determination of the wing pattern can shift the pattern in the Lepidoptera experimented upon with an order of magnitude corresponding to microevolutionary changes. (2) In many different families of Lepidoptera, forms exist with a hereditary norm of reaction permitting the production of patterns of the order of magnitude found in macroevolution, if a shift in the patterning processes is induced directly or indirectly by external agencies. The occurrence of this phenomenon in different families, in connection with the first group of facts, demonstrates that it is not a rare situation requiring peculiar conditions, but that the general potencies of development, as far as wing pattern in Lepidoptera is concerned, permit a shift of a macroevolutionary order of magnitude within the same genotype if the proper conditions are available. (3) We discussed above the case of the sexual polymorphism of *Papilio dardanus*, where a few Mendelian differences accounted for an immense difference in wing pattern. As a matter of fact, corresponding cases exist in which the sexual dimorphism of the females is of a lower order, corresponding to the order of magnitude of a microevolutionary difference (e.g., *Colias edusa* after Gerould, 1923; *Argynnis paphia* after Goldschmidt and Fischer, 1922). In these cases, then, a phenomenon closely related to the former ones is found to be based on a simple genetic difference. Obviously, the same type of possibility of shifting the embryonic pattern-determining processes is present as in the former cases, but the determining agency controlling the shift is a purely genetical one. But there is an additional feature: sex-controlled inheritance. This means that the patterning embryonic process can be shifted only in the female; i.e., in a definite developmental system realized only in the female. (For detailed explanation of this feature in terms of reaction velocities see Goldschmidt, 1927.) There can be no doubt that in this case the features of both former cases are united; namely, shifting of developmental processes within an alternative norm of reaction (the sexes) and an order of magnitude of the shift from micro- to macroevolutionary level. But here the control of the process is exclusively a genetical one.

Carefully weighing this set of facts we reach the conclusion that the potentialities of development, as embodied in the potentiality for shifting processes of embryonic determination with relation to each other, provide for a potential range of phenotypic changes of the order of magnitude of macroevolutionary changes, and, further, that a genetical change which is able to control the possibilities of shifting these processes all over their range is able to produce in one step a change of macroevolutionary order limited only by the extremes of the range of shifting which still permit the formation of a harmonious whole. This conclusion, derived from this first example in our analysis of the relation of the laws of development to evolution (Goldschmidt, 1920 ff.), will be borne out in all the further material to be discussed.

c. Mutants and phenocopies

The foregoing discussion has already led to conclusions which will reappear at the end of our analysis of the norm of reactivity. Let us now return once more to the simple facts regarding relative shifts of embryonic processes during development and analyze the relation of these expressions of the norm of reactivity to the developmental changes brought about by ordinary mutation. Let us recall the old experiments in which the treatment of lepidopteran pupae with extreme temperatures, etc., during a sensitive period produced definite changes in the wing pattern. In such experiments Standfuss (1896) and Fischer (1901) found that the modified specimens appeared phenotypically identical with certain geographic races (called species at that time) of the same species, a result which led to Lamarckian interpretations (see the title of Fischer's paper). The explanation which I proposed for such facts (Goldschmidt, 1920), in line with my general ideas regarding the action of the mutant gene, is of the same type as that reported for seasonal dimorphism. The typical processes of the embryonic determination of the wing pattern may be changed in their velocity differentially from the other simultaneous processes of differentiation by the action of external agencies such as cold and heat. The result is a different pattern. Mutations which have the same phenotypic effect also act by influencing the velocities of the same embryonic processes of pattern determination, directly or indirectly, during the period in which such a shift is possible. One might generalize from this explanation by saying that the mutant genotype produces a shift in the speed of developmental processes within the normal range of variation set by the norm of reactivity of the process in question. If this conception is correct, it follows that the phenotype of any possible mutant can also be produced as a nonheritable modification, provided that an experimental agency is available which produces the necessary shift in velocity, and, further, that the sensitive period of the process is known and that the experimental procedure is able to act differentially upon one of the several processes. I have introduced the term *phenocopy* for this purely phenotypical process of copying the type of a mutant. In extensive experiments on *Drosophila* I have been able to show (Goldschmidt, 1929b, 1934a) that it is possible to produce practically every known type of mutant as phenocopy by the action of different degrees of heat shock during the sensitive periods of the pupa. It has been shown since that cold shocks (Gottschewski, 1934) and X-rays (Friesen, 1936) produce the same effect. Numerous less elaborate examples have also come to light (see review and discussion in Goldschmidt, 1938). These facts, as far as they go (and further unpublished data from my laboratory), prove beyond any doubt that the interpretation which we have given above is correct.

In discussing the *levana-prorsa* case we gave a general interpretation of shifts in adult pattern in terms of embryonic reaction velocities. Now we see that the interpretation of the special case is only a part of a general law which simultaneously embraces development, nonhereditary modification, and mutation. It is obvious that such an insight is highly relevant to evolution, and we have already pointed to some consequences in the case where variation on the macroevolutionary level was concerned. The present discussion deals only with the smaller changes of a microevolutionary order of magnitude (however, see the discussion below on homoiosis). But it seems worth while to look at some of the consequences entailed by the phenomenon of phenocopy. Let us, therefore, once more return to the interpretation of development as controlled by the genotype; that is, to the theory of balanced reaction velocities, which I have developed during the past twenty years and which seems to be the only possible explanation of genic action in physiological terms. (Details may be found in Goldschmidt, 1920, 1927, 1938.) There can be no doubt that the germ plasm controls development by means of determining-stuffs (in the widest sense) which are the products of definite reactions, the speed of which is exactly controlled and properly fitted to all the other simultaneous reactions, which have to be integrated into an orderly, properly timed whole. The most easily imaginable consequence of a genetic change is, therefore, a change in the velocities of one or more such reactions which leads to the upsetting of the delicate timing process at some point of development. It is obvious that the actual result of such an event will depend upon the amount of upsetting, the time in development when the change occurs, and the interrelations of the reaction in question with other vital ones. If the shift of a single reaction or a group of reactions is so large that no more integration with other vital reactions is possible, the whole developmental system will be out of gear and we speak of a lethal mutation. If the change in a reaction occurs somewhere near the end of development, there is considerable probability that it will be too local to be fatal, and a more or less viable visible mutant may result, showing predominantly a single changed character with at best few other effects. If, however, the change occurs earlier in development, the probability is considerable that a series of subsequent processes will be hit by the one change. This will lead to more manifold effects and accordingly to lower viability. A still earlier incidence may affect vital reactions and be fatal. If there is a dosage relation between chromosomal change and the time of incidence of effect, one dose—heterozygous mutation—

may be viable, but two doses—homozygous—may be lethal, and in addition the effect will be dominant because of its early incidence in development. Finally, there is the problem of the type of process involved. An early change in a developmental reaction which holds a key position; e.g., in the general determination of body parts, will upset the whole fabric of development and will, therefore, be lethal, unless harmony is preserved by a simultaneous change of all other reactions (or by regulation; see below). But other early changes may be possible which are not fatal because the type of development precludes lethal interactions. The development of Diptera may serve as an example. Here imaginal discs from which such organs as wings, eyes, etc. differentiate, are formed early in development. Subsequent happenings in these discs will hardly ever interfere with other developmental processes and may, therefore, be changed completely without causing other consequences.

The evolutionary importance of the facts and their interpretation become visible when we consider some rather popular generalizations. In numerous cases in animals and plants it has been found that definite types of mutation recur as so-called parallel mutations in many forms, whether closely related or not. Thus, all domesticated rodents have black, white, yellow, and piebald mutants. Angora hair appears in rabbits and guinea pigs, as well as in dogs, goats, and cattle. Albinism is found everywhere; melanism occurs in many animals; dwarfism, variegation, etc., in many plants. These facts, especially those on closely related forms, have led to many discussions of an evolutionary nature. Vavilov (1922) (see further discussion, p. 187), to mention only one example, speaks of the law of homologous series and uses the facts met with in domestic plants to explain the spreading of forms from a common center. In a great many instances such parallel mutation is interpreted as proving that the species in question have those genes in common which produce apparently identical mutants, a point which was discussed above (p. 223). The facts regarding phenocopies demonstrate, however, the inherent weakness of such a conclusion.

There are numerous developmental processes which must be, in a general way, identical throughout large groups and which by their very nature cannot vary except in a few directions. Let us take eye pigmentation. The chemistry of the melanin pigments allows for the presence or absence, or presence in different quantities and at different times, of the basic ingredients, chromogens and oxidases. The absence of pigment may be caused not only by the absence of the chromogen or its precursors, or the absence of an oxidase, of a coferment or of a proper substrate, but also by a shift in morphogenetic processes which might perfect the eye too early for the pigment, or the pigment too late for the eye. Innumerable developmental upsets may exist which in the end cause an unpigmented eye. In insects white-eyed mutants have actually been analyzed in such different groups as Lepidoptera, Diptera, and Hymenoptera. The claim that this proves the presence of the same gene for eye pigmentation in the three classes is obviously absurd, though frequently advanced. We know, further, from the work of Ephrussi and Beadle, Caspari and Kühn (literature in Goldschmidt, 1938) that eye pigmentation requires the presence of a substance, most probably a chromogen-precursor,[5] which is absent in certain mutants. This substance (or substances) is identical in the different insect groups, probably because it is necessary in the chemistry of melanin (eye pigment) formation. Does this then mean identical genes for the production of pigment? The same argument might be used for hair form in mammals, or for any other comparable case. But we always reach the conclusion that the phenomenon of parallel mutation does not give any information about the genotype of the two or more contrasted forms, beyond the general statement that eye pigment or hair form, etc., is inherited. But it does give information as to the embryonic, morphogenetic, and physio-

5. This is not the opinion of the authors named, who call the substances hormones. I am sure that the substances will turn out to be chromogen precursors which can be hydrolyzed into the real chromogens. Whatever chemical information is thus far available points to such an interpretation.

logical processes and their intrinsic limitations as regards possible aberrations from the typical line of events.

d. Norm of reactivity and hormones

The last-mentioned examples, and others which will be mentioned in the following chapters, show that the range of the norm of reactivity is directly dependent upon the possibilities of shifting the relative speed of some simultaneous developmental processes. There is another group of facts concerning a large range of developmental potentialities, a range which on the phenotypic level is comparable to macroevolutionary changes. We mean the effects caused by an utterly simple change of internal environment; namely, a change in the features of production of active substances, especially hormones. The determining stuffs produced in development and responsible for orderly serial differentiations may be of two types: different substrates for the embryonic differentiation (embryonic segregation, fields, stratification; see Goldschmidt, 1927, 1938), and active substances controlling morphogenetic processes. The latter vary considerably in significance and bear different names, the merits of which have been discussed by J. Huxley (1935). But their type of action is in a general way very similar to that of hormones, in that definite substances induce definite and often complicated morphogenetic processes if brought into contact with a definite substrate. Wherever these determining substances of a hormonic type (using the term hormone in a generalized sense) control differentiation, a change in differentiation may be brought about by changes in the quality, quantity, time and place of formation, direction, and speed of transport of the substances in collaboration with an otherwise unchanged general developmental system. Since these substances produce an immense morphogenetic effect when called into action (see the so-called organizer of the amphibian egg), small changes of the type just indicated may lead to large results, provided that the general harmony of differentiation is not interfered with. This simple argument shows that any hereditary change in the production of these substances may have an immense effect of a macroevolutionary order if it leads to the development of a viable and properly balanced whole. Expressed differently, if the norm of reactivity of embryonic development is such as to allow for changes induced by a changed activity of hormones (that is, hormones in the widest sense, including all determining substances), a single hereditary change affecting these hormones in one of the different ways indicated above may produce an immense evolutionary effect.

Discoveries regarding the active substances of the organizer type, so important for experimental embryology, cannot yet play an important role in our present analysis, except in the way of rather generalized information. What we mean by this may be shown by an example. Embryonic undetermined skin transplanted from an amphibian donor species which has typical larval structures, like a horny beak, to the prospective mouth region of a host species devoid of such organs is induced by the inductive substances of the host to oral differentiation, which, however, assumes the characteristics of the donor; i.e., a horny beak, etc. The host inductor substances then control differentiation at a definite point, but the genetic constitution of the material controls the specific type of the differentiation. This shows that a genetical difference in the reacting system may produce a huge departure without a change of the inductive materials which initiate differentiation (Spemann, Holtfreter, Schotté; see Spemann's Silliman Lectures, 1938). The inductive substances are known to be rather unspecific and perhaps are even identical over large taxonomic groups. We know nothing about taxonomic differences in inductor material and therefore cannot discuss their eventual origin

A more concrete insight of evolutionary significance may be derived from a study of the effects of that group of determining stuffs which are called hormones proper. The gist of our argumentation becomes easily visible if we look at the well-known case of metamorphosis in Amphibia. In the classical studies of Gudernatsch (1912) it was shown that metamorphosis in frogs is controlled by the thyroid hormone. Experimental administration of this hormone produces metamorphosis long before the normal time of onset; absence of the thyroid prevents metamorphosis. In addition, the absence of the hypophysis prevents normal metamorphosis and produces giant neotenic larvae.[6] The innumerable details which have since been added to these basic facts are not of importance here. One of the extensions of this line of work is the study of the role of the thyroid in the metamorphosis of urodele amphibians. The classical case is the Mexican axolotl, which reaches maturity without metamorphosing, a hereditary condition which is absent in its nearest relatives. Feeding with thyroid makes the axolotl transform into the *Ambystoma*, a transformation which involves an immense morphogenetic change from gills to lungs, with all the concomitant changes in all systems of the body. (The complete literature is listed in Marx, 1935.) Now, it is generally known that a whole group of Amphibia, the Perennibranchiata, remain in the axolotl stage of development and do not metamorphose. It has not been possible to force them into metamorphosis by hormone treatment (except for minor changes: Noble), and it seems that they are genetically unfit for complete metamorphosis. This is best demonstrated by the experiment of grafting Proteus skin onto an axolotl and inducing metamorphosis. The typical metamorphotic skin changes do not extend to the graft (Schreiber, 1939). We shall not discuss here the old problem as to whether the perennibranchs are phylogenetically primitive or whether they are derived as neotenic larval forms from metamorphosing Amphibia. The point of our argumentation is independent of such speculations. The

6. For curiosity's sake I might mention that I was the first to realize this fact. I had obtained giant neotenic frog larvae in an experiment and on dissecting them could not find the hypophysis. Comparing this observation with those on certain human abnormalities based upon hypophyseal action, I concluded that it was reduction of the pituitary which had led to the neotenic growth. I suggested, therefore, to my student Adler that he extirpate the hypophysis in tadpoles, an experiment which produced the expected result (1914), later elaborated by Allen, Klatt, and others.

facts demonstrate that there are Amphibia in which a hereditary norm of reaction permits metamorphosis, which is a morphological change of an order of magnitude characteristic of the taxonomic difference between families or orders. The reaction takes place if thyroid hormones are produced in the proper concentration, which normally happens at a particular time. There are other Amphibia (axolotl) in which the same norm of reaction is present, but some unknown hereditary threshold condition prevents the reaction from taking place under natural conditions.[7] There are other Amphibia in which the genetic norm of reaction

FIG. 44. *Periophthalmus schlosseri* ♂ in fighting position. (From Harms.)

makes metamorphosis impossible. Whether this means that the tissues, the substrate, are unable to undergo the morphogenetic changes of metamorphosis because they have not yet acquired this faculty or have lost it again; or whether it means that the genetically determined threshold conditions for the action of the hormones are not fulfilled, we do not know. But one conclusion we may safely derive from the known facts: a single genetic change controlling either of these possibilities can produce the immense morphogenetic differences of a macroevolutionary order between nonmetamorphosing Perennibranchiata and metamorphosing Urodeles, whatever the direction of evolution may have been

7. According to recent investigations by Blount (1939) the decisive condition is the lack of production of thyreotropic hormone by the hypophysis of the axolotl.

Given a condition of hormonic control, then, large evolutionary steps are imaginable as a consequence of a small change in the genetic norm of reaction involving either part of the developmental system; i.e., reacting substrate or reaction-producing hormones, or any constituents of the whole system. We are facing a situation similar to that derived from the study of phenocopies, the main difference being that in that case simple shifts in processes of differentiation were involved, based either on genetic or on phenotypic action, whereas here the controlling agency is the production of hormones and a reactive substrate.

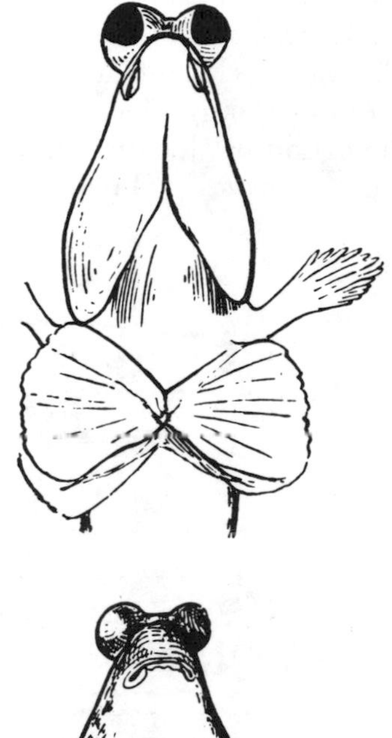

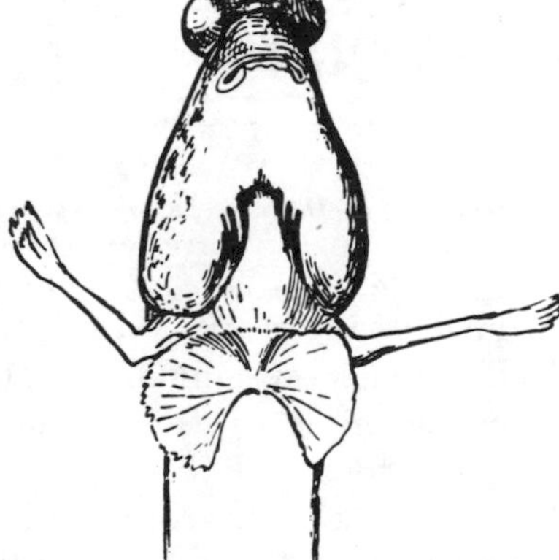

Fig. 45. *Periophthalmus variabilis.* (*a*), normal; (*b*), after thyroxin treatment. (From Harms.)

There is a very interesting set of facts of related significance available in fishes. One of the most remarkable groups of fishes are the Gobiids, because they have undergone strange adaptive changes in connection with specific modes of life. Some forms have largely given up swimming activities and live upon the sandy bottom of the sea. Their fins have changed in structure and enable them to hop upon the ground. Another group (*Periophthalmus*, etc.) has assumed an amphibian life. These are able to live out of water for a considerable time, to move on their fins as on legs, and even to climb trees. Harms (1934) has made an extensive study of this group and has described in detail the morphological changes of all organ systems in connection with the life out of water, structures which are perfected during metamorphosis of the water-adapted gobiiform larval fish. The facts relating to hormonal control of metamorphosis in Amphibia suggested in this case as well a relation between the origin of the adaptational traits and specific changes in the endocrines. (Harms explained the whole set of adaptive

Fig. 46. *Periophthalmus vulgaris* treated with thyroxin; use of anterior fin as an extremity with three lever arms. (From Harms.)

evolutionary changes on a purely Lamarckian basis; we refrain from discussing his arguments, which may be used as a fine example of the fallacies of that doctrine.) Harms actually found that during the metamorphosis of *Periophthalmus* the thyroid undergoes changes which closely parallel those occurring during amphibian metamorphosis. This suggested an experiment paralleling that of Gudernatsch, and it was actually found that metamorphosis can be induced in the larvae by feeding thyroid. Most remarkable were the results obtained by continued treatment with thyroid. The efficiency of all adaptations to life out of water was enhanced in the treated animals. The exophthalmus in creased; the skin became drier; the animals walked away from the water and even remained for considerable time on dry soil; the form of the mouth changed, and the teeth increased in size; the operculum changed, the gill cavity narrowed, and the gills became reduced in size; all the fins changed in shape, color, and structure of the skin. An extreme change was observed in the anterior fins: they narrowed and transformed into functional three-levered extremities (figs. 44, 45, 46). Simultaneously the thyroid was transformed from the diffuse structure found in the fish into a compact one of amphibian type (figs. 47, 48).

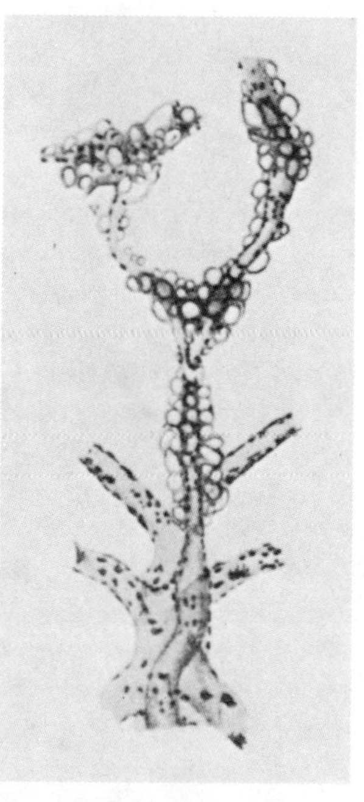

FIG. 47. *Periophthalmus vulgaris,* normal thyroid. (From Harms.)

These remarkable facts are very suggestive in connection with our present discussion. We can see that a phylogenetic differentiation within a group, associated with definite adaptations involving macroevolutionary morphological and physiological changes, is connected with the presence of specific endocrine conditions which are not present in close relatives living under usual conditions. We see that the administration of hormones enhances the same adaptational traits to an extreme degree. As the same cannot be accomplished with any other fish, these Gobiids must possess an hereditary ability to respond to thyroid action by metamorphosis. The situation, then, is very similar to that in Amphibia: a change in the reactivity of the substrate to hormones and a change with regard to the production of the hormones must have been the initial features of this evolutionary line. Again we see that simple genetic changes, which can be conceived of as occurring in a single step, may lead to macroevolutionary changes when endocrine control is the intermediary between genotype and morphogenesis. I may add finally that Harms quotes Sklower as having shown that during the metamorphosis of eels and flounders, the thyroid undergoes changes of the same type as are found in Amphibia and the Gobiids.

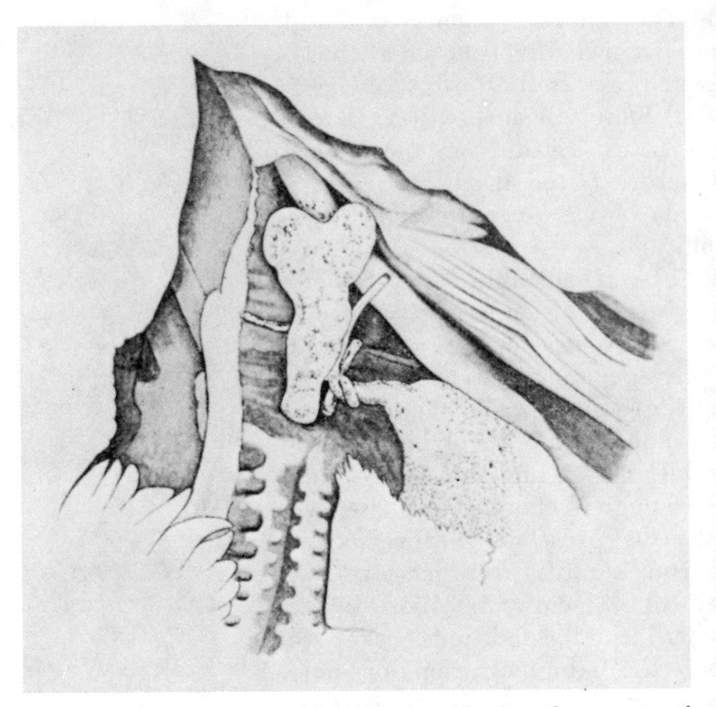

FIG. 48. *Periophthalmus koelreuteri*, thyroid after three years of administration of thyroxin. (From Harms.)

This fact certainly indicates the wide range of applicabilit of our argumentation.

There are many facts which illustrate the importance o the argument, but we need mention only a few. An importan hereditary trait in many animals is the typical behavior o the cycle of sexual propagation, which so often acts as a important adaptive trait with regard to seasonal cycles i nature. For example, the sexual cycle is controlled in Amphibia by the hormones of the pituitary glands in a definite hereditary way, in harmony with the seasons. But experimental treatment with the proper hormones permits us to shift this cycle at will. The developmental stage at parturition is a major taxonomic feature of macroevolutionary significance: compare the newborn marsupial, rat, man, and lamb. But by daily injections of thymus extract into pregnant mice, Rowntree (1935) succeeded in changing the normal rate of differentiation of the embryo so that in newly born animals eyes and ears were open, the vagina was developed, the descensus testis in males was speeded up, and the teeth were present.

One more example taken from vertebrates may suffice. The influence upon human growth of the hormones of the thyroid and pituitary is well known, and the conclusions derived from human pathology are substantiated by the experiments on mammals. The different types of giants and dwarfs show the strange morphogenetic effect upon all types of organs, including the brain, which is exercised by the presence, absence, insufficiency or hyperproduction of certain hormones, or by a change in the coördination of the whole endocrine system. It has been frequently emphasized that similar types occur as hereditary monstrosities in animals and that, therefore, in the latter cases it may be assumed that the genetic change (mutation) acts via a changed condition in the hormonic equilibrium. (See literature and discussion in Mohr, 1934; Stockard, 1931.) As an example of pathological mutants of this type we may mention the achondroplastic Dexter calf, which, according to Crew, is due to a hereditary hypophyseal defect. On extension of this argument it has been claimed that the human racial types are based upon hereditary endocrine differences; the pygmy races have been suggested as a hypothetical example. The best material for our discussion is found in Stockard's (see 1931) studies on the races of dogs. He pointed out that a considerable number of breeds of dogs are of a pathological type which closely resembles the type of well-known ab-

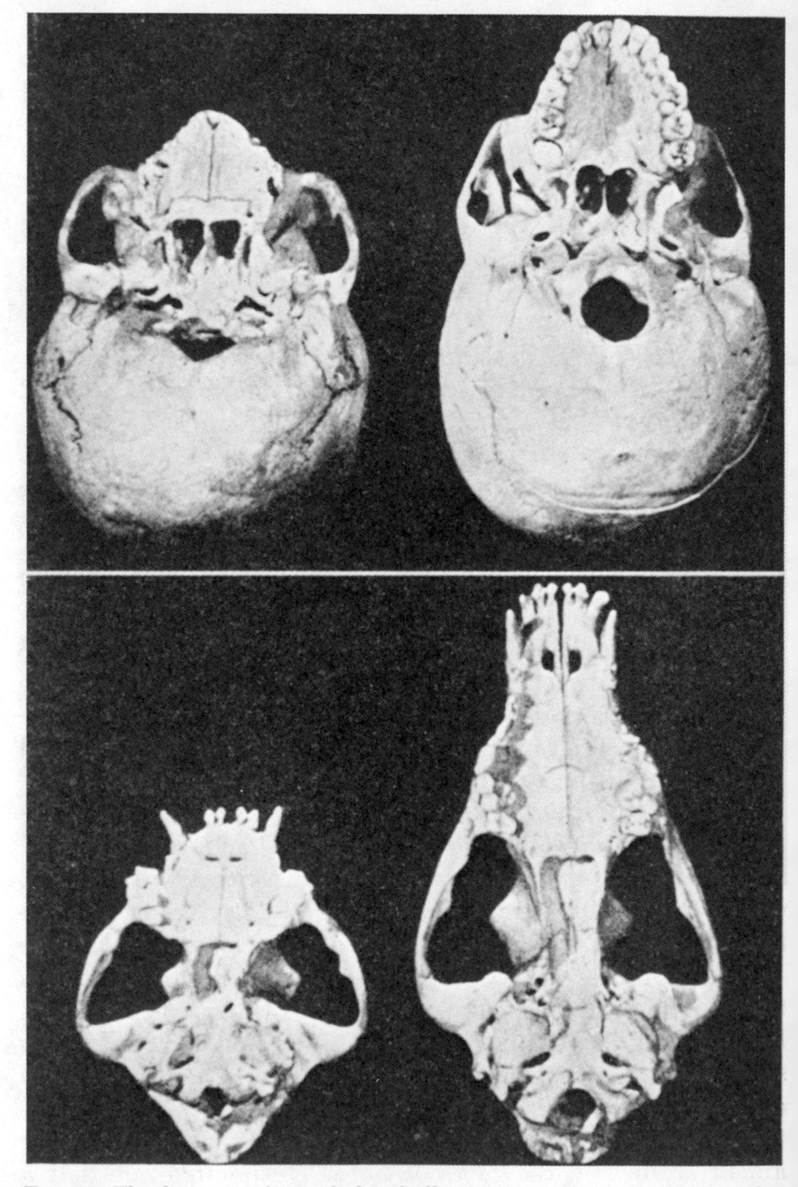

Fig. 49. The lower surface of the skull of a normal man (upper right) and an achondroplastic dwarf (upper left). Below, German shepherd dog (right) and English bulldog (left). (From Stockard. Courtesy Norton Co.)

normalities caused in mammals and man by hormonal insufficiencies or unbalance. Such forms as the St. Bernard, Great Dane, bulldog, and dachshund fall into this category, showing in growth habit, skeleton, and instincts the conditions known in pathology as achondroplasia, dwarfism, gigantism, acromegaly, all caused by abnormal endocrines (figs. 49, 50). These racial traits in dogs are certainly hereditary and they are based, as far as information goes, upon relatively simple Mendelian conditions. A study by

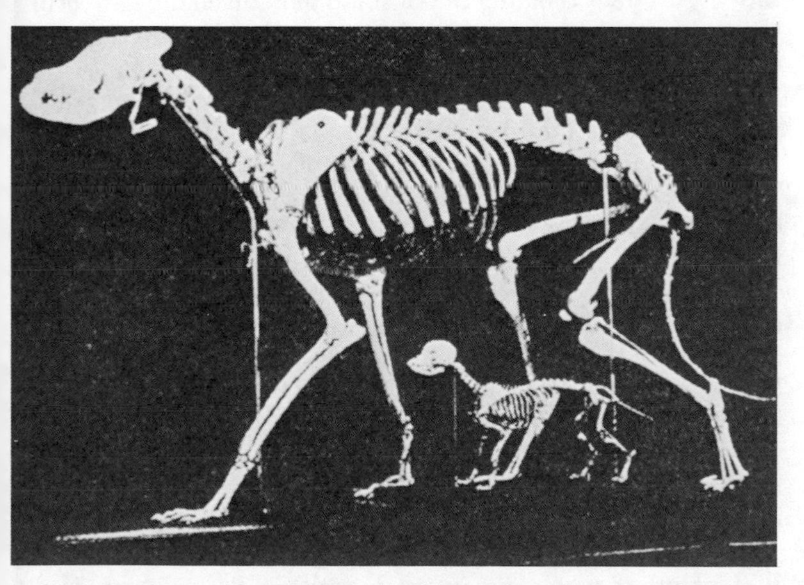

Fig. 50. Skeleton of St. Bernard dog and, below, that of a dwarf toy poodle. (From Stockard. Courtesy Norton Co.)

Stockard of the endocrines of these races revealed them to be abnormal in many different ways, so that the conclusion seems justified that the mutational changes act via endocrine disturbances. Of the many details of the situation one ought to be mentioned in connection with our problem. Some of the abnormalities which characterize the breeds involve the whole organization of the body and therefore point to "some hereditary disturbance of pituitary gland secretions causing abnormal pituitary-thyroid-parathyroid gland coordinations." In other cases, however, the abnormal growth conditions affect only one organ, say, the legs (dachshund). Stockard assumes that in such cases the glandular disturbance is of short duration and acts only during a critical moment in the origin of the embryonic limb skeleton. An alternative explanation [author] would be that a change occurs in the threshold conditions of response to hypophyseal stimulation in the limb bud. Whichever is the correct concept, the decisive fact remains that a small genetic change affecting the endocrine system may lead to general or localized growth changes of a huge order of magnitude. The skeletal differences existing between a wild dog, a Great Dane, and a poodle would certainly suffice for establishing generic differences if found among extinct forms. Actually differences of just this type must have played a considerable role in evolution, and I have not been the only one who has pointed to such facts.

Stockard himself points to a phylogenetic argumentation in this connection. Among comparative anatomists the fetalization theory of Bolk has been much discussed. This anatomist drew attention to the fact that the human head and brain retain more immature and foetal proportions than do those of any other mammal. This applies also to the relative proportions of cranium and face. Stockard recalls in this connection the fact that simultaneously the postnatal growth in man has been drawn out considerably, far beyond that of other mammals up to the apes. There is no doubt that this peculiarity of man is somehow controlled by the endocrines, as is proven by the cases of abnormal growth in infancy and sexual maturity in early infancy, all of which are connected with glandular disturbances. (See the experimental production of similar conditions in mammals by Rowntree, *loc. cit.*) Stockard further points out that we might therefore surmise that the delay in human maturity has arisen from a mutation affecting the usual mammalian coördination and balance among the endocrines. He assumes that these mutations had not yet arisen in early man, e.g.,

Sinanthropus. An evolution from this hominid to *Homo sapiens* may therefore be conceived of as having been perfected in a single genetic step, an event which is possible on the basis of endocrine control of growth and differentiation. This is certainly purely speculative. But the important point is not whether these specific conclusions can be proven. We are only interested in demonstrating that the norm of reactivity of developmental processes in regard to their hormonal control furnishes evidence in favor of our main thesis; i.e., that the potentialities of development permit changes of a macroevolutionary order of magnitude, involving the whole body or parts of it, to occur in a single genetic step.

Stockard's case may well serve as a model for all comparable situations in vertebrates. Similar evidence may also be found in invertebrates. We know nowadays, since the pioneer experiments of Wigglesworth (1934), that molting and puparium formation in insects are controlled by hormones produced in a gland near the brain. Extirpation of the gland prevents the process of molting, and injection of the hormone precipitates it. Time and number of molts are hereditary traits, which, as we saw in *Lymantria*, may be different in different races, the difference being based upon a simple Mendelian mechanism. But the occurrence of a special type of molt, pupation, makes up the essential macroevolutionary difference between ametabolous and holometabolous insects. If we try to visualize that difference in general terms of development, the formation of a pupa means that the developmental processes connected with the evagination of the imaginal discs for legs, wings, antennae, etc., are shifted from embryonic to late larval time, and, further, that a special molt which occurs at the same time enforces a period of rest within the cuticle, which is not shed as quickly as in other molts. The actual working of this timing mechanism can be inferred from cases of so-called prothetely, where a single larval organ, e.g., wings or antennae, metamorphoses alone. In a case which I studied (Goldschmidt, 1923) a group of *Lymantria* caterpillars failed to pupate at the proper molt. However, the antennae pupated, as figure 51 shows. The caterpillars lived for weeks beyond their normal time and grew to an immense size. Simultaneously their ovaries grew to the size of late pupal ovaries. Finally they died or were preserved. In this special case there is no doubt that an unknown hereditary change had produced the prothetely, which occurred in many individuals of a definite cross. (Harrison [1920] obtained a four-winged caterpillar; i.e., another type of prothetely, in a species cross.) It is not possible to give a definite explanation of this occurrence. But, since it is known that the time of molting, including the pupal molt, is genetically controlled, and, further, that the proper coincidence of the steps of pupation—evagination of discs, puparium formation, etc.—is controlled by hormones, we must conclude that the genetic change in question upset both the timing and the hormonic process, except for a single pair of imaginal discs. Such an abnormal situation, based on a small change in the genetic background, suggests that in evolution as well a small change involving the production of definite hormones in relation to the general timing mechanism of development may lead directly to a macroevolutionary step of huge magnitude. (*Vide* the preceding paragraph upon the evolution of man.)

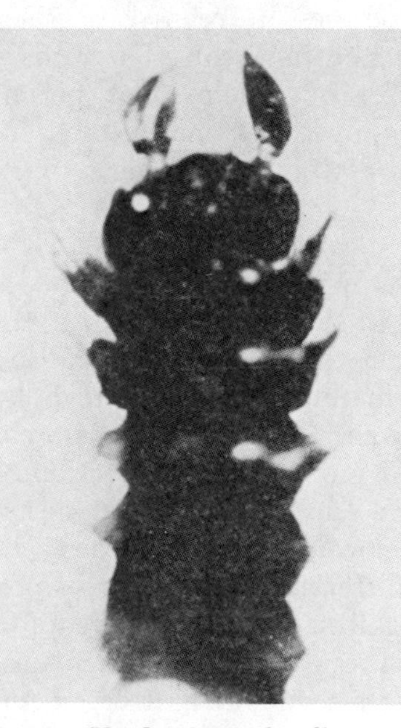

Fig. 51. Head of prothetelic caterpillar of *Lymantria dispar* with pupal antennae. (From Goldschmidt.)

The mention of hormone-controlled reactions in amphibians and prothetely in insects leads us to a short discussion of a closely related phenomenon, hysterotely, which is analogous to neoteny, though no hormonic control of the phenomenon has as yet been established. Whereas in prothetely a later step in development is anticipated, in hysterotely an earlier developmental feature is retarded and appears, therefore, in later stages. (If the earlier feature is considered to be one present in the ancestors, a hysterotelic feature is usually described as an atavism.) A good example of hysterotely has been presented (Goldschmidt, 1923) with reference to the gonads of *Lymantria dispar*, and has been discussed in the sense of our present analysis. In primitive insects (Thysanura, Japyx, etc.) the gonads are segmented structures, as in some annelids. In the development of higher insects cases are known (e.g., *Blatta*, according to Heymons) in which the earliest primordia of the gonads are still segmental. In *Lymantria* the gonad consists of four individual compartments and, as far as is known, does not show a segmented stage. But occasionally larvae are found in which the individual compartments of the gonad are separated and located in different segments. The obvious explanation of this and similar cases (many of which have been described by Schulze, 1922) is that a shift in the velocity of differentiation of the hysterotelic organ has taken place relative to the velocity of general development, though the details may be different in each case and may include additional features, especially conditions of hormonic control. The evolutionary meaning of such a situation is obvious, as here again the possibility of a large departure in a single step is offered. (A discussion of this point in line with our ideas is found in de Beer's book, 1930.) I do not know of any good case of mutational hysterotely except those such as harelip and coloboma, which cannot be used for conclusions upon evolution. But in a general way we may illustrate the correctness of the interpretation by the following example. Many mutants are known which are called regressive, because the phenotype is less complete than that of the norm. In Lepidoptera many wing-pattern mutants consist of an incompletely formed pattern, a kind of diluted condition. In some experiments on wing pattern (Goldschmidt, 1920c) I was able to slow up the development of the pigmentation process in one wing (the other wing serving as control) by operating on the pupa, with the result that a stage of pigmentation of the same type as might be produced by mutation (fig. 52) was present in the imago. This case may then serve as a model for the explanation of typical cases of hysterotely.

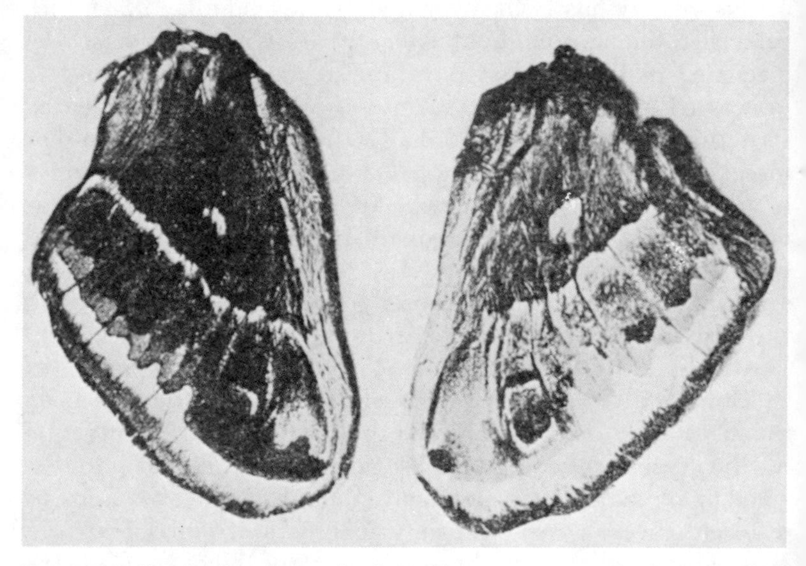

Fig. 52. Fully developed pupal wings of a cecropia moth. Left, control; right, development retarded by operation. (From Goldschmidt.)

As already indicated, neoteny is closely related to hysterotely insofar as sexual maturity is anticipated at an earlier stage of development. The meaning of neoteny for phylogeny is a much discussed topic, and in a great many cases zoologists have come to the conclusion that whole groups of animals must have been derived from neotenic larvae (see de Beer, 1930). A recent discussion of the case of the two species of *Polystomum*, *integerrimum* and *ocellatum* (Le Gallien, 1935) is very suggestive. The latter species resembles the larva of the former. Hubbs's (1926) discussion of the relation of developmental rate to adult differentiation in fishes, leading to retention of juvenile characters or extreme expression of later ornamentation, may also be mentioned. I shall not enter here into the phylogenetic argument, but want to point to the problem as another illustration of the general principles under discussion.

We have repeatedly confronted experimental facts with appropriate examples of phylogenetic features, appropriate because they may find their explanation in simple genetic changes causing huge morphogenetic consequences as a result of hormonic control. We may be permitted to add a few facts which point to the usefulness of our argument in explaining some features of macroevolution. There has always been a big rift between the Lamarckian viewpoint of evolutionary change and the Weismannian point of view. Although geneticists, with extremely rare exceptions, have accepted the Weismannian doctrine and believe that they have proven it, a great many zoologists and most paleontologists hold fast to the Lamarckian explanation. It is of little use to scoff at such convictions or to ascribe them always to insufficient knowledge, nor is it of any use to deride such ideas as purely speculative. Genetical analysis leads only to a certain point in the analysis of evolution. Beyond that point conclusions have to be based upon synthesis of all other available facts, experimental or not, and this is speculation. To erect barricades against synthetic thinking at the point where a definite type of experimentation ends, and to denounce as sheer speculation an analysis which goes beyond the possibility of direct experimentation, are indications of an attitude of snobbery which is typical of a newcomer, in this case genetics. Though I am convinced of the fallacy of the Lamarckian doctrine as thoroughly as is any geneticist, I think it necessary to try to understand why so many zoologists of broad knowledge and understanding cannot get away from the Lamarckian conception of evolution. I am sure that the reason will be found in the existence of a huge body of facts which exclude an understanding on the basis of selective accumulation of small haphazard mutants. A further reason for such an attitude is that the same facts frequently reveal such extreme conditions of adaptation that a Lamarckian, or even a psycho-Lamarckian, explanation appears preferable to the nongeneticist. It is, therefore, the duty of the geneticist to find out whether and how such facts can be explained on a genetic basis, not only without Lamarckism, but also without the improbabilities of accumulation of micromutations. As no direct experimental

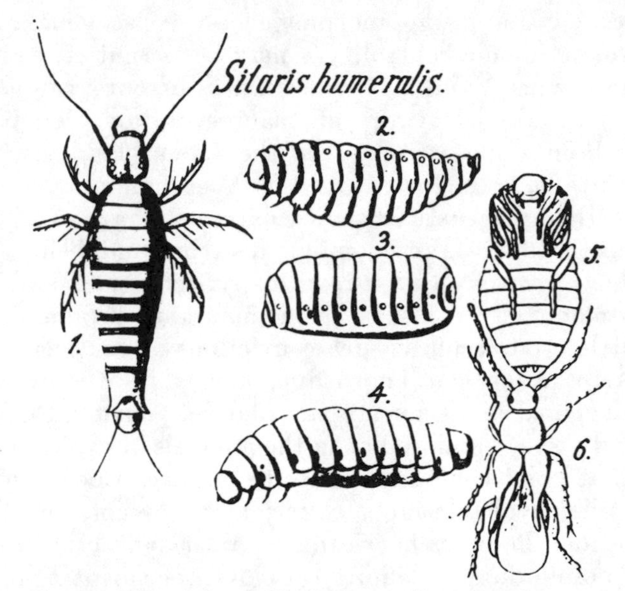

Fig. 53. Life cycle of *Sitaris humeralis*. 1, first larva of coleopteran type; 2, secondary maggot stage; 3, pseudochrysalis; 4, last maggotlike stage; 5, pupa; 6, imago. (From Fabre, after Harms.)

attack is possible, he will have to base his conclusions upon synthesis of all relevant facts. Two examples closely related to the general content of this chapter will show the direction in which we may find a solution of the old difficulties.

Sitaris humeralis (fig. 53) is a herbivorous beetle. From its eggs a typical coleopteran larva (1) hatches which will not continue development unless it succeeds in getting attached to a bee, which carries the larva into the hive. Here the larva molts and, losing eyes, legs, etc., emerges as a primitive maggot, feeding on honey (2); after some time, a resting stage (pseudopupa) occurs (3) from which another maggot emerges (4). This one finally pupates regularly (5) and from the pupa the beetle hatches (6). This strange life cycle is certainly adapted to the special ecological features, and its origin cannot be understood on the basis of selection of haphazard micromutations. Our previous discussion of embryonic reactivity to hormones indicates that there is a possibility of origin of this type of adaptation which requires neither Lamarckism nor selection of small steps. We saw that insect metamorphosis is governed in a definite way by hormones which control growth by molting as well as the processes of metamorphosis, especially the behavior of the imaginal discs. The special feature of our case is that after transfer to the hive the next molt returns the larva to a level of organization which otherwise would have been definitely passed in the embryonic stage. Though we do not know much more—here is an interesting field for experimentation—we must assume that a definite feature of hormonic regulation controls the aberrant growth and differentiation. This permits the conclusion—of only a very general nature at the present stage of our knowledge, to be sure—that a single genetic change affecting the mechanism of hormonic control may have been responsible for the initiation of the whole series of adaptational changes. There is no need to try to work out the possible details; we want to demonstrate only that a single genetic change of a definite type may entail a large departure which defies explanation in neo-Darwinian terms.

The second example which we want to mention is taken from the hermit crabs. We have already referred to Przibram's experiments on the asymmetry of shell-inhabiting pagurids. The primitive species of this group are symmetrical, like other crustacea. But the hermit crab has a completely asymmetrical abdomen, which is hidden in the shell of a snail, and an enlarged pincer which serves as a shutter for the entrance of the shell. This asymmetry begins to appear during development and increases even if no shell is available. On the other hand, crabs of extreme asymmetry, if deprived of shells, become almost symmetrical in subsequent molts. Further, the inborn right-handed asymmetry may be changed into a left-handed one, if only left-handed shells are available. Thus there is present an inherited trend to asymmetry and a wide norm of reactivity to outside conditions which permits of considerable adjustment.

One of the near relatives of this group is the coconut-thief *Birgus latro,* a huge, powerful crustacean, well known for its habit of climbing the coconut palm and of throwing down the nuts. *Birgus* has developed special features for breathing air and many other adaptations needed for his mode of life. These have been described in detail by Harms (1932). Typical pagurids undergo development in sea water. Rather early, after the last larval molt, they take to an empty shell, where they metamorphose into typical hermit crabs. Strangely enough, *Birgus* carries its eggs when these contain the zoëa stage into the sea (see Harms, 1934), where development takes place to the so-called glaucothoë stage, which already shows some specific *Birgus* characters. Still completely symmetrical, these also enter a shell which they leave occasionally at first, that is, as long as they are still symmetrical and have a normal abdomen capable of swimming movements. But soon their development becomes paguruslike, the abdomen becomes asymmetrical with all the ensuing morphogenetic consequences, and a true hermit-crab stage follows. After some time the shell is left again and symmetry is restored. The whole organization now changes into that of the typical symmetrical air-breathing land animal, with all the concomitant changes of structure and use of the different legs.

Phylogenetically, then, the hermit-crab type must have been produced by the origin of a tendency to asymmetry under conditions of a rather labile norm of reactivity to pressure (?) on the abdomen. The further steps toward *Birgus* must have been initiated by a changed growth rate beyond the size of available shells, a change in the inclination toward asymmetry and probably also in the lability of the norm of reactivity. It is, then, conceivable that the whole group of changes was primarily initiated by a genetic change with regard to the growth-controlling hormones.

These particular examples were mentioned because both have been used to demonstrate the necessity of a Lamarckian interpretation (by Harms), because both defy a purely Darwinian explanation by slow selection of micromutations, and because both suggest an explanation in terms of a single change (at least initially) in the hormonic control of growth and molting, producing an immense effect upon the whole organization at once and thus preadapting the new form to completely new ecological niches.

e. Norm of reactivity and regulation

At different points in our discussion we have met with the important phenomenon of regulation without mentioning it specifically (except in one instance). Actually this phenomenon is just as important in a discussion of the evolutionary significance of the potentialities of development as it is paramount in the experimental analysis of individual development. Regulation may be defined as a purposive response of the organism to changed conditions. Regeneration is the most typical form of regulation; the transformation of an experimentally produced half embryo into a whole is another type; the rebuilding of the spongiosa of a femur bone after a break, resulting in a new arrangement of the trajectories, is still another type. In one of our first examples of the inherited alternative norm of reaction, the air and water leaves of *Limnophila*, we observed regulation on the basis of a hereditary condition; most other plants do not show such a regulation after immersion. This example leads to an important point in our discussion. If a definite genotype is necessary for the production of a regulatory response, the ability to respond will have arisen as all genetic differences do, by some kind of mutation (or by an accumulation of mutations if the neo-Darwinian thesis is accepted), and its perpetuation under selection. Regulation, therefore, would have to be treated like any other hereditary trait. But this is not always the case. There can certainly be no doubt that whatever regulation may occur in individual cases, it is possible only within the limits set by the hereditary norm of reaction. One insect regenerates limbs; another is unable to do so. But within this obvious restriction there are different possibilities. The following is one involving a special genetical background. The older generation of biologists may recall the discussion about the regeneration of the lens from the dorsal margin of the iris in Amphibia. Wolff, the discoverer of this phenomenon, and his follower Driesch, emphasized that Darwinian principles could not explain the origin of this regulatory ability, and used the case as a proof for vitalism. Weismann, however, tried to show that this hereditary ability might very well have arisen under the influence of selection, just as might be argued for the regeneration of a limb, or the tail in Amphibia and Reptilia. Whether we agree or disagree with Weismann's conclusions is not relevant for this discussion. The purpose of the example is only to show that regulations exist which are explicable on neo-Darwinian lines if one is willing to accept the argumentation, which is bound to be often rather crude.

But there are innumerable cases of regulation in which the genetical side of the problem, and therefore the selective aspect, does not come into play, except for the obvious fact that everything happening in an organism must have its potentiality in the genotype. The following is a good example. The mechanism of walking in man consists, among other things, of the lever system of the foot with the heel-bone as the shorter arm. To this is attached the tendon of the gastrocnemius muscle which moves the lever. Marey (1887) has shown that in the white races the short arm of the lever is relatively short. In connection with this the muscle is compact and powerful, showing a characteristic featherlike arrangement of the fibers, and the tendon is long. In black races, however, the short lever arm of the heel-bone is much longer, and correspondingly the muscle has a completely different structure (absence of a pronounced calf), and the tendon is short. This is, of course, a hereditary difference. The leg of the cat is built like that of the Negro. If the cat's heel-bone is shortened by operation, regulation sets in after some time and the muscle of the calf assumes the type of structure found in the white man. Nobody can claim that this potency for regulation in the cat is the result of a selection of mutants. The same holds for many other regulations, especially all the regulations of early embryonic development known to the experimental embryologist. It is also true for a very different type of regulation usually described as atavism. If Kammerer's *Proteus* experiment is to be accepted, the formation of a normal eye under experimental conditions involves a complicated set of regulations, which cannot have been provided for by a special set of genetical conditions under the stress of selection. The potency for eye development must be provided in the genotype, but once this potency is realized, the concomitant regulatory processes are automatic; i.e., they are based upon a general ability of embryonic processes to adjust themselves for the sake of the production of a harmonious whole. It is this ability which appears over and over again in experimental embryology. Witness, for example, the numerous types of chimeras produced not only from different individuals but also from different species, even families

We must confess frankly that this power of regulation is not yet completely understood. Otherwise it would not be the favorite haunt of vitalism and its disguised variants. But experimental embryology has already furnished sufficient material at least to realize the direction in which an explanation has to be sought. Some of the most important facts have been presented by Holtfreter (1938), and his discussion of the situation is the most up-to-date statement of the problem available, though it is restricted to a special type of embryonic regulation. He describes regulation in a general way as the ability to repair a disturbance in the material construction or in the symmetry of a developing organism, in the direction of a harmonious whole. This repair can be accomplished by a reshuffling, increase, or decrease of the material without change of its character. Or it may be accomplished by a change of the original determinative character of the material. Or, finally an assimilation of parts, not primarily involved, into the new whole may be accomplished by a shift in the formative capacity, or by reshaping or resetting the material. (It is needless to say that the regulations under discussion here follow operative disruptions of normal development, which produce the tendency for regulation.) What actually will happen depends upon many conditions of the internal and external environment. Important among these are the condition of the regulating material, its location, and the time at which regulation starts. There is, first, the condition of determination of the material, which regulates only if not yet finally determined, though secondary regulation within the limits of the already determined potency is still possible. The time factor enters insofar as the time of determination is a different one for different primordia. Thus there is a definite time of still labile determination during which regulation is possible. Many examples of these rules may be found in Holtfreter's brilliant experiments. Environmental factors are mentioned, especially the mechanical conditions of the system, first exemplified in Roux's famous half embryos. The environment of the isolated blastomere, in this case contact with the other dead one, prevents regulation, which takes place if the contact-stimulus is not present (Spemann). Another extremely important factor is the intrinsic ability and tendency of the embryonic cells to move and to unite with other cells in the formation of new tissue adapted to the new mechanical conditions.

This latter point is of paramount importance in our present discussion, because it deals with a phenomenon which is not based upon a special genetic condition which might have arisen slowly in the course of evolution, but is the expression of a primary property of living cells. We might, therefore, interrupt Holtfreter's argumentation for a moment to point to the important phenomenon of morphogenetic movements (Gestaltungsbewegungen) aside from embryonal experimentation. The most convincing examples of this cellular potency are found in such cases as the following. Cells of an adult hydroid or sponge, if isolated and completely mixed by straining through fine gauze, come together again and build up a perfect new organism (Wilson; J. Huxley; Föyn, 1927—bibliography here). In the propagation of some Myxomycetae, individual, isolated amoeboid cells come together and build up by appropriate morphogenetic movements the complicated toadstoollike structure (Arndt, 1937). Arndt's film showing this development is a most stupefying sight to a thinking biologist. Similar examples are found in the behavior of cells in tissue culture, where isolated cells may come together to form tissues, the structure of which is determined by the conditions of the medium. I have found some fine examples in tissue cultures of Lepidoptera (Goldschmidt, 1916, and much unpublished work). From an isolated group of spherical spermatocysts cells migrate and finally connect all the cysts floating free in the medium into a characteristic unified tissue. Migrating cells from other organs may form, when attached to the surface of the cover glass, a flat epithelium with definite internal pattern arrangements, in response to mechanical conditions. A comparable group of facts is found in Steinmann's (1933) experiments on regeneration in planarians. The removal of the head, which ordinarily results in regeneration, leads to a formation of a structureless, wildly growing blastema of tissue-culture type, if the regeneration takes place in Holtfreter solution instead of water. This result may be described as inhibition of regulation without inhibition of cell movements. These examples may suffice as illustrations of regulative potency by morphogenetic movements.

The morphogenetic tendencies in the aforementioned cases of isolated cells in tissue culture are increased to actual regulation if embryonic parts are grown in tissue culture. Holtfreter, who performed numerous such experiments, reports, for example, that an isolated piece of chorda-primordium may produce musculature, nervous tissue, etc. It is amazing that these tissues, if given a chance, are not mixed in an irregular way, but may arrange themselves into a kind of bilateral, orderly pseudoembryo. (An example of regulation which is unsuccessful because of mechanical hindrance is, I suppose, the embryoma.) Holtfreter emphasizes in connection with these experiments that some facts indicate that the direct cause of the initiation of regulation is a rather simple chemical one. Finally he discusses experiments in which the regulative ability of a given embryonic tissue is not determined by its immediate chemical or physical environment, but by some action of the whole germ, whereas normal embryonic induction is a localized one. He adds this important statement (original in italics): "It may be assumed that these strange processes of determination surpass the usual principle of action by contact, and occur not only in experimentation, i.e., after material disturbances, but probably are important also for normal development." And further: "One thing is sure, that here a mutual relation between the parts is involved, and not only a one-sided one, as is the case with induction. The system as a whole is here taking a part in controlling all partial processes." I may add to the last statement that I have derived the same conclusions with regard to one regulative process, regeneration, from a general analysis of genetic control of development (Goldschmidt, 1927), by explaining regeneration in terms of redistribution of all determining substances, according to the given physicochemical conditions of the system as a whole.

We have gone into some of the details of regulation because they are of utmost importance in a discussion of the potentialities of development with regard to evolution. We have discussed many cases in which a hereditary change (a mutation) has produced exactly the same shift in the processes of individual development as has an experimental disturbance of development. In the experiments on regulation it is always an operative disruption of continuity between embryonic parts which sets regulation in motion. A comparable disruption may also be produced by a genetic change. We have reason to believe that many genetic changes result in a relative shift of the rate of interlocked developmental processes. Such a shift, if produced in early developmental stages, at the time of still labile determination, may act in the same way as an experimental disruption by operation; except that there is no disruption by crude separation of the parts, but a disruption by separating interlocking processes through the shifting of one integrating process. (Simile: the disruption of the function of a motor by breaking a shaft, as against dislodging it from the synchronizing mechanism.) What will be the consequence of such a mutation? In many cases the result is an upsetting of the developmental mechanism; i.e., lethality. In other cases a certain amount of regulation takes place and the result is some kind of monster. However, effective regulatory processes may be induced if the change occurs under proper circumstances (see the conditions for regulation, above). In this case the single genetically produced change of an embryonic feature results in a whole series of changed developmental processes—in other words, in a completely new type of development; i.e., a departure of a macroevolutionary order of magnitude. To take a fictitious example, a genetic change in vertebrate development which shifts the differentiation of the gill arches will lead to regulation of the developmental processes of the aortic arches, the gill pouches, and many other cephalic structures. The actually existing series of large anatomical differences between taxonomic groups does therefore not require an evolution by simultaneous selection of numerous small mutants of the determiners for every single organ, a necessary hypothesis on the basis of the neo-Darwinian view, or the current theory of the genes. A single mutational step affecting the right process at the right moment can accomplish everything, providing that it is able to set in motion the ever-present potentialities of embryonic regulation. It is needless to say that this statement also contains the explanation of atavism as well as of the positive and negative features of embryonic recapitulation. We shall have to return again to this important problem of embryonic regulation as a phenomenon of evolutionary significance.

We may conclude this chapter by stating that it has been shown before that the potential range of effects of single mutational steps coincides with the range of the individual norm of reactivity as determined largely by the range of shifting individual developmental reactions. Now we may add that this range is immensely enlarged if the norm of reactivity includes also the power of regulation.

f. The sexual norm of reactivity

In discussing the alternative norm of reactivity as well as the relation of hormones to determination, we did not mention the sexual alternative and its relation to hormones. This subject will now be discussed, as we consider it to be of great significance in the present connection. We have reported upon a number of cases which demonstrate the range of developmental potencies based upon the general type of developmental processes (we mean such types as permit a certain amount of shifting without interfering with the harmony of the resulting organism). The sexual alternative furnishes a case in which the developmental processes within a species may become so different that the resulting organisms, the two sexes, may exhibit differences of a macroevolutionary order of magnitude. As I pointed out in the essays (1920) to which I have repeatedly referred, a morphological difference of the magnitude found between the female and male genital armature in Lepidoptera (see fig. 25) would suffice for at least generic distinction if found as a somatic character distinguishing two different forms. This argument could be easily extended all over the animal kingdom. This sexual difference, however, is based upon a genetic difference—if we take only the most frequently found situation—i.e., the mechanism of the X-chromosomes, which creates within the same species two different genetic situations which determine differences in development. But these developmental differences are also based upon a definite norm of reactivity of the embryonic primordia, an alternative norm of reaction. Let us explain the situation by means of some examples. The anlage of the tissue on either side of the cloaca in mammals has an alternative norm of reaction. Under the influence of female determination it develops into labia majora; under male influence, into a scrotum. In Lepidoptera a group of cells in a certain abdominal segment develops into an ovipositor (labia) if the individual is genetically female, and into a clasping hook (uncus) if it is genetically male (see fig. 25). In both cases it can be demonstrated that the corresponding groups of cells in both sexes have an alternative potency of development. The decision over the alternative, which is usually brought about by the action of genetic determiners, according to female or male chromosomal constitution, may also be enforced within the same genetic constitution either as a consequence of a special genetic situation (zygotic intersexuality) or as a consequence of environmental influence, including action of hormones. The sexual difference, then, furnishes examples of developmental potencies of a large range within the same species, of a genetic control of these potencies via the existence of an alternative norm of reaction of the embryonic primordia, of the realization of these potencies within the same individual either by environmental influence of a simple nature or by genetic determination, and, finally, of the realization of both these potencies within the same individual in the special cases of intersexuality, both on a genetical and an environmental basis. This shows that the sexual alternative exemplifies within a single organism different aspects of the developmental norm of reactivity which are otherwise found in different forms, and that this happens sometimes with realization of an extreme range of morphogenetic possibilities. We may therefore expect to find in this field good models of large morphogenetic (as well as physiological, chemical, psychological) changes of the type occurring in macroevolution.

As we emphasized just now, the morphogenetic range of the sexual alternative is based upon the alternative norm of reaction of the primordia of the sexually different organs.

It might appear that this is a special genetical condition which cannot be used as a model for ordinary processes. However, we have already met with similar situations outside of the sphere of sex: the alternative norm of reaction of the leaves of *Limnophila*, grown in dry or moist air, the alternative norm of reaction in butterflies with seasonal dimorphism. In the latter case we could show that this norm of reaction was only a special situation within the general norm of reaction of lepidopteran development. In the case of the sexual alternative we meet with all imaginable differences of alternative reactivity. Within the same taxonomic group we may have extremely dimorphic forms as well as those in which the sexes are hardly distinguishable, which means that the tissues in one case react sharply to the genetic sexual difference (1X versus 2X), and in the other case, little or not at all. Where hormones are involved, which amounts in the sense of our discussion to a reaction to environmental conditions (internal environment), closely related forms may have an extreme alternative reaction to hormones (plumage of fowl), or a slight one, or none at all (pheasants, passerine birds). Such genetic differences cannot be considered to be different in principle from ordinary, i.e., not sexual, differences in the norm of reaction of embryonic primordia which are encountered in closely related forms. The literature on experimental embryology contains a tremendous amount of data showing that the same experiment does not give identical results if performed on different species (see, for example, lens induction). The sexual alternative, with its special features, is therefore to be considered as furnishing the same type of model of large morphogenetic changes caused by a single step as the other examples studied before and those to be considered below.

A group of facts which are of importance in our discussion is found in cases in which a definite genetical situation, definite features of alternative development, and certain taxonomic facts may be linked together. Such facts have been found in our work on intersexuality in *Lymantria dispar* (literature in Goldschmidt, 1934), facts which are extremely instructive if considered apart from the special problem of sex. We will select for our argument two topics, certain features of the genital armature and of the antennae.

In the male gypsy moth the dorsal posterior part of the armature consists of a characteristic hook, the uncus (fig. 25). The female homologues, developed from the same primordia, are the sheathlike labia flanking the body openings (fig. 25). The uncus primordia are paired like the primordia of the labia (Kosminsky). In the development of the uncus the paired anlagen unite dorsally and grow into the single uncus. Thus, strictly speaking, the basal part of the uncus is the homologue of the labia. The genetic sexual difference

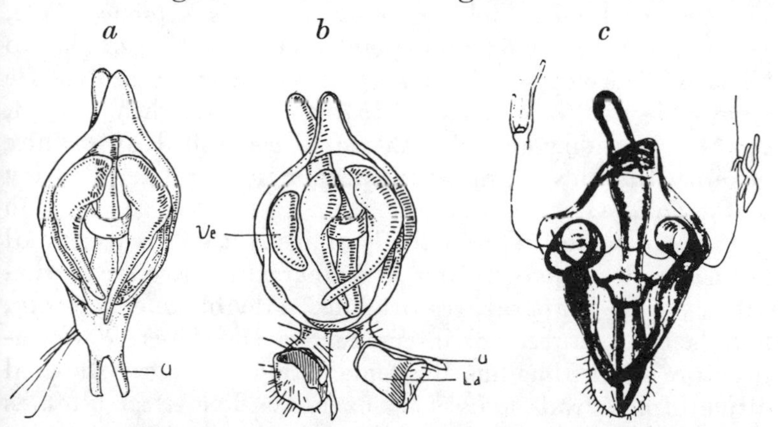

FIG. 54. Genital armature of male intersexes of *Lymantria dispar* in three different grades, *a, b, c.* La, labia; U, uncus. (From Goldschmidt.)

controls the decision whether the primordium remains paired or not, and whether it differentiates into one or the other of the two very different structures. Intersexuality is caused by a definite genetic situation (not to be discussed here) within a normal diploid female or male chromosomal constitution. Let us consider here only male intersexuality, where the male chromosomal constitution is present, but intersexuality is produced by special genetic features. The embryological consequence is that development proceeds first in the male direction and changes sooner or later in the direction of female differentiation, thus leading to different degrees of intersexuality. Therefore, parts which have not yet been finally determined when the turning point occurs continue their growth according to the female alternative. Figure 54 shows this situation for the uncus in three different degrees of intersexuality. Since the form and structure of the uncus are determined early, as can be proven by many facts, but the concrescence of the paired anlagen is determined later, the lowest degree of intersexuality leads to complete but paired unci (figure 54*a*). An earlier turning point (fig. 54*b*) finds the determination of the uncus completed, but the basal part (the zone of growth) is still undetermined and therefore grows in breadth, forming labiumlike structures (La) to each of which an uncus-point is attached (U). With a still earlier turning point, determination of the alternative has not yet been accomplished and complete female labia are formed (fig. 54*c*). The series, then, demonstrates the embryonic alternative as controlled by definite genetic situations acting at the proper time of determination of the anlage.

It has now been shown by Kosminsky that treatment of normal male pupae during a temperature-effective period with extreme temperatures produces a double uncus exactly like the one pictured for intersexuality (fig. 54*a*). The temperature shock thus has the same action as the genetic condition in low-grade male intersexuality. The effect belongs, then, to the category of phenocopies. I may add that in the case of low-grade intersexuality the position of the still normal uncus is frequently changed, as it is bent forward at its base; the same condition can also be obtained by temperature action. (It is not known how far the musculature is involved, but a normal male never shows this infolding of the uncus.) To complete this set of facts, there are near relatives of the lymantriids in which a paired uncus, exactly like the one described, is a typical taxonomic character; and others in which the single, bent uncus also occurs as a specific feature. In these instances, then, we have an embryonic anlage which has in one case *(Lymantria)* an alternative norm of reaction, which may be steered either by the normal genetic sex difference, or by an abnormal genetic constitution within one of the sexes, or by a crude change of environment. The same partial process, paired growth versus concrescence, is determined in other cases by the specific genetical constitution of a different taxonomic unit, independent of the sexual alternative; i.e., uncus vs. labia.

A very similar case can be made out for the antennae of *Lymantria dispar*. The female antennae have small side

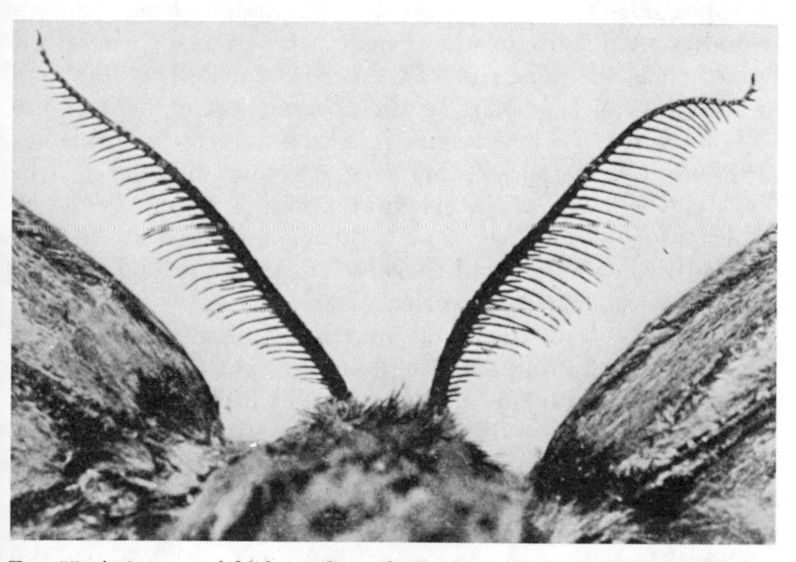

Fig. 55. Antennae of high-grade male intersex of *Lymantria dispar;* one row of branches has assumed the female condition. (From Goldschmidt.)

branches; the male antennae, very long ones. Differences in histological structure, especially of sense organs, are combined with these external differences. The sexual difference of the antennae is obviously connected with the mode of life. The females are sluggish, hardly ever fly, and are not in need of elaborate sensillae, whereas the males are energetic fliers and use their antennae in the search for the females. The embryonic determination of the general size of the antennae is fixed at the time of pupation, but the details of structure, especially the length of the side branches, can be shifted after pupation. Just as in the foregoing example, we can make a comparison of conditions in the normal sexual alternative with those found severally in intersexuality, in phenocopic experimentation and in natural taxonomic units. In female intersexuality the branches of the antennae elongate and all degrees of transition to the structure of male antennae are found. We shall not present the details of the developmental procedure in this case, as they are rather complicated. We shall confine the discussion to male intersexuality. A definite grade of male intersex is characterized by antennae with one row of branches of male type and one of female type (fig. 55). In the normal development of the male antennae a stage occurs in which one row of branches differentiates first, the other row starting only later. Obviously, then, in the intersexual antenna the male determination of the first row is already accomplished at the time of the turning point, and therefore only the second row develops in the female direction. The details, which can be understood only if the whole process of development is described, are not of importance in our present discussion. Just as is the case with the uncus, antennal differentiation may be affected by action of extreme temperatures during a sensitive period (Kosminsky, Goldschmidt). A female antenna may thus be changed into one of the intersexual type by starting the male type outgrowth of the side branches (actually not a growth but a stenciling process). Finally, the taxonomic parallel is also available: There are moths of different families, related to the lymantriids, in which the female antennae are branched more or less like the male ones. In one genus *(Orgyia)* closely related to *Lymantria*, the female antennae are like those of *Lymantria*, but the male ones are characterized by one long and one short row of branches, exactly as in the intersexual males of *Lymantria!* The importance of this set of facts to our discussion of evolution is the same as that of the foregoing example.

One more interesting point should be added. Normal differentiation of the antennae involves mainly three integrating processes (details and literature in Goldschmidt, 1934):

development of the shaft by a gradual shrinkage of the broad epithelial sac from which the antenna is modeled; formation of the side branches by a strange combination of cutting and stenciling of the branches from the sac, combined with growth at the growing point of each branch; chitinization of the whole, which ends differentiation. Shifts in the relative timing and in the detailed procedure of these processes (in addition to the size of the epithelial sac to start with) account for the normal as well as the abnormal sexual

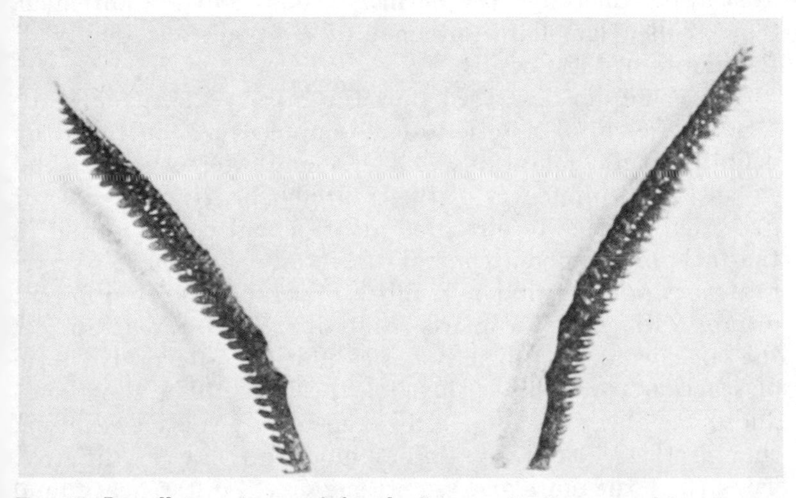

Fig. 56. Lamellate antenna of female *Lymantria dispar* resulting from a racial cross or, identically, from temperature experiment. (From Goldschmidt.)

differences mentioned. In certain racial crosses as well as in temperature experiments the type of antenna shown in figure 56 is frequently produced. Here the normal branches of a female antenna are replaced by platelike structures. In some individuals part of these plates are not developed and instead a continuous chitinous blade is found which stretches over a varying number of antennal segments. This blade or membrane is nothing but a chitinized part of the antennal sac which has failed to develop branches. This structure is connected through a complete series of intermediate stages with the type of platelike side branches which are present in the antenna in figure 56. The details clearly indicate that this strange type of antenna is produced by a chitinization of developmental stages of the antenna. In other words, an upset in the proper timing and integration of the three processes of differentiation described above accounts for the pathological type of antenna. But what is in this case a pathological feature, caused by a genetically or environmentally conditioned shift of developmental processes, is (at least approximately) the normal structure of the antennae, the so-called lamellate antennae of a far-distant family of Lepidoptera, the Cossids.

The examples described thus far involve the sexually alternative norm of reaction of development as controlled by definite genetic conditions (1X–2X and intersexuality) and as subject to influences of the environment. Another case of great importance to our present argument can be found in the facts of hormonal control of sexual characters in vertebrates. Again we find a number of primordia of different organs with an alternative reactivity. In different species the same organ may or may not exhibit this alternative norm of reaction (probably a threshold problem). The sex glands and ducts always belong to this type of alternative reaction, though their reactivity differs quantitatively in different cases (*vide* the more or less extreme sexual differences found in these organs in different groups). The copulatory organs may show an extreme alternative reactivity (mammals) or almost none at all (birds). The so-called secondary sex characters exhibit all degrees of alternative norm of reaction, from absence to presence, with innumerable modifications based upon different hereditary constitutions (example: the reaction of the plumage to sex hormones in fowl, pheasants, passerine birds). This statement includes all the metabolic and psychological features distinguishing the sexes.

In normal development the sexual alternative of differentiation of a group of cells is decided by the genetic constitution (1X–2X). This acts in vertebrates via the production of the sex hormones (aside from the disputed existence of special embryonic hormones). This feature permits us to isolate the action of the internal environment (the hormones present) in those experiments in which only a single genetic basis is involved; i.e., in experimental administration of hormones to one sex only. The decisive facts may be stated in a few words, if we neglect all details and restrict the discussion to the more important primary sex characters. By injection of sex hormones into avian or mammalian embryos all degrees of intersexuality may be produced. This means that the decision of the sexual alternative can be shifted without any genetic change in one or the other direction by the simple means of hormonic action, of course only in those organs in which the genotype provides an alternative of differentiation; e.g., cortex-medulla, Müllerian-Wolffian ducts. (See especially the recent work of Wolff, Dantschakoff, Willier, etc.)

Let us now look at the morphogenetic side of the situation and appraise the order of magnitude and range of morphogenetic differences induced within the same genetical and developmental system by the presence of a single chemical compound. Let us assume that a female is masculinized. Müller's duct will show all degrees of rudimentation, a phenomenon of the same order as any type of rudimentation occurring in evolution. Certain pronephric tubules, which would have become rudimentary in the female, grow into *vasa efferentia,* which is a functional change of the type found so frequently in evolution. The Wolffian duct, which would have become rudimentary in the female, becomes a functional organ and enters into specific associations with other parts. This is again a morphogenetic process paralleling happenings in evolution. In the cloacal region a series of concrescences, outgrowths, shifts of position, etc., take place in the production of the male genitalia. In mammals the descensus testis, with the concomitant rebuilding of the inguinal region, is an additional feature. These are morphogenetic features of an order of magnitude and of a range of concomitant correlational effects found in many evolutionary processes in the vertebrates; e.g., in the transformation of the urogenital apparatus from that of reptiles to that of mammals, which is hardly of a larger morphogenetic latitude than the hormonic effects just described. It is not necessary to continue this enumeration, as it is obvious that the norm of reactivity of development, in the case of hereditary disposition to alternative development, permits a morphogenetic shift of the order of magnitude of large evolutionary changes, and this under control of a single chemical compound, the respective sex hormone.

In order to prevent misinterpretation of the purport of this discussion, let me emphasize once more that the genetic difference between the basis of evolutionary change and of sexual change is found in the presence of the alternative norm of reaction in the latter case. But this difference is not one of principle but only of degree. In the chapter on regulation we saw that under definite experimental conditions many groups of embryonic cells may change their prospective fate. Though this is not the same thing as the sexual alternative, the latter involving a special genetic provision for different types of differentiation, the former the general potency of regulation, it is certainly an indication that the sexual alternative is a specialized case of the general potency of development to proceed in a different direction if the proper stimulus is provided. In both cases different types and degrees of reactivity are found, and in both cases the stimulus has to work on a system which is genetically capable of reacting more or less completely. Therefore I think that the facts relating to the sexual alternative may serve as a model to demonstrate the ability of developmental processes to change on a large scale as a result of a single event which may be compared, with regard to evolution, to a single mutational change affecting major features of development.

B. Mutation Affecting Early Development

One of the important points in the evolutionary discussions of my essays of 1920 was the following: I had come to understand the action of the genes in controlling development in terms of relative velocities of the integrating processes of differentiation. I had found that certain conditions of the genes, which I interpreted as different quantities of that material (for discussion, critique, and changed outlook, see Goldschmidt, 1938), were linked with an action occurring at a definite time in development. This suggested the idea that a single mutation of the type considered to involve the quantity of the genes might act upon an early embryonic process by changing its rate relative to the rates of the other integrating processes of differentiation. If at all viable, such a mutation could accomplish in a single step a huge evolutionary departure. I pointed out briefly that facts taken from the field of comparative anatomy of vertebrates, as, for example, the history of the visceral skeleton, could thus find an easy explanation. The same applies to orthogenesis and the law of recapitulation. I called this conclusion obvious and did not go into further details. Later I returned briefly to the same point (Goldschmidt, 1923, 1937), using a few other examples by way of illustration.

Originally my idea had been that evolution generally proceeds by the accumulation of micromutations, but that occasional mutations affecting early embryonic differentiation via change of rates may account for some major evolutionary changes which could not be accomplished slowly. This viewpoint was accepted by others and enlarged upon in the writings of Haldane (1932a), Huxley (1932), and especially de Beer (1930), who elaborated it in detail. But when my own work on geographical variation later led me to the conclusion that geographic races are not incipient species, and that the origin of the higher categories cannot be explained in terms of micromutations (Goldschmidt, 1932, 1933), I began to realize that the large departures, produced in a single step by what we call systemic mutations, offer the only feasible method of macroevolution on and above the specific level. I have since found out that the general idea of evolution in large steps based upon early embryonic changes has been proposed before; but I think that only the linking of such an idea with the facts of genetics and physiological genetics can raise it from the status of a hint to that of a theory. This is still more the case in view of the fact that older as well as newer discussions of such ideas have been for the most part linked with all kinds of mystical or formalistic discussions. F. Müller (1864), Kölliker (1864), Mehnert (1897), and the paleontologist Cope (1887) had long ago discussed the idea, more or less clearly, which also appeared hidden behind speculations, in the writings of Jaeckel and others. To these early authors must be added Sedgwick (1910), Naef (1917), Garstang (1922). A full elaboration of the idea was given by Severtzoff (1912) in Russian (quoted from Severtzoff, 1931). After I had discussed the problem (1917, 1920, 1923) it was taken up by a number of paleontologists acquainted with my work, Beurlen, Wedekind, and especially Schindewolf (see below). Among morphologists it was Severtzoff (1931) who pushed the idea further and furnished important materials, which will be discussed below. The value of his discussion was unfortunately impaired, because he was not acquainted with the genetic side of the problem and frequently hid important conclusions behind purely formalistic theories. Finally, the general idea has recently been "rediscovered" by the anatomist Boeker (1935), who buries it, however, under mystic speculations of the psycho-Lamarckian type, and by the botanist Ungerer (1936), who is inclined toward a similar philosophy. We shall describe later some of the material used by these authors and insert it into the general line of our argument, without stopping to discuss their speculations, which cannot be reconciled with the elementary facts of genetics.

We shall now elaborate our original thesis by analyzing relevant facts, old and new. Again we shall speak of mutation in a rather general way without discussing whether the ordinary type of mutation or only what we called systemic mutation is involved. The nongenetical material of these chapters does not convey information on this purely genetical problem, though we shall meet with a few facts which point in the direction of systemic mutations.

a. Growth and form

The repeatedly quoted essays of 1920 conclude with the following sentences: "The quantitative conception [quantitative genic control of velocities of differentiation] permits us to understand extraordinary changes in the final result, i.e., evolution over a rather considerable range on the basis of very simple and minute causes. One may read in . . . Thompson's brilliant book how most forms of shells of Mollusks can be understood as caused by very small changes in the algebraic value of the individual terms of the equation for the underlying curve; or how complicated differences in the form of body, skull, skeleton can be related to shifts in comparable systems of coordinates; how a mass of morphological differences may be reduced to the action of differential relative growth within simple mathematical laws. Add to this the proof that highly specific differential growth may be initiated by the production of specific hormones [determining stuffs] at a definite time and one will be able to visualize the host of evolutionary processes, which can be caused by small quantitative changes of the basic genes with its consequence, the host of shifts in the interplay of timed coordinations." On later occasions the same idea was expressed more specifically (Goldschmidt, 1923, 1927). These former statements contain the gist of the present chapter, and, for that matter, also the gist of much of the work which has been done by others since the above lines were written.

Faced with the necessity of linking the facts of relative growth with definite conceptions of physiological genetics and of deriving from such studies conclusions with regard to macroevolution, I found important material in D'Arcy W. Thompson's then newly published work on growth and form (Thompson, 1917). Thompson starts with the statement that in morphology our task is to compare related forms. This will lead to a recognition that one form may be a deformation or definite permutation of another. This deformation may be described exactly by relating the form to a system of coördinates. This can be done in a simplified form by a projection in two dimensions, which results in an outline which corresponds to a curve in a system of coördinates and may be described in general terms as a function of x and y. The rectangular system of coördinates can then be submitted to deformation, for instance by altering the direction of the axes, or the ratio of x/y, or by substituting more complicated expressions for x and y. In this new system of coördinates the inscribed figure will be a different one, or rather the old figure under strain, just as the new system of coördinates is the old one under strain. (The procedure is a similar one to that used in a rectangular Mercator projection of the earth, derived from the actual globular surface.) If it is possible to derive different forms from each other through such transformations, we may conclude that very simple "forces" (Thompson) suffice for an explanation of the morphological changes. As a matter of fact, the differences between related forms are those of growth and proportions (the Aristotelian scholar Thompson points to Aristotle's words, "excess and defect"). It is obvious that this method of graphic transformation is devoid of heuristic value if the several constituent parts of the body represent so many independent variants. This seems to be the opinion of many evolutionists. Thompson, who does not use genetical terms in his analysis and, therefore, does not mention the neo-Darwinian interpretation, makes it clear, however, that he is opposed to such conceptions as far as the change of form is concerned. He points out that the morphologist usually compares organisms point by point, character by character, and therefore falls into the habit of thinking and talking of evolution as though it had proceeded on the lines of his own descriptions, point by point, though a certain amount of correlation is admitted. Thompson himself prefers to assume that a comprehensive law of growth pervades the whole, and therefore that the phenomenon of correlation loses its complexity and becomes the expression of very simple conditions. The differences in form then "might have been brought about by a slight and simple change in the system of forces to which the living and growing organism was exposed." If we replace the energetic assumption of a "system of forces" in this statement by the genetical conception of single mutational change affecting early processes of differential growth, we arrive at our thesis as stated above, which we may then assume to be implicit in Thompson's less concrete statement in terms of energy.

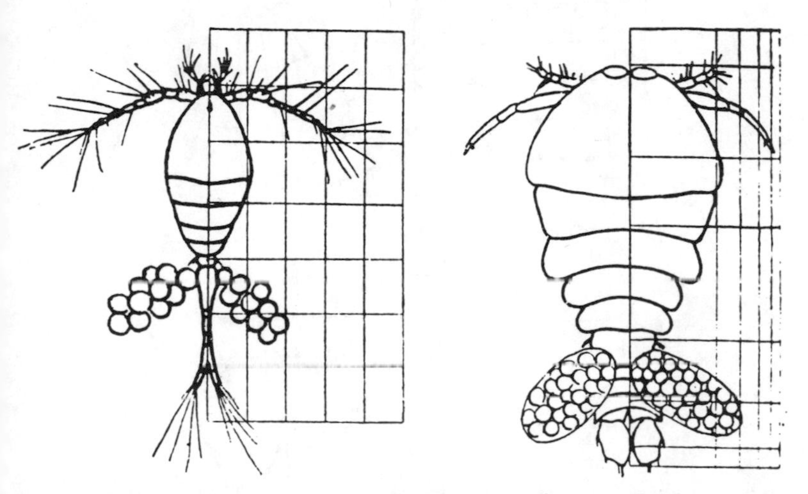

FIG. 57. Left, *Oithona nana;* right, *Sapphirina,* with the grids showing the Cartesian transformation of one form into the other. (From d'Arcy Thompson.)

Thompson proceeds to demonstrate his idea with several brilliantly conceived illustrations. He develops a number of different types of changes of a simple system of coördinates up to very complicated changes and inscribes the outlines of many different forms into such a system, showing that after deformation of the system of coördinates another related form appears through the corresponding deformation of the inscribed form. A few of the many beautiful examples will adequately illustrate the point. Figure 57 shows how the very different forms of the copepods *Oithone* and *Sapphirina* may be derived from each other by such a rectilinear transformation. A little more complicated is the example of figure 58 showing the outlines of the carapace of different crabs. Here the transformation of the coördinates (mathematically not so simple) even reproduce very unexpected details of form. The third example (fig. 59), the transformation of the outline of the porcupine fish *Diodon* into that

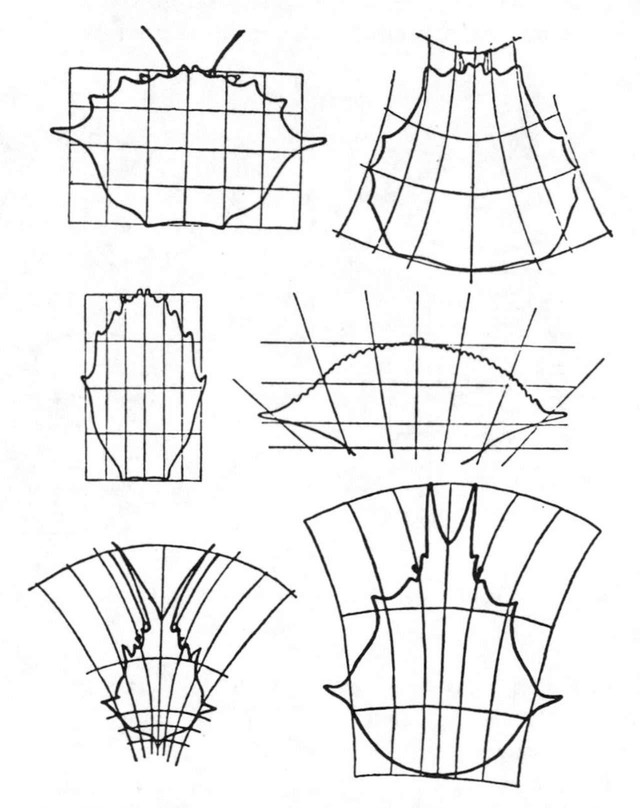

FIG. 58. Outline of carapaces of different crabs: 1, *Geryon;* 2, *Corystes;* 3, *Scyramathia;* 4, *Paralomis;* 5, *Lupa;* 6, *Chorinus;* and the Cartesian transformations needed to derive them from each other. (From d'Arcy Thompson.)

of the sunfish *Orthagoriscus*, is also very striking. Very remarkable are the diagrams which show the type of Cartesian transformations which account for well-known phylogenetic changes of form in the series of crocodiles, horses, etc.

Thompson does not fail to realize that this manner of approach is only a first approximation which is too simplistic, especially since it neglects the third dimension. But this does not detract from the immense importance of this analysis. I cannot find any statement in Thompson's book to the effect that he is ready to conclude that macroevolution must have

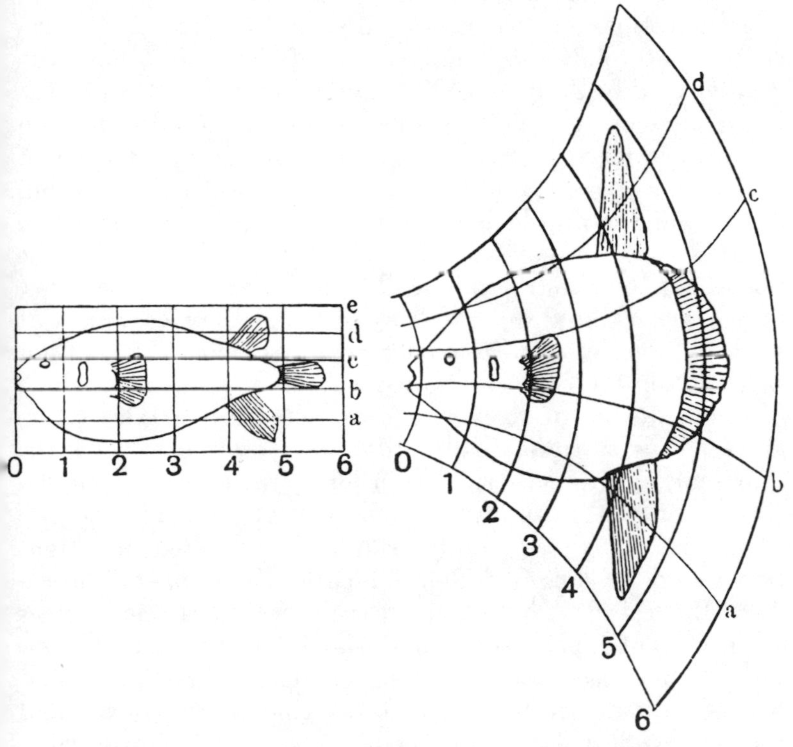

Fig. 59. Derivation by Cartesian transformation of the outlines of *Orthagoriscus* from those of *Diodon*. (From d'Arcy Thompson.)

proceeded not by accumulation of micromutations but by single steps, small from the standpoint of genetics, large, however, in their consequences because they affect primary processes of relative growth beginning in early development. But his unexpressed conclusions cannot have fallen far short of those voiced by us, as he states his general standpoint in the quotation of the rule of parsimony: *Frustra fit per plura quod fieri potest per pauciora,* a maxim which I appropriated as a motto for my essays of 1920.

I am afraid that both Thompson and I were too far ahead of our time to make an impression upon evolutionary thought, and especially upon the reasoning of the geneticists who were unable to turn their minds from the idea of micromutations and their accumulation by selection. But recently the evolutionary importance of these ideas has been rediscovered by de Beer (1930) and by J. S. Huxley (1932) in his work on relative growth. Huxley succeeded in finding the proper mathematical expression for the usual type of heterogonic growth, and this led him to consider the meaning of this law for problems of evolution. Though realizing the great general importance of Thompson's derivations, he points out their shortcomings. The graphic method can give only a general and qualitative picture of the mechanism at work, in place of a specific and quantitative one. It does not account for the change of proportions with growth, which is a consequence of relative growth following Huxley's formula. "It is static instead of dynamic, and substitutes the short-cut of a geometrical solution for the more complex realities actually underlying biological transformation." (Huxley). This is certainly true, and a detailed investigation would no longer follow Thompson's simplistic procedure. But this does not change the fact—and Huxley agrees to this—that Thompson's analysis shows that macroevolution can proceed by single small changes affecting the system of growth gradients as far as change of growth and form is involved. Huxley describes Thompson's transformation of the form of *Diodon* into that of *Orthagoriscus* (see fig. 59) in dynamic terms in the words: "From the figure it will be immediately obvious that the essence of the transformation, considered biologically, and not merely as an exercise in higher geometry, must have been the origin of a very active growth-center in the whole of the hind region of the body, whence the intensity of growth diminished regularly towards the front end. In other words, superposed on whatever growth-mechanisms may be necessary to generate a form similar to that of Diodon, there has arisen a steep and unitary postero-anterior growth-gradient extending throughout the entire body, with high-point almost or quite at the extreme hind end." To this is added later, in complete agreement with our antecedent conclusions, that the actual evidence, as well as a priori reasoning, indicates that a single mutation can act on a growth gradient as a whole, thus simultaneously altering the proportions of a large number of parts.

This dynamic interpretation of Thompson's idea is derived in Huxley's work from a fine, detailed analysis of numerous instances of heterogonic growth in terms of growth gradients controlled by a simple mathematical law. We shall not enter into the details of this analysis, as only the general features are essential to our present argument. However, some of the evolutionary consequences stated by Huxley are of great importance. There are first the cases in which species of different sizes show an increase of size of organs with heterogonic growth far beyond the increase in general body size (example: horns of rhinoceros beetles). The phenomenon is shown to be an automatic consequence of the working of the laws of heterogonic growth. The excessively growing parts are not independent genetically: given the difference in general body growth between small and large species, the rest follows automatically. This is, then, clearly an illustration of our claim of huge macroevolutionary changes based on small genetical differences, in this case a mutation in time or velocity of growth, which automatically involves all heterogonic growth. Huxley points out correctly that here a whole class of evolutionary phenomena of a nonadaptive type is apparent, which is not in need of natural selection in order to account for the details of correlative variation. "The burden on natural selection is also lightened" (the same idea as is contained in the Latin motto of our old essays) by the demonstration of growth gradients, as the factual evidence indicates that a mutation can act on a growth gradient as a whole, thus simultaneously altering the proportions of a large number of parts. This, then, greatly simplifies the picture of the genetic and selective processes at work. Thus Huxley derives from his detailed work exactly the same conclusions as I have derived from a general analysis, and he also accepts finally a general interpretation in terms of mutants controlling different rates of embryological processes, a conception from which my discussion of the subject had originally started.

In Huxley's discussion of the explanation of actual phylogenetic series on the basis of relatively simple changes of relative growth, he mentions, besides the examples used by Thompson, the famous evolutionary series of the Titanotheres. This example has been worked out in a very ingenious way by Hersh (1934) and may, therefore, serve to demonstrate the principle as applied to a special case, especially as the general conclusions which Hersh derives are in complete accord with the viewpoint which I have developed in my papers and books from 1917 to 1927. Hersh uses Huxley's formula of relative growth, $y = bx^k$, in which x is the size of the animal, y is the size of the differentially growing organ, and b and k are constants. The constant b denotes the value of y if $x = 1$; i.e., the fraction of x occupied by y when x equals unity. The constant k means that the ratio of the relative growth rate of the organ to the relative growth rate of the body remains constant, this ratio being the value of k. The relative growth rate is the actual absolute growth rate at any instant divided by the actual size at that instant (Huxley). This function is derived from, and is applicable to, ontogenetic data, as we have seen. But it may also be applied to measurements in genetically diverse but related groups, thus changing the emphasis from ontogeny to phylogeny. Thus Hersh measured the typical features of the skulls found in Osborn's phylogenetic series of Titanotheria, beginning with small, not horned forms and ending with giant horned types. He found that the formula fits the data, which means that the increase in size of the animals accounts for the differential increase of the horns (and probably other organs) in the phylogenetic series, so that all major features of the development of this group in the Tertiary are accounted for by an evolution of body size alone, given the basic conditions for growth according to the formula. To this basic feature many important details are added. Within a single genus the species differ primarily in size; i.e., the value of x alone is changed genetically. But between the genera differences in the values of b and k exist. This means, then, that smaller steps in the evolution of the animals required only a simple genetical change in the control of size, whereas large and very large evolutionary steps are accounted for by additional simple genetical changes controlling those ontogenetic processes which are expressed in the values of the constants b and k. In some cases a positive relation was found between the values of these two constants, one being related to the other by a definite function, which still further simplifies the genetic change needed for this type of evolution. An interesting implication, already mentioned by Huxley, is that the formation of a horn may have been implicitly present in the hornless ancestors of the Titanotheres, but could not find expression under the genetical conditions which determined the values of b and k before a certain size was reached. This relieves selection of the burden of explaining the inception of horn formation, which could hardly have had a selective, adaptive value. In other words, the whole phylogeny could occur without selection of innumerable small steps involving different mutants for all organs concerned, exactly as I had derived it in principle from the theory of gene-controlled rates. Hersh expresses this by saying that there is no reason to assume that the last *Brontotherium* of Oligocene times had more genes in its germ plasm than there were in ancestral *Eotitanops* of early Eocene times.

Hersh further tries to correlate these facts with facts of ontogenetic determination in the same sense as I have indicated in a general way for relative growth. He points to the fact that hereditary size differences in certain animals are caused by differences in the rate of cell division, which begin early in development (Castle and Gregory, 1929, for rabbits; Goldschmidt, 1933a, for *Lymantria*). A study of the relation of organ size to body size in rabbits (Robb, 1929) shows that the same formula of relative growth applies (identical b and k) and that, therefore, the constants of the group coincide with those of ontogeny. From this it is concluded that in the Titanotheres also the ontogeny of different species occurred with identical constants, b and k. This means that only a size difference is involved, without a difference in embryonic determination. As stated before, Hersh found generic differences based upon changes in the values of b and k. This, he points out, may also be understood in ontogenetic terms if hereditary differences in the time of determination of primordia are present; in other words, as I may add, if my explanation of genetic control of development in terms of balanced reaction velocities is accepted. This means reactions controlling the production of determining substances which produce final determination of a primordium when a definite threshold is reached. (Hersh refers to this conception as Brandt's "typological principle." Brandt himself has acknowledged in his first publications that this "principle" is an application of my "physiological theory of heredity" to the cases which he has investigated.) Going into further details, Hersh assumes that the character of relative growth between width and length of skull is established by an early embryonic determination. This event will occur in different genera at different relative times in ontogeny, which amounts to a change in the phylogenetic relative growth constants. In other words, the generic differences in b and k are indicative of different times in the ontogenetic determination of skull dimensions.

We have reported this analysis in more detail, because it demonstrates the whole argument by a definite case accessible to quantitative analysis. The cogency of the conclusions seems the more evident as they have been drawn, without the knowledge of my own general deductions, from the facts of relative growth and the general explanation of development in genetical terms. The important point in our present discussion is, again, that small genetic changes may account for major evolutionary divergence.

At the beginning of our discussion of development we pointed out that we would speak in a general way of genetic changes and mutations without discussing whether ordinary Mendelian mutations or chromosomal changes of the type which we called systemic mutations are meant. Most of the facts under discussion do not convey any information regarding this special point. But here and there hints, at least, may be found as to the underlying genetic situation. Let us recall our discussion of polyploidy and trisomics, where it was shown that one of the characteristic effects of chromosomal changes without gene mutation is an effect upon the growth habit, in both plants and animals. We pointed to the special case reported by Blakeslee and Sinnott (see p. 238) of change of relative growth after induced tetraploidy. We may consider these facts as, at present, barely hinting that macroevolutionary steps based upon a change in relative growth might be based genetically upon systemic mutation.

In favor of such a conclusion another set of facts and their as yet purely speculative interpretation may be mentioned. In our earliest discussion of the facts and principles analyzed in this chapter, we mentioned their importance for an understanding of the phenomenon of orthogenesis. Many authors have subsequently done the same. Orthogenesis, that is, evolution in a single straight direction, is a fact of which the paleontologists have assembled innumerable examples. Indeed, there can be no doubt that in frequent cases evolution is of the orthogenetic type. (Data are found in all technical books on evolution, from Eimer to the present day, especially in the books written by paleontologists.) The facts have frequently received a Lamarckian interpretation, which has been rejected by the geneticists. Another explanation with which volumes have been filled is the mystical one, the assumption of an existing urge toward improvement, or a similar transcendental principle. The explanation preferred by geneticists is the assumption of a selection of mutants which deviate in a definite direction. The orthogenetic series of changes in the evolution of the horse is considered to have received its direction by means of the selective advantage conferred to mutations in the direction of reduction of toes, etc., by a change in climate producing the new ecological niche of steppe. In the Titanotheres the analysis on the basis of the relative growth function leads to the conclusion that a selection of mutants for size accounts for the rest of the orthogenetic line of evolution. If we look at the problem from the standpoint of ontogeny, we realize that an orthogenetic line of evolution requires changes in ontogenetic determinations which occur in some rectilinear order. If the whole organism is affected by such a change, for example, change of body size, only a change in the plus and minus direction is possible, and it is probable that selection will be the controlling agency. If, however, only definite organs are involved, the situation is more complicated. The development of any primordium is closely interwoven with that of all other primordia and, therefore, a local change caused by a mutation affecting an early embryonic process with regard to time of onset, speed of occurrence, or time of determination, cannot lead to a viable result if the embryo is not able to carry out the proper regulations. The selection of the direction in which genetic changes may push the organism is therefore not left to the action of the environment upon the organism, but is controlled by the surroundings of the primordium in ontogeny, by the possibility of changing one ontogenetic process without destroying the whole fabric of development. (One may call this a modernized restatement of Roux's "Kampf der Teile im Organismus.") Thus what is called in a general way the mechanics of development will decide the direction of possible evolutionary changes. In many cases there will be only one direction. This is orthogenesis without Lamarckism, without mysticism, without selection of adult conditions. This conception of the ontogenetic side of the problem, however, does not explain why the underlying genetic changes do occur in a series of steps. We know of so many dwarf and giant mutations in animals and plants, both of the whole and of parts, that we would expect a rather irregular behavior of mutations with regard to growth, forward and backward, small steps and large. Therefore, we must assume either that the embryonic selection just described permits only small changes in addition to prescribing their direction, or that the genetic mechanism does not permit of another type of procedure. The latter would be the case theoretically if characters showing orthogenetic evolution were always based upon multiple and additive factors. The difficulty would be that in such a case selection could not be dispensed with, as such mutations actually occur in all directions, and ontogenetic selection could hardly work only in one direction in the case of small steps. Large mutational steps, however, are not a part of a multiple-factor system. But there is another possible mechanism for directed mutational changes without selection available. We may recall the work of Sturtevant and Dobzhansky on inversions in the chromosomes of *Drosophila.* Here we found that the intrachromosomal pattern was rebuilt, and we discussed the facts in relation to what we called systemic mutations and their emerging action on the phenotype after a definite threshold has been reached (see p. 242). In this case it was demonstrated that the mechanical features of inversion necessitated a certain seriation of events which must have taken place because the complicated new patterns could be formed only by an orderly series of inversions. We may call this an orthogenetic series of pattern changes, controlled in their direction by the mechanics of the chromosome. We can imagine that here a model for directed genetical change has been found, combined with the possibility of large steps, the systemic mutations. We shall not indulge in further premature speculations, but I think that we are justified in having at least intimated the interesting possibilities of future advances in this direction.

b. Homoeosis and segmentation

We turn now to a group of facts which link genetics, development, and evolution in a way which furnishes important insight into the evolutionary significance of single changes affecting early embryonic differentiation. In two of the most important phyla of animals, in arthropods and vertebrates, one of the major features of evolution is the progressive specialization in segmentation of the body. In the primitive forms all segments are practically alike, each metamere containing a nephridium, a gonad, a muscle segment, a ganglion in arthropods, a neuromere in vertebrates, and an identical pair of appendages. In the course of evolu-

Fig. 60. Normal antenna of *Drosophila* (right) and mutant *aristapedia* (left). (From Balkaschina.)

tion homomery is changed into a heteromery. In arthropods the appendages differentiate into mouth parts, legs, gonapophyses, disappear in some segments, change their function in others. In vertebrates the comparative anatomy of muscles, segmental nerves, and vertebral column demonstrates the changes from the considerable homomery of Amphioxus to all types of heteromery. Among these evolutionary steps there are many of a type which preclude an evolution by slow accumulation of micromutations. The mouth parts of

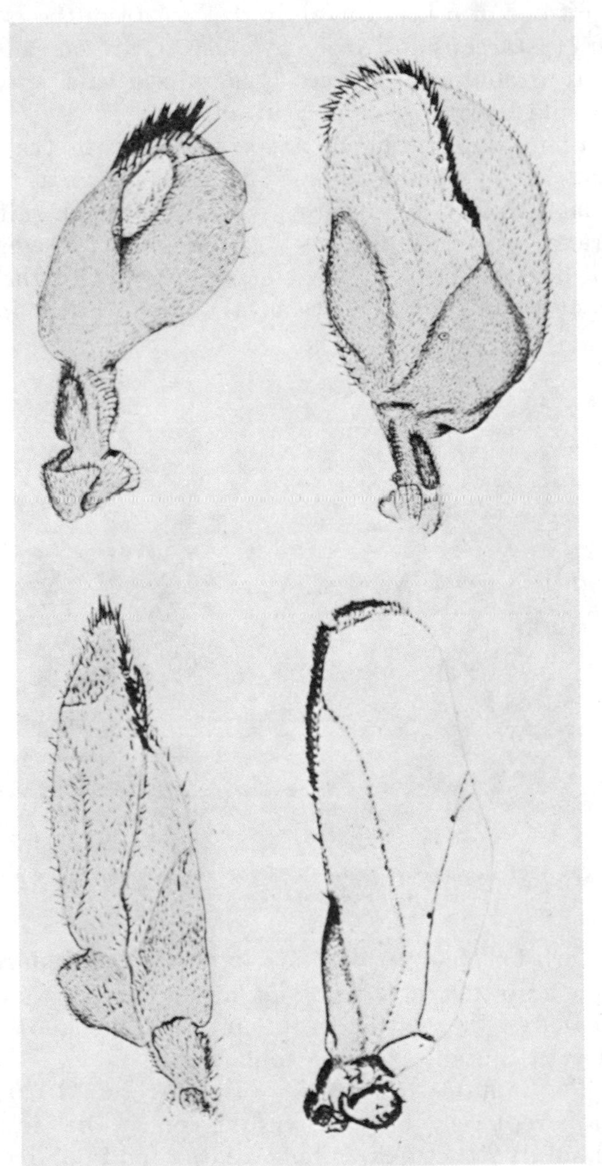

Fig. 61. Different transitional types between haltere and wing in *Drosophila* mutant *tetraptera.* (From Astauroff.)

a mosquito or of a bee, certainly derived from the primitive type of crustaceans and primitive insects, are an example in question: gradations between generalized and specialized types would have died of starvation.

For a long time the phenomenon of homoeosis (called heteromorphosis by some authors) has been known as an occasional monstrosity in arthropods. The term signifies the appearance of a homologous appendage in a segment to which it does not belong. The classical example is the regeneration of an antenna after removal of the eyestalk in Deca-

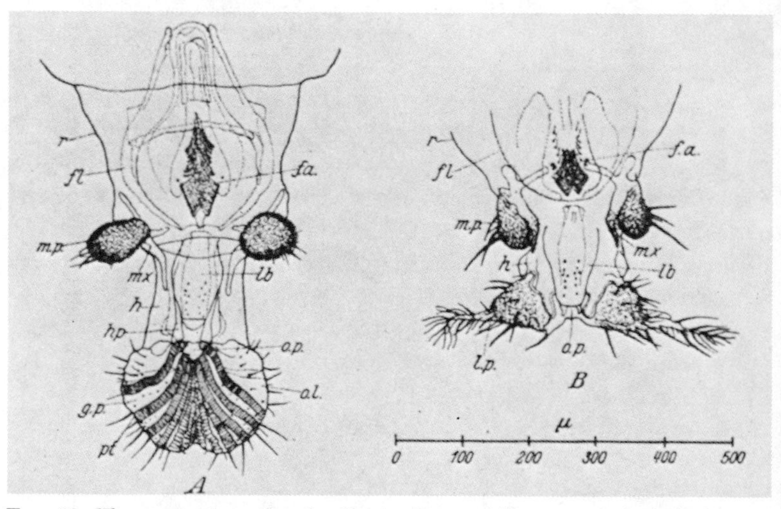

Fig. 62. The mutant *proboscipedia* in *Drosophila* compared with normal proboscis (left). (From Bridges-Dobzhansky.)

pods (Herbst). Homoeosis is now known to be produced also by simple mutation in *Drosophila*, an occurrence which permits an analysis in relation to our problem. The known types of homoeotic mutants are the following:

(1) The mutant *aristapedia* (Balkaschina, 1929). The antenna is replaced, to a varying degree, by the tarsus of a leg with all its structures. Intermediate conditions are found (fig. 60).

(2) The mutant *tetraptera* (Astauroff, 1929). The hal-

teres, i.e., palpuslike sensory appendages, replacing the hind-wings in Diptera, are transformed into wings, again to a varying degree (fig. 61).

(3) The mutant *proboscipedia* (Bridges and Dobzhansky, 1933). Here the oral lobes which form the proboscis of the fly are bent together to form a labrumlike structure with appendages exhibiting the structure of an antenna or of a tarsus (or something in between). Simultaneously the other mouth parts are so modified that they resemble somewhat the chewing mouth parts of lower insects (fig. 62).

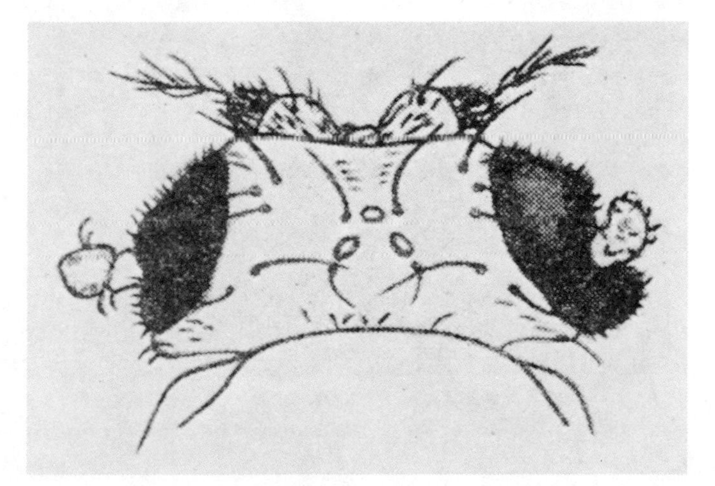

Fig. 63. Palpus in the eye of *Drosophila*. (From Valadares.)

(4) Instead of an eye, or combined with a rudimentary eye, an antennalike structure appears which shows all transitions from a knob with hairs to a segmented, palpuslike structure of 3–4 segments. I have found this mutant twice (unpublished). It characterizes a varying number of individuals in a high *kidney* allele, which I found many years ago and which, it seems, has been found again by Valadares (1938). It also characterizes a high allele of *Lobe*, which I found more recently and which will be studied in detail (fig. 63).

(5) A new homoeotic mutant was found by me years ago (unpublished, now being analyzed) in which the fore-wings are replaced by halteres, with all stages in between, thus producing a fly with, in extreme cases, four halteres (fig. 64).

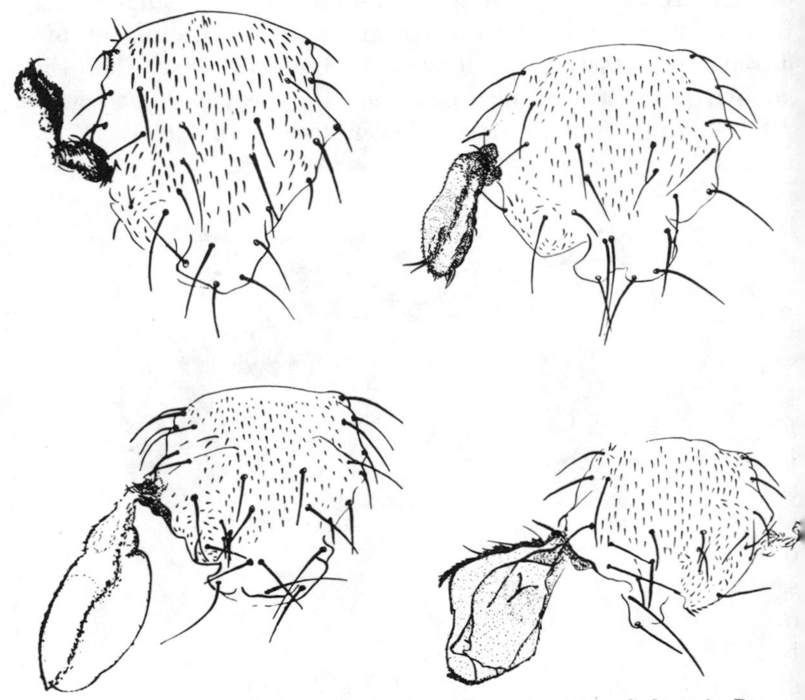

FIG. 64. Different conditions of wing transformation into a haltere in *Drosophila* mutant *tetraltera*. Only thorax and wings represented. Normal right wing not drawn where present. (Original.)

Before analyzing the meaning of these occurrences we must point out that there is a general feature which is common to all the cases, apart from their special features, and which applies to dorsal as well as to ventral appendages (antennae, mouth parts, wings). This general point is that the homoeotic organ has primarily the appearance of a palpus; i.e., a primitive appendage of a few segments, which may specialize, as the case may be, into an antenna, a tarsus, or a haltere. An explanation of the action of the mutant loci in question in determining the homoeotic change was derived (Goldschmidt, 1938) from the facts of development of *aristapedia* as described by Balkaschina. The latter found that normally the imaginal discs of the legs develop and segment prior to the differentiation of the antennae. In *aristapedia,* however, the antennae begin differentiation simultaneously with the differentiation of the legs and develop from the beginning like legs. This fact suggested the following explanation. In normal development the genetically controlled processes of determination take place in a definite order which is controlled, according to our general theory, by the production of determining substances with a definite velocity which leads to the proper threshold condition at a definite time. One such determining process results in the determination of a segmentation process in growing imaginal discs. If this inductive process takes place, all imaginal discs which are in the proper stages of growth will be stimulated to segment. The normal relative rate of growth and differentiation of the different discs is such that the proper discs are ready to receive induction at the proper moment; i.e., there is a definite timing of the disc growth with relation to the moment of induction. The seriation of the inductive stimuli as attuned to the seriation of the growth of the discs is then the simple system which takes care of the differences in local development. This system works on the basis of identical potencies of the discs before their determination by the inductive stimulus (stimulus of course in the chemical sense). A mutant, then, which shifts the time of determination of the antennal disc so that it becomes coincident with that of the leg discs, automatically produces a leg instead of an antenna. The same argument applies, *mutatis mutandis,* to the other cases of homoeosis. We have attempted to prove this interpretation experimentally. W. Braun (1939) tried in our laboratory to affect the determination by transplantation of discs between normal and mutant types, using the Ephrussi–Beadle method. But the time of determination turned out to be too early for a successful operation. However, an indirect proof could be found. If the mutant *aristapedia* was combined with other mutants affecting legs and antennae, respectively, the leglike antennae (*aristapedia*) showed the effect of leg mutants only.

The above explanation of the situation is derived from a logical application of our general ideas on genetic control of development (the physiological theory of heredity, 1920, 1927) to the special case and, therefore, fits in with many other points discussed in former chapters. It simultaneously opens an important vista into the evolutionary side of the problem. The authors who studied homoeotic mutants before (Balkaschina, Bridges, and Dobzhansky) did not fail to emphasize that in these cases a single mutational change produces in a single organ a deviation of a macroevolutionary degree: two wings and two halteres instead of four wings characterize the order of Diptera. But the real importance of these facts for a general analysis of evolution appears only in the light of our interpretation. If an embryological system of the type described underlies the process of segmental differentiation of appendages, and if this system is controlled by the genotype in the way described in the theory of balanced reaction velocities, a system obtains in which very small genetic changes in that part of the genotype which controls the speed of differentiation, the gradients of segmentation (see Seidel, 1936, on the experimental embryology of arthropods), or the time of the different inductions, may lead to sudden macroevolutionary steps in all details of segmental divergence. Let us look at the series of intermediate steps between a haltere and a wing, all produced by a single mutant. A haltere is morphogenetically nothing but a variant of the most generalized type of arthropod appendage, a palpus consisting of a few segments. The similarity of the series of intermediate conditions found in the homoeotic change from haltere to wing and vice versa makes it possible for us to conceive how one major and a few minor genetical changes in phylogeny may have produced dorsal appendages and their transformation into wings. Again, we do not want to indulge in detailed phylogenetic speculations. Our only point is to demonstrate that a proper evaluation of the facts of genetics in terms of development points in the same direction as have so many facts previously discussed; i.e., that the facts concerning the range of potential changes of development caused by a single or a few genetic steps, which are small from the genetical point of view but large in the morphogenetic result, demonstrate that it is possible, and even probable, that macroevolution takes place without accumulation of micromutations under the pressure of selection.

At many points in our discussion we were able to base our argumentation in favor of macroevolution by single large steps upon parallels between these groups of facts: A definite morphogenetic departure produced by a single mutation could be duplicated phenotypically by experimental change of development (phenocopy), or could be found to exist in nature as characteristic of a higher taxonomic category. The same triple parallel may also be drawn for the phenomenon of homoeosis. In the *Drosophila* experiments in which temperature shocks and X-rays were used to produce phenocopies no cases of typical homoeosis were found. But in a recent note by Enzmann and Haskins (1939) it is reported that homoeotic changes have actually been produced as phenocopies by neutron bombardment (the authors do not mention either phenocopy or homoeosis). Among the effects described is an exact parallel to my homoeotic *Lobe* and *kidney* mutants. Not only are all the details of eye abnormalities recorded which we found in these mutants, but also the homoeotic antennae inside the eye and their different modifications. Further, intermediate conditions between halteres and wings were also observed. Simultaneously there appeared a note by Rapoport (1939) demonstrating the production of the type *aristapedia* as a phenocopy after treating *Drosophila* larvae with certain chemicals! Thus once more the correctness of my claim that each mutant type can be copied as an experimental phenocopy if only the proper stimulus is found is

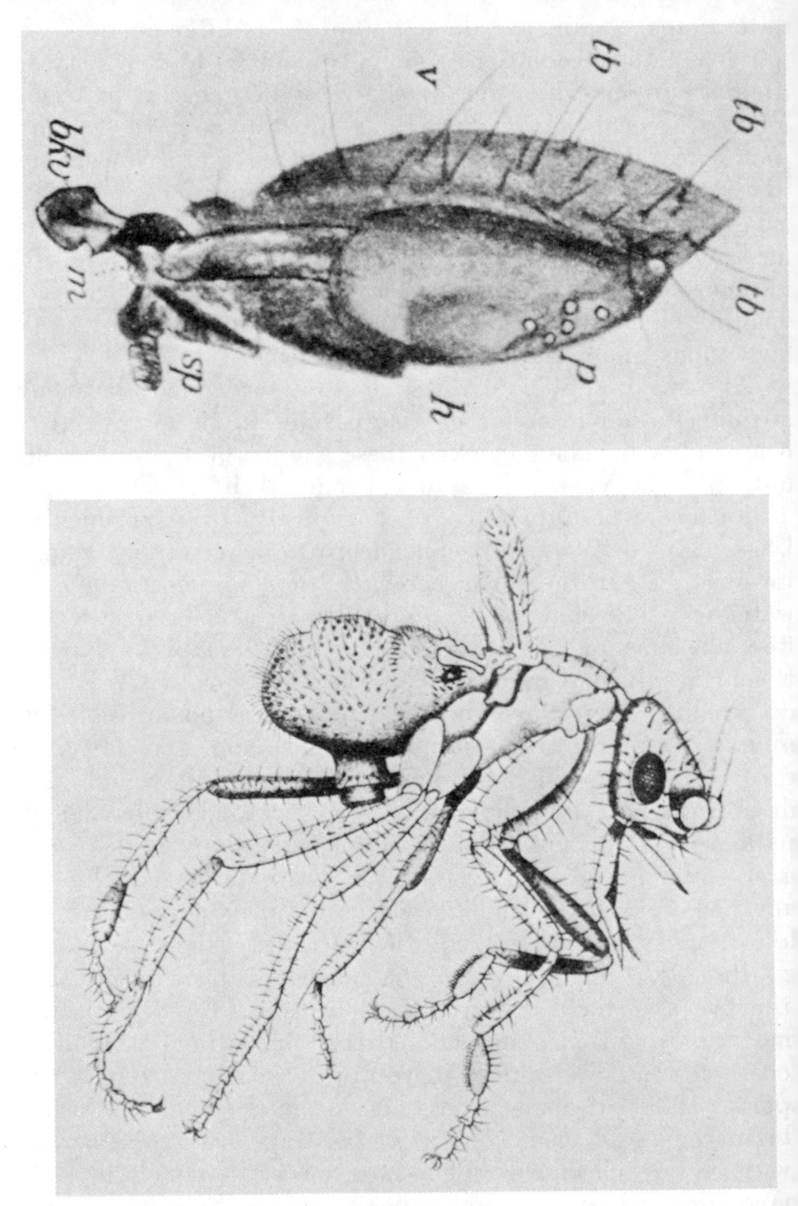

FIG. 65. Wing of *Termitoxenia* (after Wasmann). Below, *Termitoxenia* (after Assmuth).

demonstrated. Now to the third parallel, the taxonomic one, which also can be drawn in the case of homoeosis. The very aberrant termitophile fly *Termitoxenia* has minute rudimentary wings of a very peculiar type (fig. 65). A comparison with the homoeotic wings intermediate between halteres and wings, both in the mutants *tetraptera* and *tetraltera* (see figs. 61, 64), shows identity of structure! Homoeosis thus becomes a phenomenon of greatest importance in our discussion. This importance is still further enhanced when we learn (see Bezzi, 1916) that in a number of Hymenoptera, Neuroptera, Strepsiptera, and Coccids the hind-wings are reduced to an appendage resembling a haltere; and that in parasitic Hymenoptera, Orthoptera, and Homoptera the fore-wings may be halterelike. Unfortunately, a proper analysis of these facts has not been made as yet.

Again we may point briefly to the question discussed before. Do the facts under discussion contain any suggestion as to whether the macroevolutionary changes contemplated here may be based upon ordinary Mendelian mutations or upon what we called systemic mutations? Most of the homoeotic mutants which we discussed are simple recessive mutants (only the *Lobe* allele is dominant). But they all have one point in common: the great variability of their expression. In my *Lobe* and *kidney* cases hardly two individuals are alike, and only a small number exhibit the palpus in the eye. In my four-haltered mutant *tetraltera* most individuals are normal, and the abnormal ones show all transitional stages and relatively low right-left correlation, with, in addition, a considerable dependence of the expression of the character upon environment. To a more or less considerable degree the same applies to the other cases. This suggests that the genetic effect of the mutant locus affects the developmental processes in question only within a very small range of time, so small that the normal fluctuation of developmental speed caused by external and internal environment may easily shift the decisive event below the threshold. Expressed in more general terms, the genetic effect is not balanced or completely integrated with the whole of development, but is too strictly localized, whereas a comparable evolutionary process has to be completely integrated to lead to a viable and constant whole. The neo-Darwinian geneticist would explain this difference by assuming, in the case of evolution, a primary mutation and its subsequent stabilization and integration by the selection of numerous modifiers. I have serious doubts as to this explanation, as it always meets with the old difficulty that the original mutant would be wiped out before it could become established by selection of modifiers. This difficulty would, however, disappear in the case of our systemic mutation where the trial and error take place within the chromosome without external effect until the new pattern emerges which, as a completely new system, controls the emergence of a complete and, therefore, integrated change of morphogenesis. Unfortunately, no experimental attack upon this problem is at present apparent, but unbiased synthesis of existing facts seems to favor our solution. Some of the older evolutionists who are acquainted with classical literature may find that such a view, based upon recent genetical developments, savors of Weismann's theory of germinal selection, which was founded upon now obsolete ultra-atomistic conceptions of the germ plasm and its action in controlling development. But like so many other ideas of Weismann, in whom the great morphologist, ecologist, and experimentalist were combined with a great analytical thinker, this one has also been resurrected in modern guise more than once. Fisher's theory of the selection of dominance modifiers is one such case. We discussed above Ford's interpretation of the *Papilio dardanus* case in the light of Fisher's ideas, and compared it with our interpretation in terms of systemic mutation (p. 241). The discussion shows that both viewpoints may also be called germinal selection, one in terms of genes and micromutations, the other in terms of intrachromosomal pattern.

In introducing this chapter we mentioned segmentation in vertebrates. Actually the facts of segmentation furnish a considerable amount of information regarding our problem. Let us select from the innumerable details which fill the volumes of the literature on comparative anatomy a single topic which permits of clear conclusions, because the connections among morphology, taxonomy, embryology, genetics, and physiological genetics can be established (just as we tried to do in our previous examples). The most typical segmented organ in vertebrates is the vertebral column. Looking over the taxonomy and therewith the phylogeny of vertebrates, we realize that one of the major features in macroevolution has been a shift in the vertebral column—with regard to the number of vertebrae within the region, the relative position of the regions, expansion or rudimentation of regions, concrescence, and formation of various transverse processes and their regional differentiation. The problem is again as before: Are all these evolutionary steps the result of accumulation of micromutations?

Again we may base our conclusions upon genetic information brought in line with the facts of embryology. There has been a considerable amount of recent work upon the genetics and development of abnormalities of segmentation in the vertebral column. The most detailed study has been made by Kühne (1932) in man. He analyzed and studied the inheritance of abnormalities in number and position of sacral vertebrae, presence of supernumerary ribs, and similar variations frequently found in man.

The decisive results derived from an immense amount of material are the following. The abnormalities are not inherited individually. What is inherited is a tendency toward serial changes in an antero-posterior or postero-anterior direction. This change is based upon a single mutation; the dominant condition produces a cranial shift, the recessive a caudal shift, of the limits of the regions, and therewith of the behavior of the segments with regard to formation of ribs, sacral fusion, etc. Further, the same mutant controls simultaneously the type of variation of the brachial and caudal nerve plexus and of the muscles of the back and the position of the diaphragm. E. Fischer (1933) realized that these facts must be explained on the basis of a developmental system of balanced reactions, as elaborated in my theory of genic action. A simple shift in the velocity of one of the integrating processes relative to the others will account for the primary change with all the later unavoidable consequences during subsequent development (see also Fischer, 1939). Fischer also pointed to the phylogenetic significance, which is rather obvious after our discussion. If a single mutation can shift the regional relations of the vertebral column, including innumerable details of individual vertebrae, and also the pattern of other organ systems, there is no reason to expect a different procedure in phylogeny. In this special case there is not much known regarding the details of embryogeny, though it can be inferred that the decisive feature must have occurred rather early in embryonic differentiation. But in a number of closely related phenomena involving the segmentation of the tail of mammals, we have the details necessary to complete the picture (see the review of the literature by Steiniger, 1938, and the discussion of the phenogenetic aspects in Goldschmidt, 1938). Numerous mutants are known which control different types of abnormal segmentation of the tail; the developmental causation of the abnormality is known to be an early embryonic disturbance of segmentation leading in many cases to a resorption of already formed segments. And, finally, it is also known that some of the types may be produced without genetic change by experimental treatment of young embryos (phenocopy), indicating that the decisive feature is a change in rate of one of the integrating processes of differentiation. As the general line of the argument is always the same, we do not need to discuss it further.

The reliability of our conclusions is considerably enhanced by the fact that an entirely anatomical analysis of the same problem, without knowledge of the genetic side and without knowledge of our genetic theory of development, has led to exactly the same results. Severtzoff (1931) has assembled the results of his anatomical and embryological work (partly published before in Russian) which include the phylogeny of the vertebral column. As the result of a very minute analysis, he states that the macroevolutionary changes in this case do not start with small quantitative variations but with large departures, sudden changes of one organ into another. He also realizes that the sudden change must have been one affecting early development.

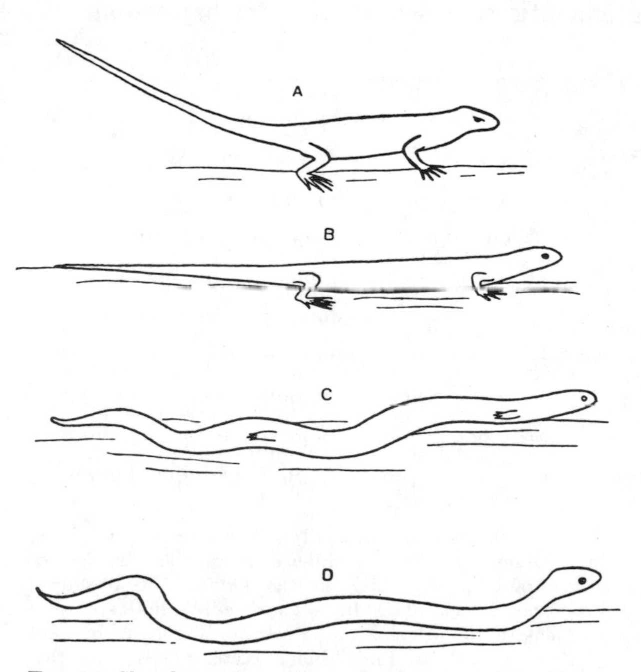

Fig. 66. Sketches of reptiles showing transition from walking to crawling locomotion. A, *Agama*. B, *Ablepharus*. C, *Seps*. D, *Vipera*. (From Severtzoff.)

His book contains many other examples from the comparative anatomy of vertebrates which lead to the same conclusions. Geneticists, who frequently are not sufficiently acquainted with material inaccessible to genetic experimentation, would profit much from a study of the facts analyzed by Severtzoff, even though they would have to overlook his general notions, which are conceived in the spirit and couched in the terminology of phylogenetic speculation, and which have remained uninfluenced by the facts of genetics as well as of physiological genetics.

Let us examine one of Severtzoff's examples. Figure 66 represents the well-known case of the assumption of snake-like form by saurians through increase in vertebral number and rudimentation of the extremities. Figure 67 gives a diagrammatic representation of the happenings in a series

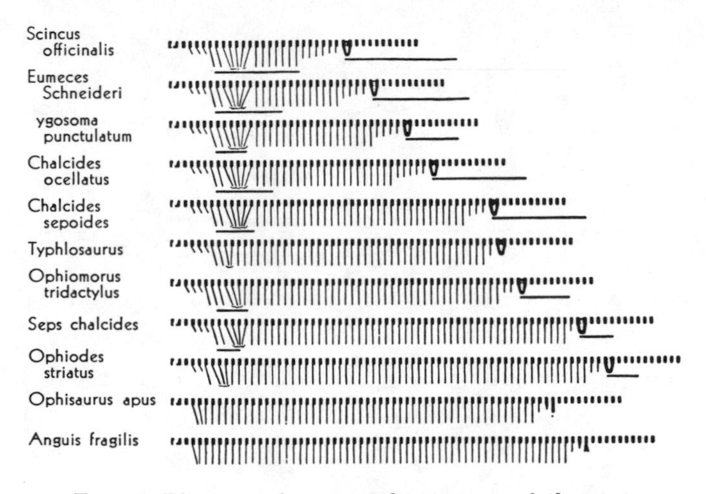

FIG. 67. Diagram of segmental structure of the vertebral column and its appendages in Sauria. The species are arranged according to the number of presacral vertebrae. Thick black lines (horizontal) indicate relative length of anterior and posterior limbs. A bracket marks sternal ribs. Small black rectangles = vertebrae. Long perpendicular lines = thoracic ribs; the same, short, = lumbar ribs; the same, short and diagonal = cervical ribs. V = sacral vertebra. (From Severtzoff.)

of such forms, indicating the individual vertebrae, their appendages, and the length of the limbs. Severtzoff assumes that the change has occurred in evolution through a series of steps, each one involving one vertebra. His idea, typical of the comparative anatomist, is that a caudal vertebra is transformed into a sacral one, and that in this way the vertebral column is elongated and the position of the hind extremities shifted backwards. On the basis of our knowledge of experimental embryology and the genetics of segmental changes, we would interpret the facts in a different way, by saying that the mutation which started the evolutionary process changed primarily the process of segmentation itself by altering its embryonic gradient and rhythm so that a larger number of segments was produced to begin with. The localization of the limb buds and, therewith, the setting of the limits of thoracic and lumbar segments is a determinative process independent of the primary segmentation. It is this process which singles out a definite segment; i.e., the one in the proper position, for regional differentiation, whatever its number in the series may be. A mutational increase in segments in the rump will find the extremities still in the same relative position, and therefore in another segment. One vertebra has not changed into another one, but primarily equipotential segments are determined as to their further fate by their topographic relations to neighboring organs. This is a conception which is in harmony with all the pertinent facts of experimental embryology. From this it follows that all the different degrees of segmentation represented in figure 67 need not have originated in an orthogenetic series, vertebra by vertebra, as Severtzoff assumes, but each may have been produced in a single step, a mutational change of the rate (rhythm) of segmentation. Severtzoff in one place actually comes to the same conclusion. In an early chapter of his book he speaks as a comparative anatomist and thinks in terms of the change of one vertebra into another and theorizes in the obsolete terms of the "principle" of Kleinenberg and the "principle" of Dohrn, to which he adds a "principle" of his own. But in a later part of the same book the embryologist Severtzoff realizes that only a change in the rate of segmentation can account for the facts. He pictures and discusses embryos of Sauria and Ophidia which demonstrate that the developmental differences in segmentation actually consist of different rates of segmentation if embryos in the same stage of differentiation are compared. Other examples of the same type are discussed, and there is no doubt that Severtzoff, on the basis of his great mastery of vertebrate anatomy and embryology, had in mind ideas similar to those developed by me during the same period, though his lack of knowledge of genetics and physiological genetics made him express dynamic ideas in the static language of comparative anatomy and in the language of "laws" and "principles" of the evolutionary schools of pregenetic Darwinism.

Though we shall discuss the problem of rudimentation in the next chapter, we may add here a word about the correlation between increase in segmentation and rudimentation of the extremities (as well as other concomitant changes). According to the current neo-Darwinian conception, both are adaptational traits which have been developed by slow selection of small mutations. It would be rather difficult to develop this idea in detail without getting into difficulties. But the neo-Darwinian explanation is not needed, after all. Since problems of rhythm and rate, as well as time of determination, of course, are involved, a change of rate in one series of exactly timed processes, which have to interlock properly with others, can result in a consecutive shift of all succeeding processes. (See the facts just quoted regarding vertebrae, muscles, and nerve plexuses.) If we describe the present example in a rather crude way—because of lack of exact experimental information—we may say: a change in the rate of segmentation may have the consequence that extremities start their differentiation too late, or reach their point of determination too early; i.e., have a changed growth rate relative to the whole body. In any case the change in the extremities, their rudimentation, can be the direct consequence of the change of rate of segmentation. It is a necessary consequence of the mechanics of development, once development has been disturbed at an early stage. A single genetic change affecting the rate of early embryonic features of segmentation may, therefore, have produced in a single mutational step at least the fundamental elements of the whole group of adaptations to crawling movement.

c. Rudimentation

There can be no discussion of problems of evolution without an analysis of the process of rudimentation. There is hardly a group of animals existing in which there is not some kind of macroevolution involving rudimentation of certain organs: the rudimentation of limbs and girdles in all classes of vertebrates, of wings and segmentation in insects and arthropods, and of almost any organ in the case of special adaptations to parasitism, life in caves, etc. In connection with our present discussion the analysis of rudimentation is of great importance. Here macroevolutionary processes may be linked in a rather simple way with the potentialities of development and can be shown to be in harmony with the major thesis of our present discussion; i.e., macroevolution by single large genetic steps involving a change of early embryonic processes on the basis of a developmental system controlled by balanced velocities of integrating reactions. As the application of these conceptions to the facts of evolution by rudimentation, facts which also include an understanding of the law of recapitulation, is rather obvious, I had suggested this application in my earliest discussions of the topic, pointing out—without going into the details of the case—that the proper conclusions may be easily drawn. A number of subsequent writers (Haldane, de Beer, Huxley, *loc. cit.*) have taken up the idea and have discussed it more or less at length. But I do not think that the many sources of information which fit together in the formation of a consistent general picture have as yet been properly assembled. Taxonomy, biometry, experimental biology, embryology, experimental embryology, comparative anatomy, paleontology, and genetics have furnished facts which can actually be woven into a simple pattern.

Before we take up specific examples we may enumerate the possibilities of rudimentation which may be derived in a general way from the known laws of development. A priori, three major types of rudimentation are provided for by the possibilities of development: a single organ may be formed and fail to complete development in the normal way; the primordium of the organ may fail to appear at all; or the primordium of the organ may not only fail to complete development but may be incorporated into another developmental process with a change of function. The most frequent type of rudimentation is the first one, and its occurrence may involve changes in different embryological processes. There is the possibility that the primordium is formed too early to permit its proper integration with the later developmental processes; and it may also be formed too late, thus preventing its keeping pace with the rest of development. The timing process of development may be normal with regard to the anlage of the organ, but the relative growth rate of the organ may be changed (by production of a lowered or even negative value of the constant k in Huxley's formula). The latter possibility would also include the contingency of a stopping of growth from a definite time on. Both possibilities just mentioned may be caused primarily by a change in the inductive process: time of action, localization, concentration and gradient of the inductor, whether this be of the organizer (evocator) type or of the specific hormonal type. Finally, rudimentation may be produced embryologically by a secondary destruction of parts or of the whole of an organ after initial normal development.

After this general introduction we take up, first, an example which will demonstrate the different types of facts and their interrelation, the rudimentation of the wings of insects. Every zoologist knows that rudimentation of wings occurs in all groups of winged insects, and that it is sometimes only a process of microevolutionary significance, sometimes one characteristic of macroevolution, and sometimes only a part of the life cycle. To give examples: In Cynipids and aphids, forms with rudimentary wings or wingless forms are a part of the life cycle within the species. In Diptera and Lepidoptera, mutants without wings or with reduced wings occur within normally winged species. In Orthoptera, different genera may be distinguished by the presence, absence, or rudimentation of wings. In Coleoptera whole families are characterized by definite types of wing rudimentation. Finally, there are whole orders without wings (mostly ectoparasites). In addition, in each group definite genera or species with more or less reduced wings occur in definite habitats: the island, mountain, and seashore species of Diptera, Lepidoptera, and Hymenoptera, cave and subterranean insects, and the guests (symphiles) of ants and termites. A complete list of the facts has been compiled by Bezzi (1916).

An analysis of the facts has to begin with the statement that what appears at first sight to be a simple and always identical process may have a very different embryological and, therewith, genetic basis. We have, therefore, to consider first the embryological and experimental facts and their significance for the situation in nature. Though an immense amount of work can still be done in this field, we already have a sufficient amount of knowledge from different sources to permit a preliminary analysis of the situation of sufficient reliability to warrant conclusions as to the evolutionary process. Most of the decisive material comes from studies of Diptera and Lepidoptera, to which contributory evidence on the aphids and Coleoptera may be added.

As a shift in wing development leading to rudimentation must necessarily be linked with the determination of the typical processes of wing formation, we must first consider what knowledge is thus far available upon this problem. In Diptera and Lepidoptera (and other metabolous insects) the wings are formed as imaginal discs, invaginations from the body surface forming pouches in which the wings grow as appendages which are pushed to the surface during metamorphosis. There can be no doubt that the major features of the wings are finally determined at a rather early stage of development of the wing discs. We know that transplanted larval discs in Lepidoptera may develop into normal wings. The best information, however, is derived from a study of venation, which is known to be one of the most conservative taxonomic features. In

FIG. 68. Pupa of a Japanese moth (species?) with indications of later venation by colored chitin. (Original.)

Lepidoptera we know that venation is determined in the larval disc a long time before wing veins are formed (Goldschmidt, 1920c). In a young pupa the wing contains a network of sinuses which have very little relation to the later veins. The tracheae and nerves are arranged in a fashion which is somewhat similar to the prospective veins, but which is considerably changed when the veins are formed. The epithelium of the wing does not indicate any structural feature in connection with the later pattern of venation. Nevertheless, this pattern, except for cross veins, is already completely determined. This can be shown without any experimentation in some pupae where the chitin secreted by the epithelial wing already shows the complete pattern of venation (pictured in Goldschmidt, 1920). Still more conspicuous is the case shown in figure 68, where the pupal shell shows the later pattern of venation in the form of pigmented bands, though the wing itself does not show this pattern as yet. In Diptera exactly the same situation is found (Goldschmidt, 1935b, 1937b). After pupation no genuine venation is present in the wing sac. Only later are the veins formed in a definite complicated order and slowly perfected through concrescence of the wing lamellae between the veins and through final chitinization. Nevertheless, the pattern of venation has already been determined in the discs of young larvae. This is best shown in *Drosophila* in the development of the wing-mutant vestigial, where the wing anlage in the imaginal disc begins to degenerate in a definite way in very young larvae. The result is, however, not a wing rudiment with some disturbed kind of venation but a wing stump with a perfectly normal pattern of venation (except for small regulations to the form of the wing remnant) in the part which is left, which is a stump comparable to a normal wing which has been clipped. These facts are of great importance, as they permit us to diagnose in a case of rudimentation which has not been studied embryologically what type of embryological process has been responsible.

Formerly we had occasion to discuss some of the problems connected with rudimentation of wings. We pointed out that such a rudimentation can be brought about by different embryological processes, as I could show for wing mutations in *Drosophila.* Three main types were found: (1) the type just mentioned for the mutant vestigial and a number of comparable mutants. Here an already determined wing anlage is reduced to stumps of different size by secondary degeneration and contraction of the wing area beginning distally and posteriorly and progressing toward the wing base in the different types until hardly a stump is left. What does remain finishes development and growth, including the postpupal growth and folding process, in a quite normal way. In this case, then, the general features of wing development are not changed, but a definite part of the anlage loses the ability to continue development and is histolyzed.[8] (2) The second type of rudimentation is in principle comparable to the first one. The difference is that not a stump but a whole shrunken wing is formed. In this case (type dumpy in *Drosophila*) a normal wing anlage is formed. After pupation the distal part of the wing is retracted; i.e., shrinks, a process which may involve a concomitant destruction of tissue below the surface. Thus a wing is produced which is more or less rudimentary in

8. In a recent preliminary note (*Proceedings National Academy Washington,* 1939) Waddington claims that the notched wings of *Drosophila* are already notched in the imaginal discs and that the normal appearance in the young pupa is due to the inflation of the wing after evagination of the disc (the wing disc is inflated by blood pressure to many times its size before it secretes the chitinous pupal sheath). According to Waddington, the absent parts of the wing become invisible on account of this inflation, but a secondary collapse brings them out again. The shortness of the report makes it difficult to find the source for this completely erroneous description. But it is easy to prove that this representation cannot be correct; only a few points will be mentioned here. (1) The author has forgotten that the chitinous sheath of the pupal wing is a cast of the secreting epithelium. In the notches this epithelium is missing and therefore cannot secrete if the notches are formed prior to pupation. Nothing cannot be inflated into something. (2) The photographs of my slides show clearly to what abnormal, more or less grotesque, forms the wing discs are inflated when, in the higher grades of notching, parts of the epithelium are already absent at the time of pupation. The chitinous sheath is thus a perfect cast of such a condition in all higher grades of scalloping. (3) In the next several pages facts found by other writers, both in species of Diptera with rudimentary wings and in corresponding cases in Lepidoptera, will be reported. They are in complete agreement with my description. In one case a normally formed pupal wing is actually completely destroyed before the imago hatches. (4) Cytolysis and histolysis are frequent phenomena in insect metamorphosis and it is, therefore, not surprising to find them at work. In insects which shed imaginal wings (ants) the entire musculature of the wings is histolyzed after having functioned.

addition to being changed in shape. The reduction is a rather late and secondary process. (3) The third type is of a very different embryological significance, as found by both Dobzhansky and myself (type miniature in *Drosophila*). Here the wing develops quite normally until it is

IG. 69. Normal male and saclike female of *Orgyia antiqua.* (From Paul.)

completed in the pupa. In normal development there follows a second phase of growth without cell multiplication, by increase in cell size (with regard to epithelium, hairs, etc., but not connective tissue) and subsequent folding within the pupal sheath. In the miniature wing this second phase of growth either does not occur or does not proceed very far. The result is a complete wing with the same number of cells, hairs, etc., but smaller in size.

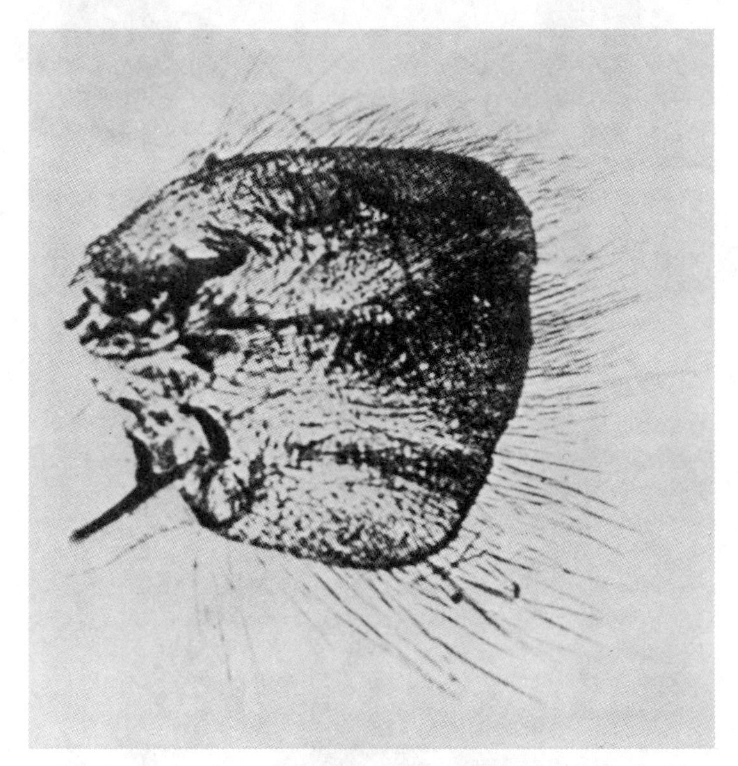

FIG. 70. Fore-wing of female *Orgyia antiqua*. (From Paul.)

There is not much material available which furnishes reliable information on the development of rudimentary wings in cases of greater evolutionary significance. But there is a little information on two cases of wing rudimentation in nature. The antarctic Chironomid *Belgica antarctica* possesses only minute wing stumps. Keilin (1913) has traced their development. In the larva the wing discs are already smaller than normal, but still good-sized. The pupal wing (and its chitinous sheath) is almost of normal size, though not perfectly shaped. Within the pupa this wing is reduced, according to Keilin, by resorption to a

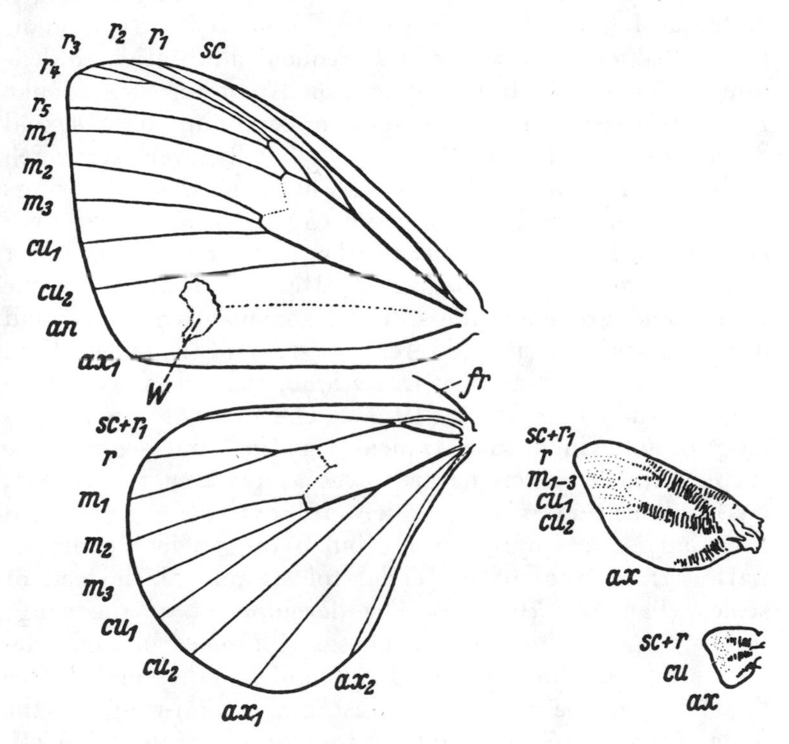

Fig. 71. Venation of fore- and hind-wings of male and female *Orgyia antiqua*. (From Paul.)

minute stump of about one twentieth of the size of a normal wing. Another remarkable case of the same type has been recorded by Hopkins (quoted from Bezzi). In *Pryxia scabiei* the pupae of both sexes have normal pupal wings, but in the females these are completely resorbed during pupal life. According to Stange (quoted from Bezzi), in *Melophagus* small pupal wings are formed but histolyzed afterwards.

These few data show clearly that the types of embryonic wing destruction which I have described for the *Drosophila* mutants vestigial and dumpy are also encountered in cases of rudimentation of a taxonomic nature. Except for *Drosophila*, complete embryological information is only available, as far as I know, in one case. In the moth family Lymantriidae reduction of wings is a frequent adaptation to definite modes of life and propagation. In flying genera like *Lymantria*, certain species show a beginning of a sexual dimorphism insofar as the females hardly ever fly, which is the case in *Lymantria dispar*. The strength of the wings does not suffice to lift the heavy egg-laden abdomen for a sustained flight. In addition, the moth never feeds but starts laying eggs immediately after fertilization by the males, who are good fliers. In the genus *Orgyia* we find different species with different degrees of rudimentation. An extreme case is that of *O. antiqua*, which has been thoroughly studied by Paul (1937). The male of this species has normal wings with typical venation and scales. The saclike female, which never moves away from the cocoon, has rudimentary wings, which are extremely small and show only traces of venation (but arranged in the proper pattern!), a trace of epithelial folds, and hair instead of scales (figs. 69, 70, 71). The development of the wings in the imaginal discs shows no sex difference in early development and follows the plan found in other moths (see fig. 72). In the last larval instar the hind-wing of the female begins to lag, and in the final prepupal growth of the wing the female fore-wing keeps pace with the male one, while the hind-wing grows very little, thus increasing the difference which became visible in the last instar. Thus, the details of morphogenesis of the fore-wing are alike in male and female up to pupation, but the relative growth rate is different. The female moth is much larger and has a longer period of growth, but the wings grow more slowly and, therefore, reach a smaller size relative to the size of the body. There is also another difference. At the time of pupation there is a stretching of the folds of the wing which became folded within the disc.

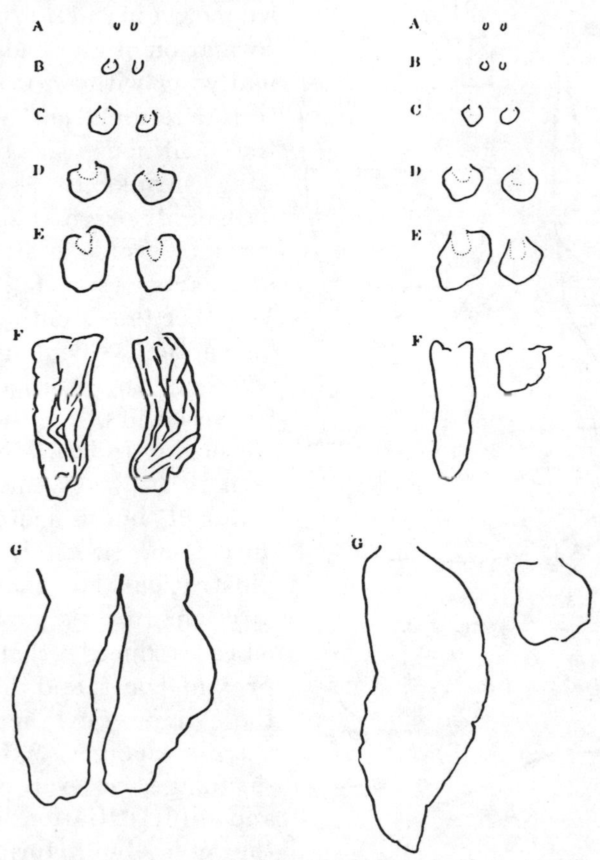

FIG. 72. Development of imaginal discs in the caterpillars of *Orgyia antiqua*. A-E, first to fifth instar. F, prepupa folded. G, after unfolding. Left series male, right series female; in each, left fore-wing, right hind-wing. (From Paul.)

The wing is subsequently inflated by pressure of the hemolymph. (The same process occurs in *Drosophila.*) This process is less pronounced in the female and thus the female pupal wing becomes absolutely smaller than the male

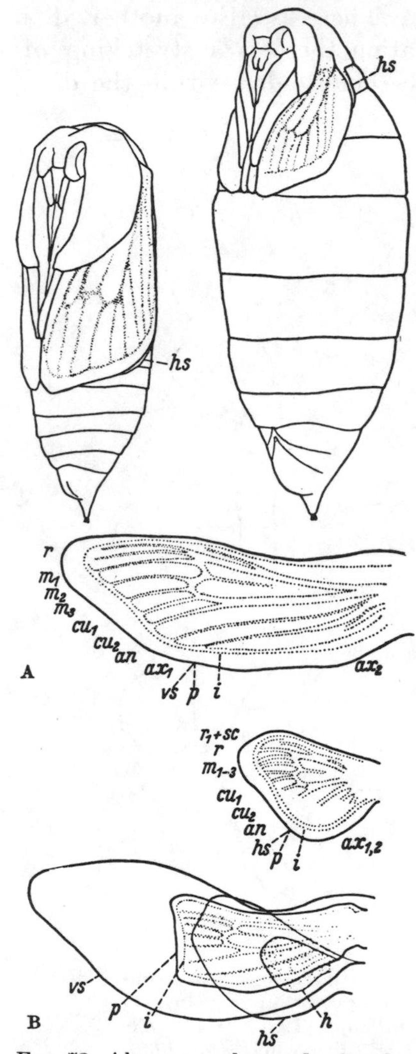

FIG. 73. Above: male and female pupae of *Orgyia antiqua* showing wing venation (*hs*, sheath of hind-wings). Below: A, fore- and hind-wing of young female pupa. B, 3 days later. *vs*, sheath of fore-wing; *p*, margin of wing; *i*, margin of later imaginal wing; *hs*, sheath of hind-wing; *h*, hind-wing. (From Paul.)

one. Only after pupation does the major process of rudimentation of the female wings begin; i.e., after the formation of an almost normal wing with normal pattern determination! Within the chitinous sheath the wing shrinks to less than half of its original size but without visible destruction of parts (see fig. 73). Whether this is nothing but an actual shrinkage, or whether an internal histolysis occurs, is not known. According to Paul, the pattern of the wing veins is not changed, but is merely reduced in size. But his illustrations show that actually the distal parts are more reduced than the proximal ones, and that also the shape of the wing margin changes. After the shrinkage, however, growth and differentiation occur: the epithelium forms the typical transverse folds on the rims of which the scales are formed. At the time of hatching, when normally these folds are leveled by the influx of hemolymph into the wings under pressure, the female wing is not stretched. Thus the rudimentation of the wing is produced: (1) by reduced relative growth of the wing discs in the larval stages; (2) by a reduced expansion at the time of evagination of the discs; (3) by the shrinkage of the wing in the pupa; (4) by absence of imaginal inflation. If we compare this rudimentation with the cases known in *Drosophila,* we find that none of the latter correspond completely to the *Orgyia* case. The shrinkage within the pupal shell in *Orgyia* is practically identical with the similar occurrence in the development of the mutants dumpy and rudimentary in *Drosophila.*

In the vestigial mutants of *Drosophila* the wing discs seem to grow more slowly than do normal discs, but only a part of the already determined surface of the wing develops, and this to its normal size, while the rest degenerates. Thus, not a complete small wing, but a stump of normal size, is formed. In Lepidoptera a corresponding process is known, but as a normal feature, without any apparent connection with rudimentation. As Süffert (1929) has found in my laboratory, in normal development of the wings of Lepidoptera the edge of the larval wing degenerates in the pupa. In cases of special wing shape, like the tails of *Papilio,* these are cut out from a pupal wing of ordinary shape by degeneration of peripheral tissue. These facts show that the vestigial type of rudimentation is potentially present in Lepidoptera, though it has apparently never occurred in evolution.

The last remarks lead to a survey of the known processes of rudimentation in the evolution of insects, insofar as they may be inferred from morphological descriptions without any knowledge of the embryology. We have already mentioned the existence in Lepidoptera of impaired flight without wing rudimentation. There is also an intermediate condition between this and the situation just described for *Orgyia antiqua.* In *Orgyia thyellina* the females of one generation have normal wings, those of another, rudimentary wings (Cretschmar, 1928). But here the rudimentation is the result of only one of the processes described for *O. antiqua;* namely, the lack of unfolding of the wing in hatching (see below).

Of great importance are the facts concerning Diptera, because here a direct comparison with the best-known experimental animal, *Drosophila*, is available. In nature many species of flies have more or less rudimentary wings. Such species are typical of definite habitats: dunes, oceanic islands, and such specialized ecological niches as are involved in ectoparasitism and commensalism, like association with ants and termites. A complete review of such

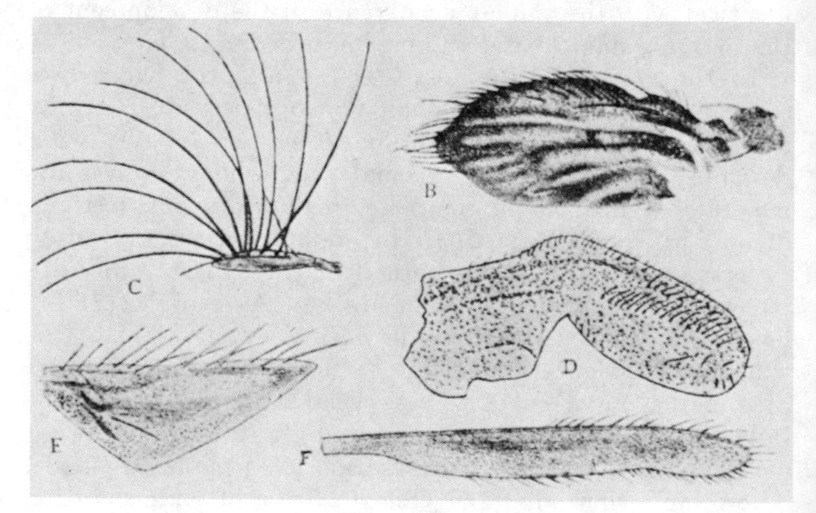

Fig. 74. Some types of rudimentary wings in Diptera. B, *Apetenus litoralis*. C, *Acontistoptera mexicana*. D, *Tipula simplex*. E, *Commoptera solenopsidis*. F, *Psamathiomyia pectinata*. (From Bezzi.)

species and their respective habitats is found in Bezzi (1916). Among the many interesting types of reduction which are of importance in our present discussion (but none of which have been properly analyzed by the host of dipterologists working with the material) a few may be mentioned. Judging from the insufficient number of pictures available, the following principal types of wing reduction are found. There are species which do not fly, though they use their wings in jumping. In these the wings show different degrees of rudimentation within a single brood. Good pictures of this varying process of rudimentation are found in a paper by Brauns (1938) on shore flies. It seems that in such cases rudimentation is confined completely to postpupal differentiation. Two processes are involved: the postpupal growth is more or less inhibited, and the formation of the veins is retarded and remains more or less unfinished in different individuals. The ectodermal dif-

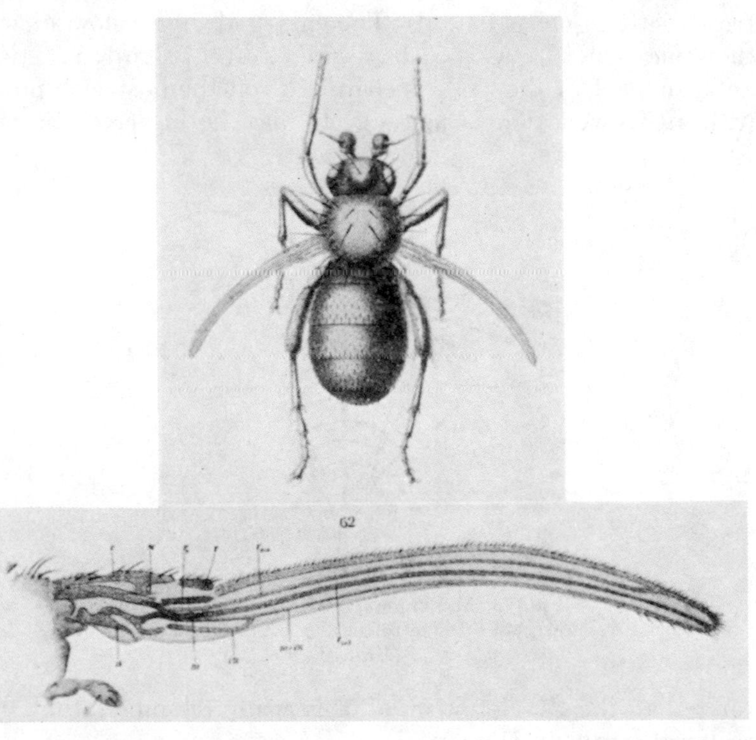

Fig. 75. *Amalopteryx maritima,* an antarctic island fly. Below, details of wing. (From Enderlein.)

ferentiation (hair, etc.), however, proceeds rather normally (see fig. 74, B). Obviously, then, an inhibition of some but not all developmental processes occurs after development has proceeded normally up to pupation. No comparable mutant is thus far known in *Drosophila,* though one wing mutant in *Lymantria dispar* is somewhat comparable. A second type looks as if wing development had almost completely stopped after pupation. The wing looks like a regular or more or less misshapen pupal wing. But some ectodermal development has taken place, as the presence of hairs shows (see fig. 74, D, E, F). A third type which seems to be rather rare is an exact copy of the vestigial case in *Drosophila.* Figure 75 represents an antarctic fly from the Kerguelen Islands. The enlarged wing below shows the same structure as the *Drosophila* mutant strap; i.e., the anterior part of the wing is complete and the posterior part with all its venation is missing. It may be inferred, there-

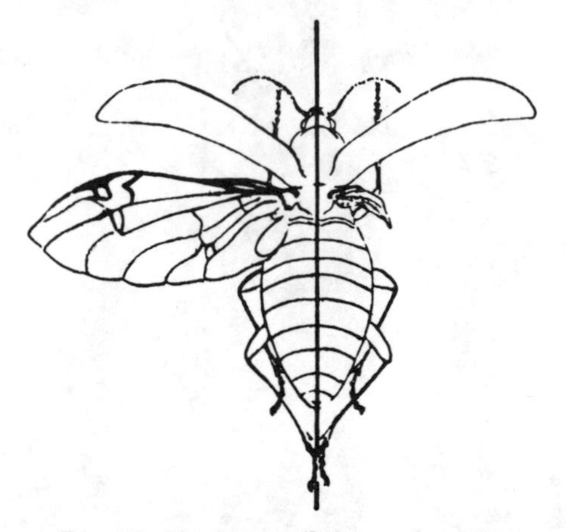

FIG. 76. Maximum (left) and minimum (right) of wing formation found in *Carabus clathratus.* (From Oertel.)

fore, that the development of this wing resembles that of vestigial-strap in *Drosophila.*

Another type shows minute stumps of different forms (see fig. 74, C). Here belong the cases in which the development is known to involve a secondary destruction within the pupa after a more or less normal development up to pupation, as described above. Some of these stumplike wings have an appearance intermediate between wings and halteres. We have already pictured (fig. 65) the wing of *Termitoxenia* and compared it with the *Drosophila* mutant *tetraltera*. Probably many more wing stumps of the type pictured in figure 74, C will turn out to be halterlike homoeotic appendages.

Finally, visible wings have completely disappeared and simultaneously the body segmentation has become as regular as in lower insects. This type of fly (resembling an ant) is an exact replica, in these general features, of the wingless beetles living in the same surroundings. Unfortunately, no information is available as to the development of such forms.

In Hymenoptera we might mention the reduced wings

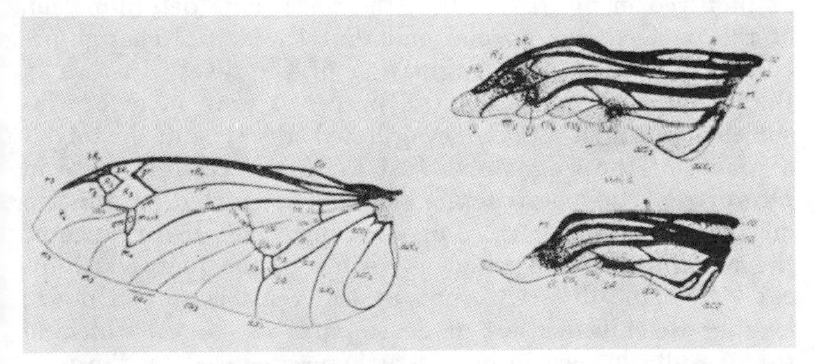

FIG. 77. Details of the same as in figure 76. (From Oertel.)

of some Cynipids. Kinsey's illustrations show completely patterned small wings, so that we must assume either a process of the type described for *Orgyia antiqua*, or one as pictured for the miniature mutants of *Drosophila*. The situation in ants and termites will be discussed later.

A rather thorough study of wing rudimentation has been made in Coleoptera by Rüschkamp (1927) and Oertel (1924). In this group, just as in others mentioned before, there are species with normal wings which do not fly and others with rudimentary wings, from small perfect wings down to completely abnormal stumps, with all intermediate stages.

In the group of Carabid beetles a few species have normal wings and fly, but a majority show wing rudimentation, even within the same genus, and some species do not have hind-wing rudiments at all. The elytra are frequently united into a solid carapace. There are species in which the amount of rudimentation varies from individual to individual and even between the two sides of the body. The range of variation is extreme, as figure 76 shows. The optimum is a practically normal wing (fig. 77), and the most rudimentary type (fig. 77) shows considerable reduction. There is nothing known about the development of these, but in the light of our knowledge of *Drosophila* and moths we may safely conclude from the structure of a rudimentary wing, as pictured in figure 77, that the embryonic determination of the pattern was normal and that the actual change occurred only after the evagination of the discs. Just as in the model cases, probably two processes were involved: insufficient growth of the evaginated discs, and histolysis of parts of the wing area. Just as in the vestigial case in *Drosophila*, this destruction may have started at the tip and at the posterior edge of the wing. The disturbance of the remaining veins, which is visible in the pictures, indicates that the differentiation of the veins is slowed down, leading to chitinization of developmental stages, which in turn results in mechanical disturbances caused by the partial degeneration. The individual asymmetry of right and left wings is also encountered in some combinations of the vestigial series in *Drosophila*.

This is not the only type of rudimentation. In another species some geographic races have normal, others rudimentary, wings (fig. 78). In such cases the rudimentary wing is a complete wing with normal venation (except for small reductions) which has changed its shape into that of a narrow band (fig. 79). Here, then, the rudimentation occurred by differential growth of the wing area followed by stoppage of growth. We know of *Drosophila* mutants (and phenocopies) with such bladelike, but not rudimentary wings, so that the type described in Coleoptera is comparable to a combination of a process of changing wing proportions with one entailing a miniature type of rudimentation.

We shall not go into more details of the variations found, but shall mention only the rudiments found in species with considerable reduction of the wings. Here very bizarre forms of rudiments are formed (fig. 80). A comparison with the figures of developmental stages of the higher grades of the vestigial series in *Drosophila* which I have published (Goldschmidt, 1935b, 1937b) shows that the destruction of an already formed wing area must have proceeded exactly as in the *Drosophila* case. Finally, in some species

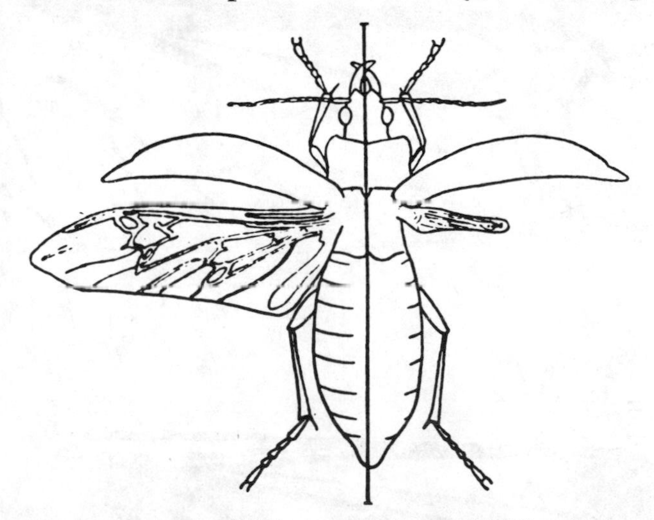

FIG. 78. Maximum (left) and minimum (right) of wing formation found in *Carabus granulatus*. (From Oertel.)

wings are found in which the two lamellae are separated by a fluid-filled space in which tracheae of a completely embryonic type (with capillaries) are suspended. In such cases, then, the other features of rudimentation were clearly combined with a stoppage of development at a stage near the time of pupation. As I said, there is not much known regarding wing development in such cases. But the correctness of our interpretation can be tested by the aspect of the wing sheath of the pupa, which is a cast of the wing at the time of pupation. This is normal in the less reduced species, indicating normal development up to pupation;

it is reduced and irregular in the extreme cases, indicating degeneration before pupation. In the only case in which some more data on development have been furnished (Dewitz, 1883), a case of complete absence of the hind-wings in the imago, the larva and young pupa contain minute rudiments, indicating a condition parallel to the most extreme mutants of vestigial in *Drosophila*. We may mention finally

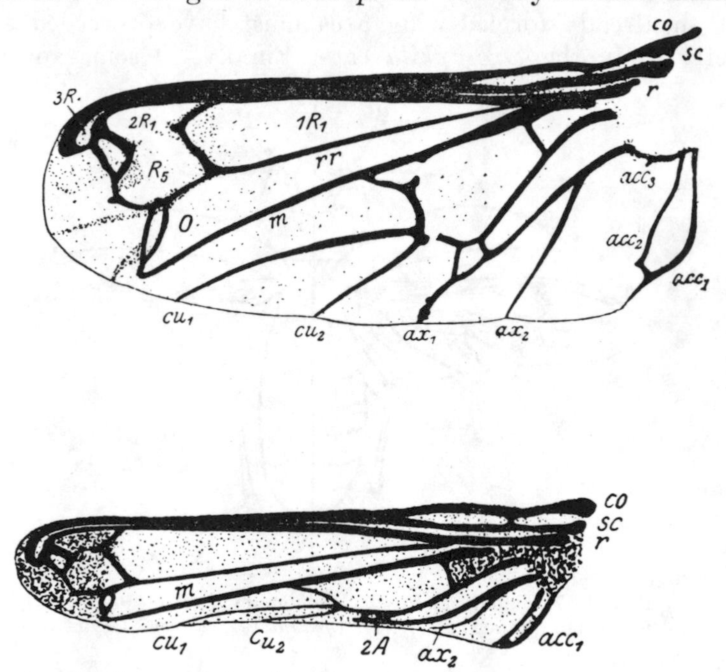

Fig. 79. Above, details of optimal wing of *Carabus granulatus* with beginning reduction at the tip. Below, minimum found in the same form, venation deformed but otherwise almost complete. (From Oertel.)

Oertel's observations upon the details of degeneration of the veins in the rudimentary wings, which disappear in a definite sequence. I should point out that in *Drosophila* the development of the wing veins proceeds in a definite order; a halting of differentiation at different times of pupal development would, therefore, change the remaining venation correspondingly. Whatever the details are, the essential point is clear: the macroevolutionary wing rudimentation in Carabid beetles does not necessitate the assumption of a gradual accumulation of micromutations, but can be explained by single steps, which change the relative

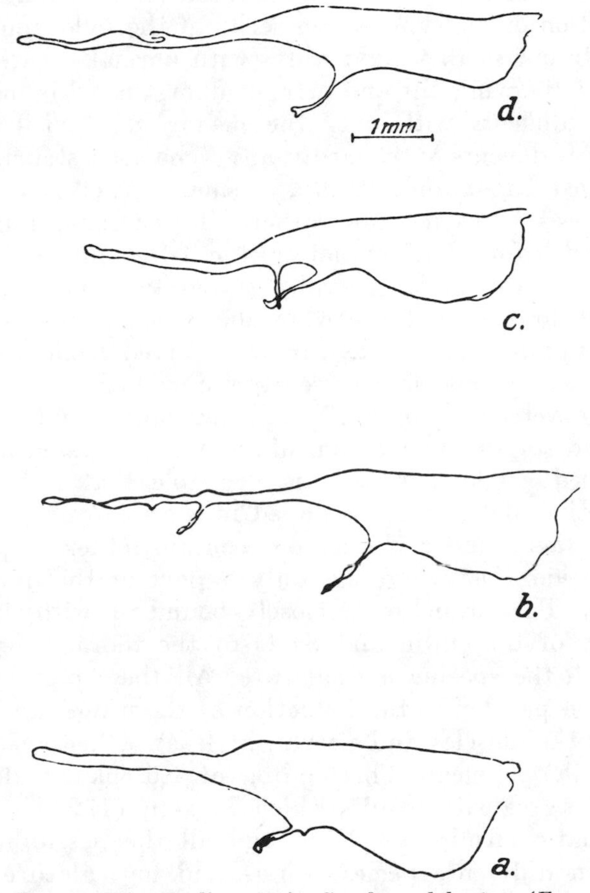

FIG. 80. Wing rudiments in *Carabus glabratus*. (From Oertel.)

rates of processes of growth and differentiation at an earlier or later time in development.

Only one more example will be given to show the identical nature of all the cases even though there are certain individual variations in the collaboration of change of growth rate, stoppage of development at a larval stage, and secondary destruction of already formed tissue. Rüschkamp has made a special study of the Chrysomelid beetles. Here again, all degrees of wing rudimentation are found. The description of the types shows series of the following order: complete but small wings; wings with normal venation, but parts of the wing tip and margin unorganized or missing; small rudiments with only the major veins visible; still smaller rudiments with hardly any venation; structureless, fluid-filled bags; and, finally, a small structureless scale. This series seems to be a rather different one from those described before. Unfortunately, the details of development are not known. It is generally stated that imaginal discs are present and that the wing shows some differentiation after pupation. As far as can be inferred from the facts, in this case the major feature of rudimentation is a halting of the development of the wing discs and pupal wings at different stages. The rudimentary wing is essentially an unfinished wing, though it is probable that a histolytic destruction of the parts is added in some cases.

The facts under discussion contain other important information. The wings are only a part of the flying apparatus. Their function is closely bound up with the morphology of the chitin and parts of the thoracic segments and with the specific musculature. All these parts show a reduction parallel to the reduction of the wings, and in the case of the muscles and nerves, at least, a histolysis after normal development. This finding of Rüschkamp does not seem to agree with results which Poisson (1924) obtained for aquatic Hemiptera. He found all the possibilities realized in different species: wings and musculature developed but not used, wings without muscles, muscles without wings, and both of them abnormal. According to Rüschkamp, however, the muscles are formed in Coleoptera but histolyzed in imaginal life; i.e., present in young imagines and absent in old ones. Poisson's first two types, therefore, may be based on oversight. Indeed, the variability actually found among different specimens favors this explanation. Where, however, wings were much reduced, Poisson found that the myoblasts were already histolyzed before emergence of the imago. A few cases are recorded in which muscles were still present in spite of considerable reduction of wings. But in the majority of cases, wings and muscles were reduced simultaneously, and the muscles were first formed and then histolyzed, either before or after eclosion. We may then say that, barring the possibilities of variation, the flying mechanism is reduced as a whole in consequence of a genetic change which affects development of the whole complex simultaneously. It must be kept in mind that in insects the widespread use of histolysis in metamorphosis furnishes a priori a ready mechanism for reducing parts. It is generally known that in the queen ant the wings are clipped after the nuptial flight. This is immediately followed by the degeneration of the flight muscles. In the weevil *Sitona*, Jackson (1933) has made a special study of these muscles. He finds imaginal degeneration of the muscles after a period of flight, as well as primary degeneration during metamorphosis leading to flightless individuals within a flying species. Obviously, histolysis of the muscles may occur at different times independently of the condition of the wing. This, however, does not exclude mutational changes of the wings from usually affecting the development of the whole flying apparatus. In a general way the latter certainly is the case. However, in the vestigial mutant of *Drosophila* the muscles are frequently but not always normal, according to Cuénot and Mercier (1923). Individual variations found in different species lose their importance for phylogenetic analysis, when such variations may occur within a single species as normal fluctuations of embryogenetic processes. Just as we said before in the chapter on embryonic regulation, the developmental system permits manifold and complicated regulations to take place after a single embryonic disturbance. This certainly applies to the present case involving the wing discs. The potentiality for regulation is well exemplified in occasional abnormalities such as those in the Procrustes (Carabid) mentioned by Rüschkamp, where the wings were reduced to small halterelike appendages. In this case the abdominal tergites, which are usually covered by the elytra and are rudimentary, had chitinized normally just as in completely wingless forms.

There is another interesting point. The facts thus far reported relate only to the hind-wings, since the fore-wings in Coleoptera are transformed into elytra. The example of *Orgyia* demonstrated the relative independence of both pairs of wings, and the halteres of Diptera showed the combination of a type of rudimentation with change of function. This is also the case for the elytra. In the series of rudimentations of the hind-wings the elytra may change correspondingly toward concrescence, or they may be reduced with the hind-wings. All possible types are found, as every naturalist knows.

Thus we may gather from the whole body of facts regarding Coleoptera that a macroevolutionary type of wing rudimentation with all its consequences for segmentation and special morphology can be based on a single primary change in the growth and differentiation of the wing discs. This change may involve both winged segments, or the segments may react independently, which, however, does not necessarily mean two independent genetic steps, as a single primary cause of the gradient type may be involved. We shall turn to this point when discussing the experimental evidence. This necessarily short review of the more important facts must suffice. A good compilation of the literature is given by Finkenbrink (1933).

We turn now to experimental evidence which has furnished sufficient insight into the embryological as well as the genetical basis of wing rudimentation to permit definite conclusions. We may mention, in passing, the earlier experiments which were intended to show that definite external conditions are responsible for wing rudimentation in nature. There are, first, the much discussed experiments of Dewitz, who claimed to have produced (of course, nonheritable) wing rudimentation by different chemical and physical agencies, all of which were supposed to act via a change of the oxidation process. His data show, however—and Finkenbrink *(loc. cit.)* has demonstrated it again—that he actually had not produced any rudimentation but only crippled conditions of normally developing wings by mechanical hindrance of the processes of pupation and hatching. But there are many exact data available which show that the process of wing rudimentation can be produced experimentally and that this is based upon the same developmental processes as those involved in the cases of hereditary rudimentation. We mentioned before the production of phenocopies in *Drosophila;* i.e., the experimental production of the phenotype of mutants by simple modification of development. In such experiments I succeeded (Goldschmidt, 1929b, 1934a) in producing by heat treatment the exact copy of such mutants as dumpy, rudimentary, miniature, and some alleles of the vestigial series; i.e., all three types of wing reduction described above were produced in normal wild stock. Since that time it was found that the same effect is also produced by X-rays (Timofeeff, 1930; Friesen, 1936), and by ultraviolet radiation (Epsteins, 1939). See also Enzmann and Haskins *(loc. cit.)* on neutron bombardment and Rapoport on chemically produced phenocopies. The detailed analysis showed (as reported above) that the effect is produced if the experimental disturbance of development acts at a definite sensitive period, preceding embryonic determination, and that it acts by stopping growth, by changing the rate of growth, or by initiating histolysis of already formed tissue. As is to be expected, the effect is stronger if earlier developmental stages are affected. The relation between general speed of embryonic differentiation and the relative growth rate of the wing anlage is best demonstrated in those cases in which the rudimentation is increased by delaying development by starvation (Braun, 1939) or by the action of ultraviolet light upon early larval stages (Epsteins, 1939). It is then obvious that the balanced system of development permits the occurrence of the known types of wing rudimentation, provided that it is possible to influence differentially the speed of some of the integrating processes. (This applies also, as was quoted above [p. 333], to the type of rudimentation which produces halterelike structures.) This effect may be produced either by an experimental slowing down of some, or speeding up of other, processes of development, or by the presence of a mutation which has a parallel effect upon development.

Though no systematic experiments of this type have been made with other insects, it may be mentioned that Larsen (1931) obtained different conditions of lengthening of the wing in an aquatic Hemipteran by changing the oxygen supply in the last larval instar. In this case the situation is opposite to the one reported from the *Drosophila* experiments: a hereditarily rudimentary wing is lengthened. A parallel to this is found in *Drosophila.* Since the work of Roberts (full literature in Goldschmidt, 1938) we know that a vestigial fly raised at a temperature near the possible maximum (29°–31°C.) will develop more or less complete wings. Again, the action requires a definite sensitive period. We may mention in the same connection the experiment, which no biologist has yet succeeded in accomplishing but which is regularly performed by ants and termites: unknown treatment of the larvae by the workers prevents or favors the development of the always present imaginal discs of the wings, thus producing the apterous or the winged castes.

In the case of the Lepidoptera with rudimentary wings in one sex, Paul *(loc. cit.)* has gained some insight into the condition of determination. Wing discs transplanted from one sex to the other retain their determination. The area surrounding the disc is capable of regenerating a wing, and a transplanted area will regenerate a wing if the wing disc is removed. Such a transplant made at an early larval stage from male to female regenerates a male wing, but a transplant from female to male also regenerates a male wing. Thus we see that the wing area of the male is finally determined at a time when it is still undetermined in the female: the rate of determination differs in the sexes. We may conclude from this that the rudimentation of the wing is caused by a special relation of the rate of growth to the rate of determination.

Another type of information concerning the developmental mechanics of wing rudimentation is derived from a study of cases in which wing rudimentation occurs in definite forms within a species, either as a constant dimorphism (the winged and wingless castes of ants and termites) or as a cyclical dimorphism (alternation of wingless and winged aphids). A kind of transitional situation between sexually dimorphic wing rudimentation and normal behavior of the wings is found in some moths. We may recall the series of rudimentation found in the Lymantriid family, from *Lymantria dispar* and *monacha* with nonflying females to *Orgyia antiqua* with saclike females with wing stumps. In between is found the species *Orgyia thyellina*. Here a seasonal dimorphism of generations is observed. According to Cretschmar (1928), the first generation has normal wings in both sexes; in the second generation, a part of the females do not expand their wings; and in the third, the majority have unexpanded wings. Cretschmar thinks that the availability of different amounts of hemolymph is responsible for the difference. But according to some experiments made by the author (unpublished) this difference is one which is very easily controlled. The rudimentary females actually have complete wings, which, however, are not spread after hatching. But if the moth is bred at a high temperature (*ca.* 25° C.), the wings spread out, and a succession of generations with normal females is obtained. Probably the decision whether female wings spread or do not spread is controlled by the temperature during a sensitive period. Such an effect, however, is already known to us to be linked with differential velocities of integrating processes of development. Probable comparable cases are not rare. A number of facts are known for Hemiptera and Neuroptera which indicate a hereditary tendency to brachyptery and aptery, the realization of which is largely controlled by temperature. The data, which are in no way conclusive, may be found in Ekblom (1928). The material does not lend itself to proper experimentation, which, however, is possible with aphids.

The determination of wing rudimentation or wing formation in aphids has been made the subject of a detailed study by Shull and his students. (We shall not mention his numerous predecessors in this field who worked without present-day knowledge of the physiology of development. See Shull, 1928.) In a general way one may say that in spring winged migrants appear which transfer the forms to another host, and are succeeded by generations of wingless forms and, finally, once again winged re-migrants. Experiments show that light and temperature are the main controlling agents. It is especially the application of intermittent light which produces winged individuals. This factor acts only if applied to the parents at a definite time, and the analysis shows that a direct action upon the embryo during a sensitive period is involved. This period occurs about thirty-four hours before birth, and just before the first visible trace of wing development begins. Intermittent light and 14° C. acting at this time induce wing formation; constant light at 24° C. prevents it. But the wingless forms also show the first beginnings of wing development, rudiments of which may still be found at advanced stages of development. Usually the hind-wing bud is absent. (The same antero-posterior gradient of wing determination is also found in Lepidoptera.) The winged aphid is distinguished from the wingless one not only by the wing characters but also by other features: the winged form has three ocelli, wing muscles, and numerous sense organs on the antennae, all of which characters are missing in the wingless forms (which have few sense organs instead of many). Shull's embryological analysis of wing determination is based upon the existence of intermediates between winged and wingless forms, which can be produced in a definite percentage by the proper variation of the controlling factors, intermittent light, and a higher temperature. Just as in all the comparable cases which we studied above, this situation suggests a simple explanation in terms of rates of determining processes. Shull has worked out in detail the possibilities based upon our general theory of integrating, balanced velocities of production of determining substances. According to this theory, a considerable number of variables are coöperating to produce the result, and a changed situation may be the product of a change in any of the variables; e.g., the rate of production of a determining substance, the threshold at which it works, the length and the position in time of the critical period, the simultaneous occurrence or the differential timing of the critical periods for the determination of the different characters involved (details in Goldschmidt, 1927, 1938). Shull (1937) discussed the different possibilities which may be involved in the aphid case and which can explain the combinations of the different characters found in intermediates; i.e., more or less intermediate wings combined with presence, absence, or partial presence of one or the other or of all the differential characters. The details of this analysis, ably illustrated by diagrammatic curves of the type used in my work on physiological genetics and sex determination, are not of importance in our present discussion. The essential point is that the experimental analysis shows that wing rudimentation and all the other considerable morphogenetic changes correlated with it, as well as possible intermediate conditions, may be produced by a single experimentally produced shift in some of the variables of the system controlling determination in development. This result agrees with all our former conclusions and demonstrates again that in evolution a single mutational step changing the rate of one of those variables can account for any type of wing rudimentation with all its obligatory morphological consequences.

Only a few words need be added here to point out the wide applicability of all these deductions. We mentioned above the castes of ants and termites as an example of immense morphogenetic differences of a macroevolutionary order of magnitude, produced upon the basis of a single genetic constitution by some experimental procedure known to the insects though not yet to the biologists. Here a difference of winged and wingless forms is also involved, in addition to which it is known that the wingless forms have wing discs in development. Though the experimental analysis of this situation has not yet led to much success, there are some experiments performed by nature itself. It is known that ants parasited by *Mermis* can develop into intermediate types, the intercastes, which have been much discussed (see Vandel, 1930; Wheeler, 1928). Though the wings are not involved in this intermediacy (never?), all other morphological characters are. Wheeler realized that the production of the intercastes must be based upon a simple principle of shift in embryonic determination, just as I had postulated for the production of intersexes. This conception brings the intercastes in line with the intermediate aphids and again confirms the general trend of our analysis from a different angle.

At this point of our analysis of the evolutionary significance of wing rudimentation, we may insert a short digression upon another feature of wing evolution in insects. It is known that the general features of wing venation characterize the different orders of insects, and that the specific features of venation are among the most constant taxonomic differentials of families, genera, and species. An understanding of the evolution of this character would help much in understanding the general laws of evolution. Further, a character is involved here which can hardly be assumed to change by accumulation of micromutations and which in addition does not lend itself very well to the assumption of selection of such mutants. There is certainly not yet sufficient information available to permit a complete analysis of the evolution of venation from the standpoint of embryonic potentialities and their genetic change. But at least a number of facts are available which permit us to show that the evolutionary principles involved are the same as the ones derived thus far. Knowledge of the embryonic determination of wing venation is far from complete, but the few points which are known suffice for a generalized discussion.

In Lepidoptera the pattern of the wing venation is determined before pupation. The young pupal wing is a flat sac filled with fluid. But the upper and lower epithelial surfaces are connected by funnellike, reinforced pillars which, seen from the surface, produce solid islands within the fluid-filled sac. The fluid is thus confined to a fine network of sinuses, which can be demonstrated by injection (Goldschmidt, 1920c). Within this network wider longitudinal sinuses are, in some cases, already visible. These correspond roughly to the main veins which develop later. If the hemolymph in these sinuses is under pressure while the chitin of the pupal shell is being formed, the arrangement of the sinuses may be visible as a surface sculpture indicating either the later veins (saturnids) or the network of sinuses *(Vanessa)* (see also fig. 68). In the pupal development of the wings, the sinuses disappear by concrescence of the two wing lamellae, and the remaining channels are the veins, which become heavily chitinized. (Details may be found in Goldschmidt, 1920c; others in Behrends, 1936.)

In Diptera *(Drosophila)* also the pattern of the veins is determined in early larval stages of the wing disc, as the process of development of the mutant vestigial proves (see above). The actual differentiation of the veins by concrescence of the lamellae of the wing occurs much later, in the pupa, where, first, wide sinuses are formed, which are later narrowed down in a definite order to form the veins (details unpublished). (Many features are visible in my photographs in Goldschmidt, 1933c. A detailed description has recently been given by Waddington; see footnote, page 346.) In *Drosophila* a number of mutants are known which influence the veins. Most of them do not affect the typical pattern: one or both cross-veins may disappear or be incomplete; other veins may be incomplete; extra veins may appear at definite places between the veins. But in some cases the primary veins are also shifted, dislocated, connected in a new way. A special study which I made (unpublished) on a series of wing conditions of the plexus type, beginning with a small extra vein at a definite point and leading to a highly plexate wing with numerous extra veins at definite points and a concomitant distortion of the primary veins, made me realize some relevant points. In all these cases the concrescence of the wing lamellae is disturbed, and the epithelium over the unclosed parts of the wing cavity secretes dark chitin, the extra veins. (Waddington describes two different types of this process.) The seriation of their appearance and their localization show that the abnormal pattern is mainly determined by the time in pupal development at which the normal concrescence of the lamellae and the concomitant narrowing of the sinuses into veins takes place. Again we face, just as in so many other cases, a complicated pattern effect produced by a stoppage of definite embryogenic processes at definite and severally different moments in development. The details are, however, more complicated and not yet fully understood. Waddington's preliminary description does not, as yet, make the process completely clear. In this case, again, experiments demonstrate that the genetic differences producing the abnormal venation act via changing rates of the decisive embryogenic processes, in this case the concrescence of the wing membranes: The same temperature shocks, etc., which produce phenocopies of other features of the wing, if applied at definite sensitive periods, also produce plexate wings. In addition, it has been shown in my laboratory (Braun, 1939) that a prolongation of the time of development changes the degree of plexation and, further, that the formation of the cross-vein can be prevented experimentally (Braun, in press). We do not need to go into further detail. The reported facts show that the developmental physiology of the wing veins is such that simple shifts in velocities (including stoppage) of the integrating processes may produce different patterns, with a larger departure from type the earlier in development the change occurs. The evolutionary conclusions are obviously the same as in former examples: a single mutational step affecting the timing mechanism of early embryonic differentiation of the wing may produce large pattern effects, the larger the earlier the genetic change begins to act. Let us suppose that in a lepidopteran wing chitinization occurred at the stage of the network-like arrangement of the sinuses. The resulting wing would resemble that of a neuropteran insect. In the same way, vice versa, the special venations of higher insects may have developed in a single step from the primitive network. We are certainly not interested here in any details of phylogenetic speculation. The only point which interests us is the demonstration that the actually existing potentialities of development, based upon the timed system of integrating processes of determination in development, permit large evolutionary steps to take place as a consequence of single mutational changes affecting the coördination of embryonic processes.

After this interruption we return again to the problem of wing rudimentation and consider now the genetic side of the problem. In the case of experimental animals such as *Drosophila* and *Habrobracon* (Whiting) the situation is simple and clear. All three types of rudimentation previously described may be caused by ordinary mutation. All the existing intermediate steps of rudimentation of either type can be produced independently of each other by single mutations of the multiple-allelic type, as well as by mutations at different loci. Each type of rudimentation may also be influenced by other mutants, which are called modifiers. The mutants for rudimentation may be dominant, recessive, or incompletely dominant, and the dominance may be shifted by modifying factors or external agencies. In addition, some mutants show the phenomenon observed in the rudimentary wings of Carabids; namely, great individual variability and low right-left correlation, increasing with the degree of rudimentation. Finally, the pattern of rudimentation is different for different mutants of the same type;

e.g., vestigial—Beadex. The vestigial type of rudimentation may also be conditioned by genetic changes other than standard mutations; namely, deficiencies and chromosomal aberrations. One type of mutation paralleling conditions in nature has not yet appeared in *Drosophila;* i.e., sex-controlled rudimentation affecting only one sex, which occurs so frequently in Lepidoptera and in some Diptera (see below). But in the beetle *Bruchus,* a case with simple Mendelian, sex-controlled inheritance is known (Breitenbecher, 1925). One should expect such mutants in Lepidoptera, especially in near relatives of naturally dimorphic forms. But *Lymantria dispar,* belonging to the same family as *Orgyia,* with its rudimentary females, has never produced a wingless mutant, though bred by the hundreds of thousands. The only wing mutant known has soft crumpling wings (Machida). In the silkworm a wingless mutant is known, affecting both sexes and behaving as a simple recessive (picture in Goldschmidt, 1927.)

It would be of great interest to know how natural rudimentation of wings is inherited. This could be analyzed by crossing winged or wingless species. Such crosses have been made by Harrison (1914–16) and by Meisenheimer (1928) in geometrid moths. F_1 females show wings of intermediate length and more or less normal structure, demonstrating that the onset of abnormality in development begins at a later stage than in the rudimentary female. Backcrosses and F_2 show a considerable variation, with occasional reappearance of the parental forms. A definite genetic analysis, however, is not possible for the following reasons (see Federley, 1925). The chromosome numbers of the species are different and conjugation at meiosis is disturbed. Therefore, complicated conditions arise in RF_2 and F_2 which cannot be analyzed in simple Mendelian terms. Furthermore, triploid intersexes are produced, and wing rudimentation, being a secondary sex character, responds also to intersexuality. The intermediate types, therefore, may be due to the degree of intersexuality and thus cannot be analyzed simply in terms of wing mutation. Therefore we are forced to derive our genetic information as to the possible range of the effects of single mutations from the mutants of *Drosophila*.

We have described the facts of wing rudimentation rather extensively, and we have gone rather thoroughly into their experimental analysis, because this phenomenon is one of the favorite topics of evolutionary speculation. One could write a history of evolutionary thought drawing all examples from rudimentation of wings in insects. Darwin discussed the report by Wollaston to the effect that most of the insects of Madeira are wingless, and proposed to explain the facts in terms of natural selection: flying insects are more apt to be blown out to sea by violent storms. This explanation was generally accepted when it became known that the fauna of most oceanic islands exhibited the same phenomenon. It was further assumed that this selection was based upon accumulation of small favorable variations. The existence of all transitions from poor fliers to apterous insects via winged nonfliers and short-winged forms within the same family was accepted as proof of such a hypothesis. Rüschkamp, for example, considered the rudimentation series found in Coleoptera to be a definite demonstration of slow rudimentation by selection. Evolutionists with Lamarckian tendencies used the facts to prove their viewpoint: the abundance of wingless forms in caves and on high mountains, the frequently observed increase of rudimentation in a northerly direction, impressed the authors with the idea that the influence of environment was at work, an idea which seemed to be justified by the results of such temperature experiments, etc., as were mentioned above. Again, the evolutionists (led by Cuénot) who based their opinion upon the results of genetics, emphasized that wingless forms occur everywhere, that the nearest relatives of wingless island forms found in the nearest continent also contain wingless species, and that the same is true for cave-inhabiting forms. They pointed to the occurrence of wing reduction by simple mutation, and held the view that already existing wingless mutants are able to occupy a niche which is unavailable to winged forms. I think that all modern biologists hold this view, with such modifications as individual cases may require. We mention these general conceptions only in order to demonstrate that the known agencies of evolution, selection, and adaptation do not necessarily entail the idea of slow accumulation of changes in the case under discussion. Actually the manifold conditions of rudimentation of wings in connection with specific adaptations are easily conceivable as based upon single mutations, with subsequent occupation of an environmental niche to which the mutant is preadapted; i.e., macroevolution in a single step. From the material which was presented in illustration of such a conclusion we need select only one case, which exemplifies one of the most extreme departures and therefore also includes the less extreme ones. The very aberrant fly *Termitoxenia*, a commensal of termites, is not only flightless (not to mention other features) but the wings have been transformed into strange organs which possibly are specific sense organs (see p. 332 and fig. 65) intermediate between wings and halteres. In *Drosophila* we found two simple mutants in which exactly the same type of wing transformation occurs, as was described and pictured (figs. 61, 64) in our discussion of homoeosis and its phylogenetic significance. In this case it is perfectly clear—see our previous discussion —how the potentialities of development permit, by a single shift in the processes of embryonic determination, rudimentation of an organ and simultaneously sudden emergence of a completely different organ capable of a different function. It may be added that the mutant in question demonstrates also (just as does *Termitoxenia*) the manifold effects upon organization which follow in the wake of a single developmental change; figure 64 shows how the form of the thorax and the details of the bristles are changed with the change in development of the wing discs. I like to underscore this beautiful example, which ought to impress every unbiased evolutionist.

Having concluded this discussion, we may return briefly to the problem discussed before, whether it is likely that the large evolutionary departures accomplished by a single mutational step have occurred in the form of ordinary Mendelian mutations, or by what we called systemic mutations. The genetic and taxonomic facts which we reported do not give any answer, as the relevant cases in nature do not lend themselves to an analysis. Therefore, only very indirect evidence is available as yet. The points mentioned in the former discussion (p. 321) shall not be repeated. We shall mention only a few facts which some day may be found to add weight in favor of systemic mutations. In Lepidoptera closely related species distinguished by wing rudimentation can be hybridized and will produce partially fertile hybrids. These species frequently have very different chromosome sets, which is otherwise rarely the case with closely related species of Lepidoptera. The domesticated *Bombyx mori*, with reduced wings incapable of flight, has one chromosome less than its wild ancestor *Theophila mandarina*. *Orgyia antiqua*, with rudimentary wings, has twenty-eight (2n) chromosomes; *thyellina*, without real rudimentation, twenty-two chromosomes. *Nyssia hirtaria* has twenty-eight, and the wingless *pomonaria*, fifty-six chromosomes. Another bit of indirect information might be derived from the following facts. In *Drosophila* wing rudimentation of one type (scalloping) is, aside from causation by different mutants, a frequent consequence of abnormal chromosome numbers (triploids, haplo IV). These facts may be kept in mind for future use.

Before leaving the subject of rudimentation in insects, we may mention very briefly a closely related topic, that of eye rudimentation in cave animals, a phenomenon which shows striking parallelism to the facts just discussed. It occurs in arthropods and in vertebrates, and has been still more widely discussed with regard to evolution. We could present an account of this set of facts which would closely resemble the discussion of wing rudimentation. To avoid unnecessary duplication we will briefly mention the essential points. Rudimentation of eyes is one of a number of features characterizing animals in certain habitats which are partly the same as for wing rudimentation: the seashore, the subterranean life zone, caves, and the dark interior of the homes of ants and termites in the cases of commensals. In all such cases pigmentation decreases more or less simultaneously with the loss of eyesight, and compensating sensory functions are developed. Just as in the case of wing rudimentation, all evolutionary theorists have claimed the facts in their favor, most conspicuous among them the Lamarckians. In the same way as reported for the former subject, modern discussions have shown that blind cave animals have near relatives with sensory organs compensating for the lack of eyes, and other adaptations to a hidden life. These and numerous other facts regarding the varying degrees of blindness, distribution of the blind among their seeing relatives, and correlation of blindness with other adaptations, have made it clear that preadaptation, occupation of new niches, and absence of selective value for eyesight are the decisive agencies of evolution in this case. Just as in the case of wing rudimentation, it has been assumed that the different degrees of eye rudimentation actually found represent stages of slow retrogressive development, orthogenetic in direction, and parallel in different groups of arthropods or vertebrates, respectively. But, again, all the facts available point out that this assumption of slow accumulation of small genetic changes is not needed. There are first the facts of natural variation. In some Isopods and fishes a population of the same species may show all variations from presence of eyes to eyelessness, without any indication that this variation may be genetic (cf. the case of wings in Carabids). Further, mutants are known in the nearest relatives of blind Isopods which reduce the pigmentation and, in part, the eyes. These mutants may have a slight or a large effect (Vandel, 1938b; Kosswig, 1935, 1936). In related animals used for experimentation mutants producing different degrees of eye rudimentation are known (Gammarus, Sexton and Clark, 1936), and in *Drosophila* there are many mutants which produce different types and degrees of eye defects, including presence or absence of right-left correlation, as well as constancy or high variability of the effect within a line, just as was the case for wing rudimentation. As in the case of the wings we must therefore conclude that a single mutation is able to produce in one case a slight effect, in another, the extreme effect, or any stage in between. There is not much known about the developmental side of the problem, but a few facts strongly suggest that the embryogenetic system which permits this rudimentation is of exactly the same type as was discussed before. The types of rudimentary eyes known in Isopods (Kosswig, *loc. cit.*) and in vertebrates (Eigenmann, 1909) strongly suggest that rudimentation means in this case a stoppage of development of the eyes at an earlier or later point in development, which may be followed by destruction of already existing parts and by correlative changes produced by embryonic regulation in the neighborhood of the organ which does not keep pace with the rest of development. Though experiments which would permit a complete analysis are thus far lacking, inferences may be drawn from the facts of development of mutants in *Drosophila*, and also of eye determination in vertebrates. There is hardly any doubt that the time of stoppage or slowing down of differentiation in development is connected with the degree of rudimentation in the simple relation, which we discussed repeatedly. A mutant causing these changes in early development, therefore, will produce the maximum effect in one step. We may recall the fact that the rudimentary eye of the blind newt *Proteus* may develop into a normal eye, just as the wingless earwig *Anisolabis* occasionally grows wings, and just as an ant or termite may grow wings or not. These few remarks may suffice to show the complete parallelism between the situations for wings and eyes. The conclusions are obvious. (The most recent discussions of the subject, to which we may refer for details and which also contain some of the facts mentioned here, are those by Kosswig (1935, 1936), Vandel (1938b), and Hubbs (1938).)

In view of the great importance of the facts of rudimentation in all theories of evolution, we shall discuss another case in which at least some of the important points are known. A classic example in vertebrates is the rudimentation of the limbs in Sauria, starting with the normal extremities of lizards and proceeding through all intermediate steps to the completely snakelike forms. We have already discussed in the last chapter that part of the problem which deals with

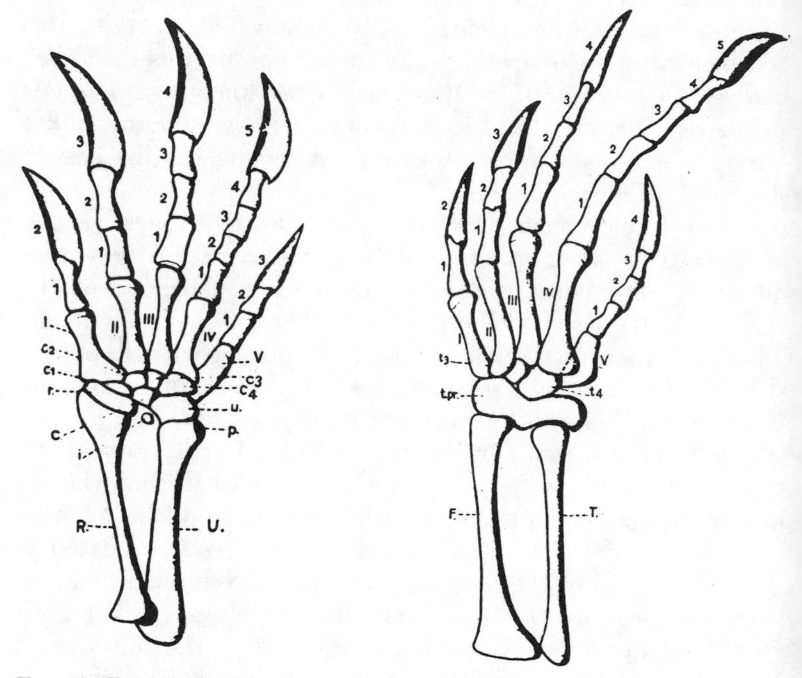

Fig. 81. From left to right: Skeleton of anterior limb of *Eumeces schneideri;* dto. posterior limb. C, centrale, cu_{1-4}, carpalia distalia; i, intermedium, p. pisiforme; R, radius; U, ulna; r, radiale; u, ulnare; I-V, metacarpalia; 1-5, phalanges; F, fibula; T, tibia; t pr, tarsale proximale; $t_{3\ 4}$, tarsalia distalia. (From Severtzoff.)

the correlation of this rudimentation with segmentation. We turn now to the problem of the limbs alone. In figures 81 and 82 we reproduce from Severtzoff a series of drawings showing the limbs of a series of forms with different degrees of rudimentation compared with normal ones (fig. 81). Without going into a detailed description we may say that the identical lettering of the bones shows how individual phalanges

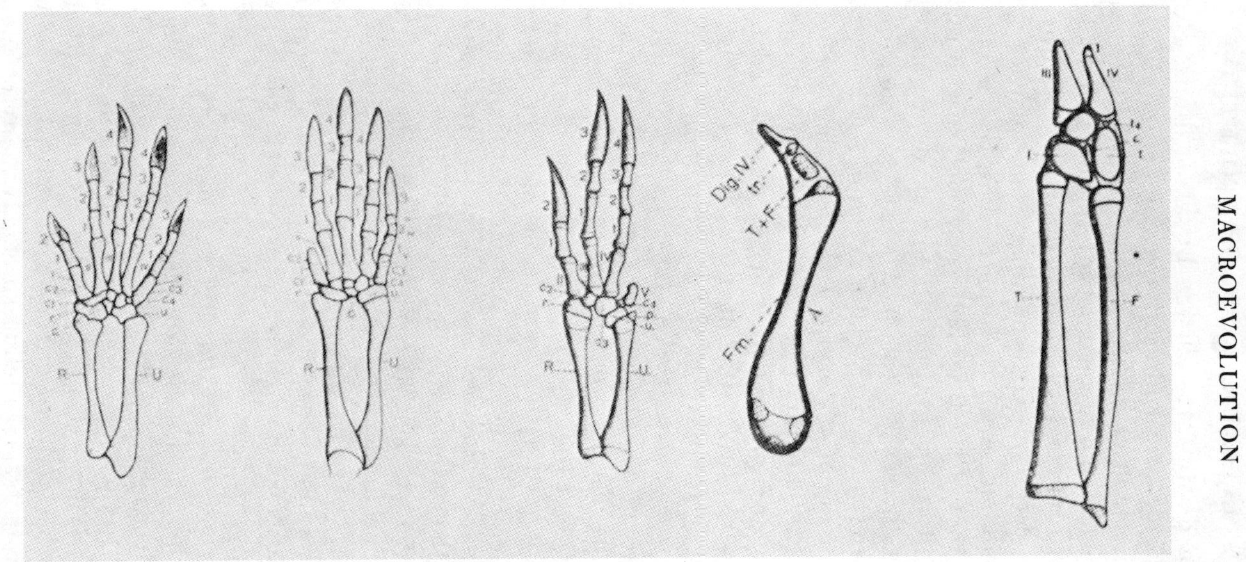

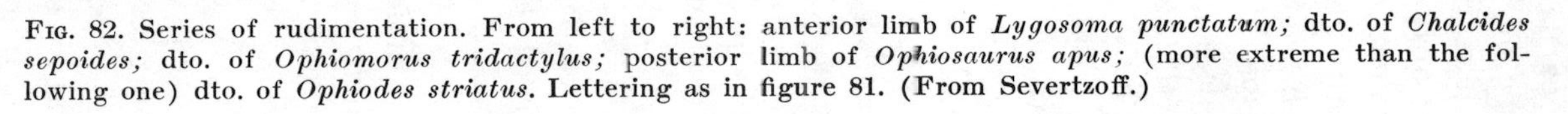

Fig. 82. Series of rudimentation. From left to right: anterior limb of *Lygosoma punctatum;* dto. of *Chalcides sepoides;* dto. of *Ophiomorus tridactylus;* posterior limb of *Ophiosaurus apus;* (more extreme than the following one) dto. of *Ophiodes striatus*. Lettering as in figure 81. (From Severtzoff.)

and carpalia disappear, followed by whole digits, etc., up to the one-rayed last rudiment present before all vestiges are gone. Severtzoff *(loc. cit.)* also investigated the development of the limbs in a normal, an intermediate, and a highly rudimentary form corresponding to figures 81–82, third and fourth from left. In corresponding embryonic stages in which the limb buds are just being formed, their size is proportional to that of the later limb; i.e., smaller, with increasing degrees of rudimentation. In further development and growth these differences remain. There is then no difference in relative growth rate but one in initial size of the primordium. But the essential feature is the following. The differentiation of the individual bones does not occur simultaneously but in a definite order, which is represented for the normal saurian in table 10. This table shows the subsequent stages in the development of the skeleton of the forelimb of *Ascalabotes fascicularis.* Seven stages are recorded (1–7); for each stage the condition found is represented in the horizontal columns. Each symbol means that the corresponding part is present in that stage. The table bears out the rule which we mentioned with regard to the order of differentiation and shows that there is a proximo-distal gradient and an ulno-radial one, though the details are a little more complicated. In the rudimentary limbs exactly the same order is observed. But such skeletal elements as are missing are not formed at all. The decisive point is—and this agrees with the facts of comparative anatomy of such forms as have not been studied embryologically—that the disappearance of the individual bones occurs in the reciprocal order of their embryonic appearances. *The last to appear in the development of a normal limb are the first to disappear in rudimentation.* In other words, the different degrees of rudimentation are produced (1) by a reduction in size of the limb bud proportional to the amount of rudimentation in the respective species, (2) by a stoppage of differentiation within the bud at an earlier and earlier time with increase of rudimentation. These facts, paralleling so closely many of the previously discussed facts of primary importance in physiolog-

Table 10

DIAGRAMMATIC REPRESENTATION OF THE ORDER OF DIFFERENTIATION OF THE LIMB BONES IN THE SAURIAN *ASCALABOTES FASCICULARIS*. (AFTER SEVERTZOFF.) c_{1-5}, CARPALIA DISTALIA 1–5. C, CENTRALE CARPI. DIG. I–V, DIGITS. i, INTERMEDIUM. H, HUMERUS. R, RADIUS. r, RADIALE CARPI. U, ULNA. u, ULNARE CARPI. v_{1-5}, PRECARTILAGES OF DIGITS 1–5. I–V, METACARPALIA 1–5. 1–5, FIRST TO FIFTH PHALANX OF DIGITS.

Stage			*Dig. I*	*Dig. II*	*Dig. III*	*Dig. IV*	*Dig. V*
1	H.R.U.					v_4	
2	H.R.U.	u.			c_3 (v_3)	c_4 v_4	v_5
3	H.R.U.	u.	v_1	v_2	c_3 III. 1.	c_4 IV. 1.	(c_5) v_5
4	H.R.U.	r.u. (C.)	v_1	(c_2) (II)	c_3 III. 1.	c_4 IV. 1.	c_5v_5
5	H.R.U.	r.u.C.	(c_1) (I)	c_2 II.	c_3 III. 1.	c_4 IV. 1.2.	c_5 V.
6	H.R.U.	r.u.i.C.	c_1 I. 1.2.	c_2 II. 1.2.	c_3 III. 1.2.3.	c_4 IV. 1.2.3.4.	c_5 V. 1.2.
7	H.R.U.	r.u.i.C.	c_1 I. 1.2.3.	c_2 II. 1.2.3.	c_3 III. 1.2.3.4.	c_4 IV. 1.2.3.4 5.	c_5 V. 1.2.3.

ical genetics, demonstrate clearly that the genetic change producing rudimentation is one affecting rates of differentiation and times of determination. A slow rate of limb-bud formation relative to the rest of the body produces a smaller limb, and the end of the period of embryonic determination (obviously controlled by gradients of the inductor substance within the bud) occurs too early, before all parts are determined, either because the supply of the inductor is too small and its flow stops, or because the moment of final determination arrives before the induction is completed. The details of the happenings can certainly be analyzed only by proper experimentation, but the general system is perfectly clear. It indicates, as in all the cases discussed before, that a single mutation influencing the decisive rate processes may produce the most extreme as well as all intermediate stages in one single evolutionary step.

Unfortunately, there is not as much genetical material available for our argument as there was in other cases, in which we showed repeatedly that the presence of a simple developmental system based on integrated, balanced rates of developmental reactions enables macroevolution to proceed in single large steps, and that the presence of such a system and its evolutionary workings can be inferred from a comparison of morphological, embryological, experimental, and genetical facts. In the present case the genetical facts are restricted to a number of data from human pathology. A considerable number of mutants are known which affect the development of the limbs, starting with absence or, in other cases, concrescence of phalanges and increasing to a very extreme rudimentation (cf. literature in Müller, 1937). I pointed out on a former occasion (Goldschmidt, 1927) that the facts of development show that these mutants act by changing the rate of differentiation in practically the same way as was just shown for the Sauria. Mohr, the geneticist who is best acquainted with this topic, has concurred in this opinion. Another group of facts which we frequently used in discussions along similar lines concerned the experimental production of phenocopies. I do not know of any such experiments involving limb rudimentation in Sauria. But, whereas phenocopies are known for comparable cases, like change in number of vertebrae in fishes, or rudimentation of the uropygial skeleton in birds, and, whereas nonheritable abnormalities of the limbs of a comparable type are known in pathology, we may consider the case for limb rudimentation just as well established as if the proper experiments were available.

In our discussion of wing rudimentation in insects we found different types, one of which involved the secondary destruction of already formed parts of the organ. In the limb rudimentation of vertebrates this type was not found (though its existence cannot be denied, in view of the scant embryological knowledge). But it is frequently met with in other cases of rudimentation and it is typical in cases of extreme rudimentation; i.e., complete loss of organs. We may mention the embryonic teeth of whalebone whales, the embryonic vertebral musculature of turtles, or the tails of Anura, all formed and afterwards destroyed. Our previous discussion showed sufficiently that in these cases also evolution must have proceeded in single huge steps, at least with regard to the major features.

I began the discussion of the present topic by pointing out that a comparative anatomist had been forced by the facts in his field to come, in a general way, to the same conclusions which we derived from a different type of analysis, though he expressed them in somewhat unfortunate terms. But after having developed his theory of "Phylembryogenesis" based on "negative Archallaxis," "Anaboly," "Aphanasy," and similar "principles," Severtzoff makes a statement in a footnote (p. 332) which we may quote to show that pure morphology finally leads to the same point which we have reached from another angle: "But I do not want to claim that this morphological theory furnishes a final explanation of the process of evolution: its importance is rather found in the demonstration of the importance of the time factor in the development of embryonic primordia for the evolution of the adult characters of animals. The question why the phylogenetically important changes of organs take place at early, medium or late stages of morphogenesis, lies, we think, outside of the sphere of competence of the morphologist and has to be solved probably by the geneticists." The writer of these lines was not aware that this solution had already been proposed ten years earlier and had since been discussed repeatedly.

At many points in our discussion we have pointed out that a genetic change involving rates of embryonic processes does not affect a single process alone. The physiological balanced system of development is such that in many cases a single upset leads automatically to a whole series of consecutive changes of development in which the ability for embryonic regulation, as well as purely mechanical and topographical moments, come into play; there is in addition the shift in proper timing of integrating processes. If the result is not, as it frequently is, a monstrosity incapable of completing development or surviving, a completely new anatomical construction may emerge in one step from such a change. This idea, which we first intimated in our essays (Goldschmidt, 1920), is based upon an insight into the action of genetic changes upon development and upon the present knowledge of experimental embryology, especially induction (evocation, organizer action). The facts discussed in this chapter furnished examples for the general idea. This assumes special importance when macroevolutionary processes in which rudimentation is combined with a change of function are studied. In my first discussions of the subject I alluded to the phylogeny of the auditory ossicles in vertebrates as a beautiful example. Neo-Darwinism (and also classical Darwinism) had to assume that a slow adaptation of some external features to changed environment took place by selection of small variations. This in turn produced a slow rebuilding of internal organization by similar steps. This was called correlation, and many types of correlation were stated and discussed in the older Darwinian literature. I wonder if anybody could ever succeed in explaining the phylogeny of the auditory ossicles in this way. But the log-

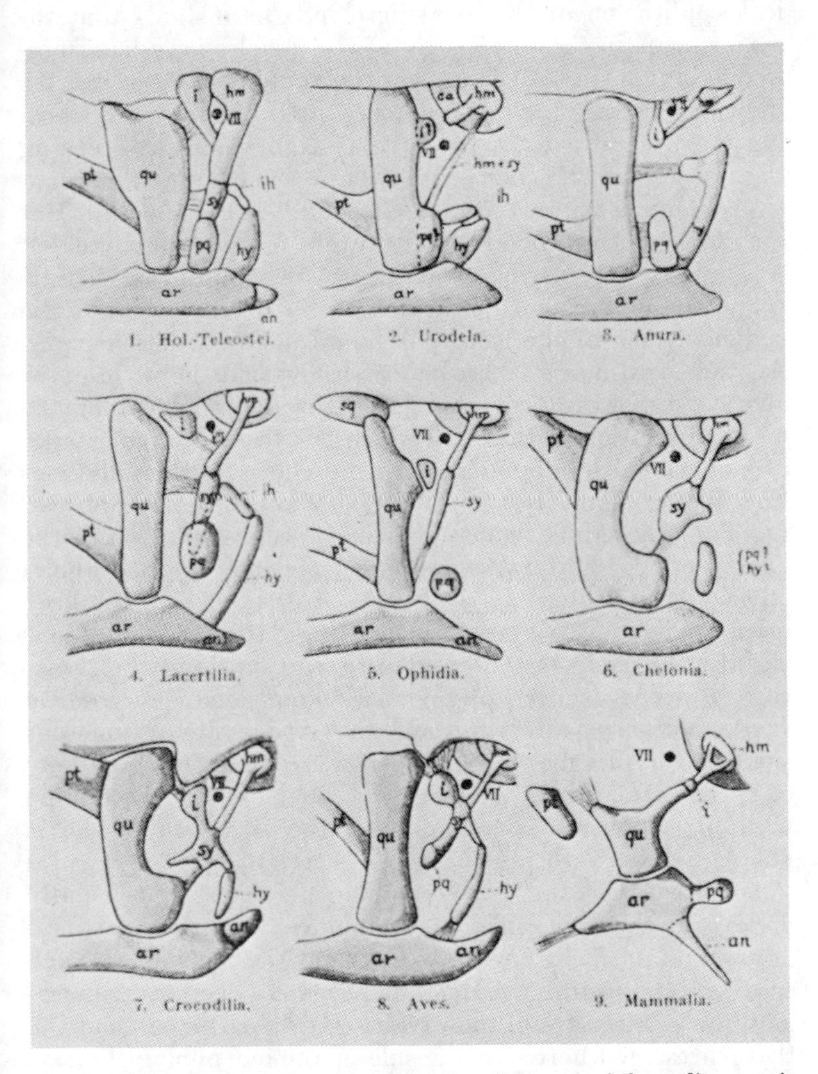

FIG. 83. Diagrammatic representation of the phylogeny of the auditory ossicles. ar, articulare; an, angulare; hm, hyomandibulare; i, intermedium; ih, int. hyoid; hy, hyoid; pq, palatoquadratum; pt, pterygoid; qu, quadratum; sy, symplecticum. (From Nauck.)

ical application of the ideas developed here shows that the major steps in the phylogeny of the ossicles may be visualized as single steps. A long chapter would have to be written to bring together the facts of comparative anatomy concerning the visceral arches, the facts of embryology concerning the ossicles, and the facts of paleontology concerning primitive mammalian precursors, and to analyze every detail. But even without this, the argument can be made perfectly clear. A single change in the relative growth rate of the dentale versus articulare and quadratum could have produced the rudimentation of the latter, as found in fossil theromorphs. Another still more extreme change might have brought about a topographical change in the arrangement of the bones in relation to the skull which left to subsequent stages of embryonic development only the choice: either destruction by secondary lysis or incorporation into the auditory region of the skull. The diagrammatic representation of the history of the auditory region in vertebrates (fig. 83) allows an easy visualization of the course of events which has been briefly indicated. We may add that regulation as well as new mutations may have contributed toward finishing the structure. But the decisive phylogenetic step must have been a single mutation affecting an embryonic rate. Numerous similar examples may be found in the comparative anatomy of vertebrates. The argument is so obvious that there is no need for further elaboration. But the evolutionist who is not acquainted with physiological genetics and experimental embryology, and the neo-Darwinian geneticist who usually disregards these two fields as well as comparative morphology, would profit by trying to work out the facts of one such case, say, the auditory ossicles, in terms of selection of micromutations. They would soon realize the heuristic value of the thesis presented here. On the side of the morphologists Severtzoff came rather near to this realization, when he wrote that heterochrony in development is a means of topographic coördination; i.e., new adaptation of the parts to each other. This is seen by the fact that the organs which primarily change in evolution are formed earlier in embryogeny and that the organs which adapt themselves to the former ones by coördination are developed later. De Beer *(loc. cit.)*, as mentioned, followed up our idea from the standpoint of development, and, among geneticists, Haldane and Huxley *(loc. cit.)* realized the importance of the problem.

Though we are discussing here only the potentialities of development as offering definite insight into the type of genetic changes available for evolution, we may at least mention at the end of this chapter the problem of recapitulation which is so closely linked with the present discussion. The views on the so-called law of embryonic recapitulation of phylogeny reflect very well the changes in general evolutionary outlook. Only compare the formulations of von Baer and Haeckel with the skeptical discussions of the experimental embryologists and the recent discussions. I think that the facts presented above, and their interpretation in terms of physiological genetics, lead to only one explanation. If macroevolutionary changes proceed by mutations affecting the rate of embryogenetic processes at a definite time in development, the ontogeny of all descendants of the mutant form must continue along ancestral lines up to the stage in development first affected by the mutant. Obviously, the mechanics of development do not permit any other course. If the mutation which changed the long tail of the Archaeopteryx, with its segmental tail feathers, into the rudimentary tail of birds with fanlike tail feathers, occurred in such a way that after formation the tail segments were made to grow together, etc., the present embryology of birds must necessarily contain an Archaeopteryx stage, which is actually the case (Steiner, 1938). If the mutation in question had changed tail segmentation primarily; i.e., before the stage of visible segmentation, no recapitulation of the Archaeopteryx condition would occur in the embryogeny of birds. The presence of recapitulation shows positively that the original mutational change in the ancestors affected development after the stage which is recapitulated. The fact that recapitulation is an ubiquitous feature of development suggests that macroevolution has progressed mainly by this type of change. The reason is obviously to be found in the relation between the genetic basis and the physiology of development: a genetic change affecting the rate, time of inception, time of determination, range of regulatory ability of embryonic processes, may occur in a single step without requiring a rebuilding of much of the genetic material. The genetic change is probably a permutation of some of the genetic elements controlling development, whatever theory of such changes we choose to accept in detail, and does not require the origination of new genetic determiners or determining systems. On the other hand, a genetic change involving a huge qualitative departure which would completely revolutionize the processes of development from their very initiation, would wipe out the possibility of recapitulation and would mean such an immense departure that it probably could rarely if ever lead to a viable product. A viable product would be a new phylum. Recapitulation, then, is an ubiquitous fact, unavoidable because of the method of evolution by large single mutational steps affecting rates, etc., of embryonic processes occurring at a definite time, and because of the mechanism of development built upon a timed system of serial processes, the order of which is unalterable.

d. The hopeful monster

In a former paper (Goldschmidt, 1933) I used the term "hopeful monster" to express the idea that mutants producing monstrosities may have played a considerable role in macroevolution. A monstrosity appearing in a single genetic step might permit the occupation of a new environmental niche and thus produce a new type in one step. A Manx cat with a hereditary concrescence of the tail vertebrae, or a comparable mouse or rat mutant, is just a monster. But a mutant of Archaeopteryx producing the same monstrosity was a hopeful monster because the resulting fanlike arrangement of the tail feathers was a great improvement in the mechanics of flying. A fish undergoing a mutation which made for a distortion of the skull carrying both eyes to one side of the body is a monster. The same mutant in a much compressed form of fish living near the bottom of the sea produced a hopeful monster, as it enabled the species to take to the life upon the sandy bottom of the ocean, as exemplified by the flounders. A dog with achondroplastic bowlegs was a monstrous mutant until man found the proper niche for it—to follow the badger (dachs) into its den—and selected the hopeful monster as a dachshund. Here, then, we have another example of evolution in single large steps on the basis of shifts in embryonic processes produced by one mutation. I think that this idea of the hopeful monster has come into its own only recently. Only now is the exact basis for an appraisal of its evolutionary significance available. This basis is furnished by the existence of mutants producing monstrosities of the required type and the knowledge of embryonic determination, which permits a small rate change in early embryonic processes to produce a large effect embodying considerable parts of the organism.

Actually, the idea expressed in the somewhat unconventional but plastic term "hopeful monster" is not a new one. We may refer back to Darwin, who pointed out that under domestication monstrosities occur which resemble normal structures in widely different animals. But Darwin did not regard them as interesting, as he believed that they could survive only under rare and special circumstances, and that they would be swamped by cross-breeding with normal forms. We know now that these criticisms are not valid. The idea under discussion has since cropped up again and again. Professor R. R. Gates has kindly drawn my attention to a little-known and in many respects rather amateurish book by Bonavia (1895) in which a whole chapter is devoted to the subject and from which Gates (1921) quoted in his book. Bonavia pointed out that monstrosities might actually have played a large role in evolution by providing specific adaptations in a single step. He even anticipated the idea of preadaptation when he declared that such monsters might have been able to occupy new habitats and there continue a special evolution. He also insisted strongly upon the possibility of single large steps in evolution for which the monsters provide the proper material, and he mentioned a number of writers on evolution who had realized this. He even had a vague idea of the embryological basis of large sudden deviations, when he wrote that "a little more atomic disturbance here, a little less there, during the embryonic stage may produce a *new compound*,[9] which then may be called a species, a genus, or even an order, as the case may be." He actually used as an example the long tail of Archaeopteryx, which he suggested might have been reduced in one generation to the short tail of the modern bird. This statement is remarkable, as the existence of an Archaeopteryx stage in the development of birds has only recently been discovered, and as the numerous tail-reducing mutants in mammals and birds, as well as the experimental production of related abnormalities, belong to present-day biology.

The facts and conclusions regarding the hopeful monster are so obvious that there is no need for a long discussion. Only a few points might be briefly mentioned. Certain types of monstrosities occur rather easily as mutants in different groups of animals of comparable architecture. For example, mutants reducing the extremities are known to occur in man, in mammals, and in birds. Hairlessness and tail-lessness occur as mutations in different species of mammals. Bulldog-head is known as a mutant in vertebrates from fishes to mammals. Wing rudimentation occurs in many groups of insects and birds. Reduced eyes occur as mutants in insects, crustacea, mammals. Telescope eyes occur as mutants in fishes. All these types of monstrosity and many others are considered in other cases as taxonomic traits characteristic of forms adapted to special conditions of life. Therefore there is no reason to assume for such taxonomic traits an origin by slow selection of micromutations instead of origin in one large step. Sometimes it is argued that the existence of different degrees of monstrous features is proof of an origin by gradual evolution. Cormorants are poor fliers. Cormorants living on the islands near the Pacific Coast of North America show slight signs of wing rudimentation (Grin-

9. Bonavia's italics.

nell). The Galapagos cormorant is flightless. But we know that mutants producing rudimentation and other monstrosities may have a partial effect in one case or a maximum effect in another. The interpretation of this case is clear. A single mutant may produce any degree of wing rudimentation. If such a mutation occurs in a hawk, for example, the resulting monster will not survive. But if it occurs in such a bird as the cormorant, which is already organized for catching its food while swimming under water, the monstrosity will not be deleterious and might even be of the "hopeful" type if it enhances simultaneously the swimming and diving capacity (by lessening friction). Whether a complete or a partial reduction of the wing can take place depends upon the habitat. Obviously a cormorant on the Pacific Coast cannot survive without flight, but needs a certain amount of it to change its fishing grounds; an island cormorant finds enough fish with very little flying; and a Galapagos cormorant can do well without flying at all. The three different degrees of mutation, all of the type of the hopeful monster, have fitted the respective mutants to three different niches characterized by the distribution of fish supply. Only one more example need be mentioned. Rumplessness is an ordinary mutant in fowl, producing deformity and shortening of the rump; it is based on a genetic disturbance in early development and may also be obtained as a phenocopy by cooling the embryo during a critical period (cf. literature in Goldschmidt, 1938). The eventual appearance of this mutant monstrosity in a bird which has to be a good flier in order to survive would certainly not fit it for survival. But the same monstrosity may enable another bird to start a line of running birds occupying grasslands or steppes. Thus, the combined facts of genetics, embryology, and taxonomy demonstrate that the hopeful monster is one of the means of macroevolution by single large steps.

C. A Few Facts from Botany and Paleontology

In our discussion of the potentialities of development in relation to evolution, the plants were almost completely neglected. The reason is obvious. Experimental embryology of plants has not yet furnished the information which would enable us to connect hereditary changes with developmental processes in a simple way. In higher plants the determinative processes occur largely at the growing points. The determination of the taxonomic features of a flower, for example, must take place at an early stage in the development of the bud. We have reason to suppose that the decisive processes are not different in principle from those in animals. Relative growth and pattern formation are still the basic features. The general type of determination is probably similar, aside from the fact that it does not occur in a closed, unified system as in animal embryos. Critical periods of determination seem to occur (Harder), and the determination itself probably involves a flow of determining substances (hormones) in definite gradients with definite thresholds. This, unfortunately, is still a largely unexplored field of experimentation. In addition, we know of the existence of phenocopies in plants; i.e., of morphogenetic changes induced by the environment and resembling hereditary traits. Many examples can be found in Turesson's writings *(loc. cit.)*. Rudimentary organs of taxonomic importance are also known (in the flower and embryosac, for example) and may be analyzed just as in animals. Monstrosities produced by single mutations are also frequent (peloria, calycanthemy); these are of the type of macromutational difference in other forms. I do not doubt, therefore, that with better knowledge of *Entwicklungsmechanik* and physiological genetics in plants, an analysis paralleling the one presented here for animals could be made and that the conclusions would be the same. These few remarks should suffice, as I do not feel entitled to discuss in detail a subject on which I do not have firsthand information

The last sentence applies even more to the field of paleontology. It is a well-known fact that biologists and paleontologists have frequently been at odds on the methods of evolution. One of the main points of difference has already been mentioned: the Lamarckian attitude of most paleontologists, who believe that evolution as they see it in their material cannot be explained in any other way. In addition, paleontologists have always been greatly impressed by orthogenetic series in the evolution of numerous groups of animals, and have frequently expressed a belief in an explanation which is based on some transcendental principle of improvement. This situation has changed considerably. The younger generation of paleontologists has tried to bring its reasoning in line with the facts of genetics and development. I need only quote Schindewolf (1936), the most progressive investigator known to me, who showed that the material presented by paleontology leads to exactly the same conclusions as derived in my writings, to which he refers. He elaborates the thesis that macroevolution on a higher level takes place in an explosive way within a short geological time, followed by a slower series of orthogenetic perfections, as exemplified in the oft-quoted evolutionary series. He realizes that the conception of preadaptation accounts completely for this type of evolution. He shows by examples from fossil material that the major evolutionary advances must have taken place in single large steps, which affected early embryonic stages with the automatic consequence of reconstruction of all the later phases of development. He shows that the many missing links in the paleontological record are sought for in vain because they have never existed: "The first bird hatched from a reptilian egg." Schindewolf and a few others also realize that the genetical and phenogenetical facts and ideas from which my thesis was derived furnish the basis for an understanding of such a process of evolution. Thus we see that the results of paleontology—see Schindewolf for references to other authors who have come to similar conclusions—vindicate the thesis which we developed here. It is gratifying that all the disciplines which furnish material for the understanding of evolution—taxonomy and morphology, descriptive and experimental embryology, static and dynamic (physiological) genetics, comparative anatomy and paleontology—supply ample and parallel evidence for a theory of evolution which is more plausible than the neo-Darwinian theory

V. CONCLUSION

The theses presented in these lectures have been derived from a large body of research in diverse fields of biology, undertaken, at least in part, with the problem of evolution in mind. They have developed and changed with the progress of my own work and with increasing acquaintance—much of it firsthand—with material studied by others. The result as it stands today, and which we have tried to base upon a large body of diversified facts converging toward a single center, may be expressed in a few sentences. Microevolution within the species proceeds by accumulation of micromutations and occupation of the available ecological niches by the preadapted mutants. Microevolution, especially geographic variation, adapts the species to the different conditions existing in the available range of distribution. Microevolution does not lead beyond the confines of the species, and the typical products of microevolution, the geographic races, are not incipient species. There is no such category as incipient species. Species and the higher categories originate in single macroevolutionary steps as completely new genetic systems. The genetical process which is involved consists of a repatterning of the chromosomes, which results in a new genetic system. The theory of the genes and of the accumulation of micromutants by selection has to be ruled out of this picture. This new genetic system, which may evolve by successive steps of repatterning until a threshold for changed action is reached, produces a change in development which is termed a systemic mutation. Thus, selection is at once provided with the material needed for quick macroevolution. The facts of development, especially those furnished by experimental embryology, show that the potentialities, the mechanics of development, permit huge changes to take place in a single step. The facts of physiological genetics and their explanation in terms of coördinated rates of processes of differentiation furnish the insight into the possibilities of macroevolution by single steps. A considerable role is assigned to such genetic changes as affect early embryonic processes and automatically entail major deviations in the entire organization. The general picture of evolution resulting from such deliberations is in harmony with the facts of taxonomy, morphology, embryology, paleontology, and the new developments of genetics. The neo-Darwinian theory of the geneticists is no longer tenable.

A theory of evolution, which in the last analysis is based upon the control of velocities of reaction by catalysts which in some way or other must make up the hereditary material, may appear to some evolutionists to be too mechanistic and too simplistic. I think that all theories of evolution tend to reflect the scientific trends of their time. I have lived to see the purely morphological period of biology with its evolutionary corollary, the construction of phylogenetic trees, invention of missing ancestors, and a philosophical outlook variously termed mechanism, materialism, monism. The following period of experimental biology was skeptical of, if not actually hostile to, evolution, as it could not be attacked in laboratory experimentation. Mechanism became unpopular and vitalistic and teleological trends invaded evolutionary thought in the form of creative evolution, emergent evolution, psycho-Lamarckism. The rise of genetics brought back a mechanistic attitude; evolution started to become an exact science. Just as there is no room for transcendental principles in experimental physics and chemistry, in the same way a factual attack upon the problems of evolution can work only with simple mechanistic principles. Genetics showed the evolutionists that evolution can be attacked scientifically only on the basis of known analyzable processes, which are by their very nature relatively simple. But, just as has been the case in chemistry and physics, mechanistic analysis of evolution will sooner or later reach a point where an interpretation in terms of known processes will meet with difficulties. In such a situation chemistry and physics have never invoked transcendental principles on the assumption that nature is so frightfully complicated that it cannot be understood otherwise. The actual developments have shown that this is not the case. The modern development of the electronic theory has shown that rather simple principles govern the most complicated phenomena of matter. Of course, there is always an unexplained residue on which the investigator may train his personal metaphysical predilection, but certainly no chemist would look to metaphysics for an explanation of a difficult phenomenon, say catalysis. In the same way the evolutionist, who meets with difficulties in mechanical interpretation at a lower level, may enjoy letting loose his metaphysical yearnings. But as an investigator he can only work under the assumption that a solution in terms of known laws of nature is possible.

We frequently encounter the idea that life phenomena are infinitely more complicated than those of inorganic nature and that they therefore cannot be understood on the same basis. Applied to evolution, this outlook would mean that one has to look for very complicated features, preferably such as require a metaphysical interpretation. I cannot agree with this. If life phenomena were not based on very simple principles, no organism could exist; if embryonic development were not controlled by a few simple basic properties and laws of matter, an organism could never be developed in a series of processes unrolling with the precision of clockwork. If evolution had not been made possible by relatively simple features inherent in the material basis of organization, it would never have occurred. I said before that evolutionary theory reflects the philosophy of the time, meaning by philosophy not the metaphysical speculations of some thinkers, but the general attitude toward the solution of the riddles of nature as based upon the results of scientific research. This philosophy is today simplistic and cannot be otherwise, in view of present-day knowledge of the constitution of matter. A few principles expressible in simple numerals govern the essentials of physics and chemistry. In biology a group of chemical substances, many of which are closely related and none too complicated in constitution, the vitamins, hormones, and Atmungsfermente, etc., control the most decisive processes of life, and heredity will have to be referred, at least in a general way, to the properties of proteins. This shows that a simplistic attitude is not a flaw but the ideal goal for a theory in science and, therefore, also for a theory of evolution. I quote again: *Frustra fit per plura quod fieri potest per pauciora.*

BIBLIOGRAPHY

ADLER, L. 1914. Metamorphosestudien an Batrachierlarven. I. Extirpation endokriner Drüsen. A. Extirpation der Hypophyse. A. Entwicklmech., 39:21–45.

ALPATOV, W. W. 1929. Biometrical studies on variation and races of the honey bee (*Apis mellifera* L.). Q. Rev. Biol., 4:1–58.

ANDERSON, E. 1928. The problem of species in the northern blue flags, *Iris versicolor* L. and *Iris virginica* L. Ann. Mo. Bot. Gard., 15:241–332.

1936. The species problem in Iris. Ann. Mo. Bot. Gard., 23:457–509.

1939. Recombination in species crosses. Gen., 24:668–698.

ARLDT, TH. 1910. Handbuch der Paläogeographie. Leipzig, 1910.

ARNDT, A. 1937. Rhizopodenstudien. III. Untersuchungen über Dictyostelium mucoroides Brefeld. A. Entwicklmech., 136:681–744.

ASTAUROFF, B. L. 1929. Studien über die erbliche Veränderung der Halteren bei Drosophila melanogaster Schin. A. Entwicklmech., 115:424–447.

BABCOCK, E. B., AND M. S. CAVE. 1938. A study of intra- and interspecific relations of Crepis foetida L. Z. ind. Abst., 75:124–160.

BALKASCHINA, E. I. 1929. Ein Fall der Erbhomöosis (die Genovariation "Aristopedia") bei Drosophila melanogaster. A. Entwicklmech., 115:448–463.

BATES, M. 1939. Hybridisation experiments with Anopheles maculipennis. Am. J. Hyg., 29C:1–6.

BATESON, W. 1914. Inaugural address. The Australian meeting of the British Association. Nature, 93:635–642.

BAUER, H., AND TH. DOBZHANSKY. 1937. A comparison of gene arrangements in Drosophila azteca and D. athabasca. Gen., 22:185.

BAUR, E. 1925. Die Bedeutung der Mutationen für das Evolutionsproblem. Z. ind. Abst., 37:107–115.

1932. Artumgrenzung und Artbildung in der Gattung Antirrhinum, Sektion Antirrhinastrum. Z. ind. Abst., 63:256–302.

BEDICHEK, S. 1939. Sex balance in the progeny of triploid Drosophila. Gen., 24:94–95.

BEEBE, C. W. 1907. Geographic variations in birds, with special reference to the effects of humidity. Zoologica. N. Y. Zool. Soc., 1:1–41.

DE BEER, G. R. 1930. Embryology and evolution. Oxford. 116 pp.

BEHRENDS, J. 1936. Über die Entwicklung des Lakunen- Adern- und Tracheensystems im Flügel der Mehlmotte Ephestia Kühniella. Z. Morph. Oek., 30:573–596.

BELIAJEFF, N. K. 1930. Die Chromosomenkomplexe und ihre Beziehung zur Phylogenie bei den Lepidopteren. Z. ind. Abst., 54:369–399.

BENSON, S. B. 1933. Concealing coloration among some desert rodents of the southwestern United States. U. Calif. Publ. Zool., 40:1–70.

BERG, L. S. 1926. Nomogenesis or evolution determined by law. Constable & Co., Ltd., London. 477 pp.

BERGNER, A. D., AND A. F. BLAKESLEE. 1932. Cytology of the ferox-quercifolia-stramonium triangle in Datura. Proc. Nat. Acad. Sci., 18:151–159.

BERGNER, A. D., S. S. SATINA, AND A. F. BLAKESLEE. 1935. Prime types in Datura. Proc. Nat. Acad. Sci., 19:103–115.

BEZZI, M. 1916. Riduzione e scomparsa delle ali negli insetti ditteri. Natura, 7. Pavia.

BIEDERMANN, W. 1912. Farbe und Zeichnung der Insekten. Handb. vergl. Phys., 3:1657–1994.

BLAKESLEE, A. F., A. D. BERGNER, AND A. G. AVERY. 1937. Geographical distribution of chromosomal prime types in Datura stramonium. Cytologia, Fujii jub. vol.: 1070–1093.

BLOUNT, R. E. 1939. Heteroplastic transplantation of the hypophysis between different species of Ambystoma. Proc. Soc. Exp. Biol. Med., 40:212–214.

BOEKER, H. 1935. Artumwandlung durch Umconstruction. Act. Biotheor., 1:17–34.

BONAVIA, E. 1895. Studies in the evolution of animals. London.

BRAUN, W. 1939. The role of developmental rates in the production of notched wing characters in Drosophila melanogaster. Proc. Nat. Acad. Sci., 25:238–242.

1939*a*. An experimental attack on some problems of physiological genetics. Nature, 144:114–115.

BRAUNS, A. 1938. Die Flügelrückbildung bei der Strandfliege Conioscinella brachyptera Zett. (Diptera; Chloropidae) und die Beziehungen zur Ausbildung der Flügelsinneskuppeln. Zool. Anz., 123:281–295.

Breider, H. 1936. Eine Erbanalyse von Artmerkmalen geographisch vikariierender Arten der Gattung Limia. Z. ind. Abst., 71:441–499.

Breitenbecher, J. K. 1925. An apterous mutation in Bruchus. Biol. Bull., 48.

Bridges, C. B. 1922. The origin of variations in sexual and sex-limited characters. Am. Nat., 56:51–63.

Bridges, C. B. and Th. Dobzhansky, 1932. The mutant "proboscipedia" in Drosophila melanogaster—a case of hereditary homöosis. A. Entwicklmech., 127:575–590.

Brown, L. A. 1929. The natural history of cladocerans in relation to temperature. I. Distribution and the temperature limits for vital activities. Am. Nat., 63:248–264.

Bytinski-Salz, H. and A. Günther. 1930. Untersuchungen an Lepidopterenhybriden. I. Z. ind. Abst., 53:153–234.

Castle, W. E. and P. W. Gregory. 1929. The embryological basis of size inheritance in the rabbit. J. Morph. a. Physiol., 48:81–93.

Cavazza, F. (1938). Dell' ibridismo di specie. 1 discendenti di una mula feconda. Riv. Biol., 24:14.

Chauvin, M. von. 1877. Über das Anpassungsvermögen der Larven von Salamandra atra. Z. wiss. Zool., 29:324–351.

Chittenden, R. J. 1928. Notes on species crosses in Primula, Godetia, Nemophila, and Phacelia. J. Gen., 19:285–314.

Clausen, J. 1926. Genetical and cytological investigations on Viola tricolor L. and V. arvensis Murr. Hered., 8:1–156.

1931. Cyto-genetic and taxonomic investigations on melanium violets. Hered., 15:219–308.

Clausen, J. and collaborators. 1937. Plant relationship as determined by experiment. Exhibition of the Carnegie Inst. of Wash. pp. 18–23.

Clausen, J., D. D. Keck, and W. M. Hiesey. 1939. The concept of species based on experiment. Am. J. Bot., 26:103–106.

Cope, E. D. 1887. The origin of the fittest. Appleton, New York. 467 pp.

Coutagne, G. 1896. Récherches sur le polymorphisme des mollusques de France. Ann. Soc. Agr. Sér. Lyon.

Crampton, H. E. 1916. Studies on the variation, distribution, and evolution of the genus Partula. The species inhabiting Tahiti. Carnegie Inst. Wash., Publ. 228. 313 pp.

1925. Studies on the variation, distribution, and evolution of the genus Partula. The species of the Mariana Islands, Guam, and Saipan. Carnegie Inst. Wash., Publ. 228A. 116 pp.

1932. Studies on the variation, distribution, and evolution of the genus Partula. The species inhabiting Moorea. Carnegie Inst. Wash., Publ. 410. 335 pp.

Cretschmar, M. 1928. Das Verhalten der Chromosome bei der Spermatogenese von Orgyia-Bastarden. Z. Zellforsch. u. mikr. Anat., 7:290–399.

Cross, J. C. 1938. Chromosomes of the genus Peromyscus (deer mouse). Cytologia, 8:408–419.

Cuénot, L. 1911. La genèse des espèces animales. Félix Alcan, Paris. 496 pp.

1935. L'Adaptation. Paris, Doin Co.

Cuénot, L. et L. Mercier. 1923. Les muscles du vol chez les mutants alaires des drosophiles. C. R. Paris, 176:1112–1113.

Darlington, C. D. 1936. Crossing-over and its mechanical relationships in Chorthippus and Stauroderus. J. Gen., 33:465–500.

1937. Recent advances in cytology. 2d ed. Blakiston, Philadelphia, 671 pp.

1939. The evolution of genetic systems. Cambridge University Press.

Darlington, C. D. and A. E. Gairdner. 1937. The variation system in Campanula persicifolia. J. Gen., 35:97–128.

Dewitz, H. 1883. Über rudimentäre Flügel bei den Coleopteren. Zool. Anz., 6:315–318.

Dice, L. R. and P. M. Blossom. 1937. Studies of mammalian ecology in southwestern North America with special attention to the colors of desert mammals. Carnegie Inst. Wash., Publ. 385. 129 pp.

Diver, C. 1939. Aspects of the study of variation in snails. J. Conch., 21:91–141.

Dobzhansky, Th. 1933. Geographic variation in lady-beetles. Am. Nat., 67:97–126.

1935. The Y chromosome of Drosophila pseudoobscura. Gen., 20:366–376.

1935*a*. Drosophila miranda, a new species. Gen., 20:377–391.

1937. Further data on the variation of the Y chromosome in Drosophila pseudoobscura. Gen., 22:340–346.

1937*a*. Further data on Drosophila miranda and its hybrids with Drosophila pseudoobscura. J. Gen., 34:135–151.

1937*b*. Genetics and the origin of species. Columbia University Press, New York. 364 pp.

Dobzhansky, Th. and J. Schultz. 1934. The distribution of sex-factors in the X-chromosome of Drosophila melanogaster. J. Gen., 28:349–386.

Dobzhansky, Th. and D. Socolov. 1939. Structure and variation of the chromosomes in Drosophila azteca. J. Hered., 30:3–19.

Dobzhansky, Th. and A. H. Sturtevant. 1938. Inversion in chromosomes of Drosophila pseudoobscura. Gen., 23:28–64.

Dobzhansky, Th. and C. C. Tan. 1936. Studies on hybrid sterility. III. A comparison of the gene arrangement in two species, Drosophila pseudoobscura and Drosophila miranda. Z. ind. Abst., 72:88–114.

Dorfmeister, G. 1864. Über die Einwirkung verschiedener, während der Entwicklungsperioden angewendeter Wärmegrade auf die Färbung und Zeichnung der Schmetterlinge. Mitt. Natur. f. Steiermark.

Drosihn, J. 1933. Über Art- und Rassenunterschiede der männlichen Copulationsapparate von Pieriden (Lep.). Ent. Rundsch., 50:

Dubinin, N. P. and collaborators. 1934. Experimental study of the ecogenotypes of Drosophila melanogaster. Biol. Zhurnal, 3:166–216.

Dubinin, N. P., N. N. Socolov, and G. G. Tiniakov. 1937. Intraspecific chromosome variability. Biol. Zhurnal, 6:1049–1054.

Duerden, J. E. 1920. The inheritance of the callosities in the ostrich. Am. Nat., 54.

Du Rietz, G. E. 1930. The fundamental units of biological taxonomy. Sv. Bot. Tijd., 24:333–428.

East, E. M. 1913. Inheritance of flower size in crosses between species of Nicotiana. Bot. Gazette, 55:177–188.

1916. Inheritance in crosses between Nicotiana langsdorffii and N. alata. Gen., 1:311–333.

1935. Genetic reactions in Nicotiana. II. Phenotypic reaction patterns. Gen., 20:414–442.

1936. Genetic aspects of certain problems of evolution. Am. Nat., 70:143–158.

Eigenmann, C. H. 1909. Cave vertebrates of America. Carnegie Inst. Wash., Publ. 104. 241 pp.

Ekblom, T. 1928. Vererbungsbiologische Studien über Hemiptera-Heteroptera. I. Gerris asper Fieb. Hered., 10:333–359.

Eller, K. 1939. Fragen und Probleme zur Zoogeographie und zur Rassen- und Artbildung in der Papilio machaon-Gruppe. Verh. VII. int. Ent. Kongr., 1938, 1.
(See Pagast, Naturw. 1939)

ENDRÖDY, S. VON. 1938. Die palaearktischen Rassenkreise des Genus Oryctes (Ill.). Arch. Naturg. 7.

ENZMANN, E. V. AND C. P. HASKINS. 1939. Note of modifications in the morphogenesis of Drosophila melanogaster occurring under neutron bombardment. Am. Nat., 73:470–472.

EPSTEINS, F. F. 1939. Ueber Modifikationen (Phaenokopien) der Flügelform nach Bestrahlung mit U.-V. Licht bei Drosophila. Genetica, 21:225–242.

ERNST, A. 1918. Bastardierung als Ursache der Apogamie im Pflanzenreich, eine Hypothese zur experimentellen Vererbungs- und Abstammungslehre. G. Fischer, Jena. 665 pp.

FEDERLEY, H. 1913. Das Verhalten der Chromosomen bei der Spermatogenese der Schmetterlinge Pygaera anachoreta, curtula and pigra, sowie einiger ihrer Bastarde. Z. ind. Abst., 9:1–110.

1920. Die Bedeutung der polymeren Faktoren für die Zeichnung der Lepidopteren. Hered., 1:221–269.

1925. Gibt es eine konstant-intermediäre Vererbung? Z. ind. Abst., 37:361–385.

1927. Ist die Chromosomenkonjugation eine conditio sine qua non für die Mendelspaltung? Hered., 9:391–404.

1928. Chromosomenverhältnisse bei Mischlingen. Verh. V. Int. Kongr. Vererb. Berlin. Z. ind. Abst., Suppl. I:194–222.

1938. Chromosomenzahlen Finnländischer Lepidopteren. I. Rhopalocera. Hered., 24:397–464.

FINKENBRINK, W. 1933. Experimentelle Untersuchungen zur Dewitzschen Hypothese des Apterismus bei Insekten. Z. Morph. Oek., 26:385–426.

FISCHER, E. 1901. Experimentelle Untersuchungen über die Vererbung erworbener Eigenschaften. Allg. Zeit. f. Ent.

1924. Die F-$_2$Generation eines Artbastardes. Schweiz. Ent. Anz., 3:53–54.

FISCHER, EUGEN. 1933. Genetik und Stammesgeschichte der Menschlichen Wirbelsäule. Biol. Zentralbl., 53:203–220.

1939. Versuch einer Phänogenetik der normalen körperlichen Eigenschaften des Menschen. Z. ind. Abst., 76:47–115.

FLANDERS, S. E. 1931. The temperature relationships of Trichogramma minutum as a basis for racial segregation. Hilgardia, 5:395–406.

FORBES, W. T. M. 1928. Variation in Junonia lavinia (Lepidoptera, Nymphalidae). J. N. Y. Ent. Soc., 36:305–320.

FORD, E. B. 1936. The genetics of Papilio dardanus Brown (Lep.). Trans. Roy. Ent. Soc. London, 85:435–466.

1937. Problems of heredity in the Lepidoptera. Biol. Rev., 12:461–503.

FÖYN, B. 1927. Studien über Geschlecht und Geschlechtszellen bei Hybriden. II. Auspressungsversuche an Clava squamata (Müller), mit Mischung von Zellen aus Polypen desselben oder verschiedenen Geschlechts. A. Entwicklmech., 110:89–148.

FRANZ, H. 1929. Morphologische und phylogenetische Studien an Carabus L. und den nächstverwandten Gattungen. Z. wiss. Zool., 135:163–213.

FRIESEN, H. 1936. Roentgenmorphosen bei Drosophila. A. Entwicklmech., 134:147–165.

GARSTANG, W. 1922. The theory of recapitulation. A critical restatement of the biogenetic law. J. Linn. Soc. London, Zool., 35:81–101.

GATES, R. R. 1921. Mutations and evolution. New Phytologist Reprint No. 12. Cambridge Press. 118 pp.

GATES, W. H. 1926. The Japanese waltzing mouse; its origin, heredity, and relation to the genetic characters of other varieties of mice. Carnegie Inst. Wash., Publ. 337:83–138.

GEITLER, L. 1938. Weitere cytogenetische Untersuchungen an natürlichen Populationen von Paris quadrifolia. Z. ind. Abst., 75:161–190.

GEROULD, J. H. 1923. Inheritance of white wing color, a sex-limited (sex-controlled) variation in yellow pierid butterflies. Gen., 8:495–551.

GOLDSCHMIDT, R. 1903. Histologische Untersuchungen an Nematoden. I. Zool. Jahrb. (Anat.), 18:1–51.

1904. Dto. II. *Ibid.*, 21:1–100.

1906. Mitteilungen zur Histologie von Ascaris. Zool. Anz., 29:719–737.

1908. Das Nervensystem von Ascaris lumbricoides und megalocephala. I. Z. wiss. Zool., 90:73–136.

1909. Dto. II. *Ibid.*, 92:306–357.

1910. Dto. III. Festschr. R. Hertwig, Jena. pp. 254–354.

1911. Einführung in die Vererbungswissenschaft. 1st ed. 502 pp.

1916. Notiz über einige bemerkenswerte Erscheinungen in Gewebekulturen von Insecten. Biol. Centr. 16:160–167.

1917. A preliminary report on some genetic experiments concerning evolution. Am. Nat., 52:28–50.

1920. Die quantitative Grundlage von Vererbung und Artbildung. Roux Vortr. Aufs. Entwicklmech. Berlin, Springer. H. 24. 163 pp.

1920*a*. Mechanismus und Physiologie der Geschlechtsbestimmung. Berlin, Bornträger (English ed. 1923, Methuen). 251 pp.
1920*b*. Erblichkeitsstudien an Schmetterlingen. III. Z. ind. Abst., 25:89–163.
1920*c*. Untersuchungen zur Entwicklungsphysiologie des Flügelmusters der Schmetterlinge. A. Entwicklmech., 7:1–24.
1923. Einige Materialien zur Theorie der abgestimmten Reactionsgeschwindigkeiten. A. mikr. An., 98:292–313.
1923*a*. Das Mutationsproblem. Sitzber. Deutsch. Ges. Vererbg. (Z. ind. Abst. 30), pp. 260–268.
1924. Erblichkeitsstudien an Schmetterlingen. IV. Z. ind. Abst., 34:229–244.
1924*a*. Untersuchungen zur Genetik der geographischen Variation. I. A. Entwicklmech., 101:92–337.
1927. Physiologische Theorie der Vererbung. Berlin, Springer. 246 pp.
1927*a*. Neu-Japan. Berlin, Springer. 304 pp.
1929. Untersuchungen zur Genetik der geographischen Variation. II. A. Entwicklmech., 116:136–201.
1929*a*. Einführung in die Vererbungswissenschaft. 5th ed. Berlin, Springer. 568 pp.
1929*b*. Experimentelle Mutation und das Problem der sogenannten Parallelinduktion. Biol. Centralbl., 49:437–448.
1931. Die sexuellen Zwischenstufen. Berlin, Springer. 528 pp.
1932. Die Genetik der geographischen Variation. Proc. Int. Genet. Congr., Ithaca, 1:173–184.
1932*a*. Untersuchungen zur Genetik der geographischen Variation. III. A. Entwicklmech., 126:277–324.
1932*b*. Dto. IV. *Ibid.*, pp. 591–612.
1932*c*. Dto. V. *Ibid.*, pp. 678–768.
1933. Some aspects of evolution. Science, 78:539–547.
1933*a*. Untersuchungen zur Genetik der geographischen Variation. VI. A. Entwicklmech., 130:266–339.
1933*b*. Dto. VII. *Ibid.*, pp. 562–615.
1933c. Gen und Ausseneigenschaft 3. Biol. Centralbl. 55:535–554.
1934. Lymantria. Bibl. Genet. 11:1–180.
1934*a*. Gen und Ausseneigenschaft. I, II. Z. ind. Abst., 69:39–131.
1935. Geographische Variation und Artbildung. Naturw., 23: 169–176.

1935*a*. Multiple sex genes in Drosophila? A critique. J. Gen., 31:143–153.

1935*b*. Gen und Ausseneigenschaft. III. Biol. Centrabl., 55:535–554.

1937. Spontaneous chromatin rearrangements and the theory of the gene. Proc. Nat. Acad. Wash., 23:621–623.

1937*a*. A critical review of some recent work in sex determination. I. Q. Rev. Biol., 12:426–439.

1937*b*. Gene and character. IV. Univ. Calif. Publ. Zool., 41:277–282.

1938. Physiological genetics. New York, McGraw-Hill. 375 pp.

1938*a*. A Lymantrialike case of intersexuality in plants? J. Gen., 36:531–535.

1938*b*. A note concerning the adaptation of geographic races of Lymantria dispar L. to the seasonal cycle in Japan. Am. Nat., 72:385–386.

1938*c*. The theory of the gene. Sci. Month., 46:1–6.

In press. Chromosomes and genes. Stanford Sympos. Centen. Cell Theory.

Goldschmidt, R. and E. Fischer. 1922. Argynnis paphia-valesina, ein Fall geschlechtscontrollierter Vererbung. Genetica, 4.

Goldschmidt, R., E. J. Gardner, and M. Kodani. 1939. A remarkable group of position effects. Proc. Nat. Acad. Wash., 25:314–317.

Goodspeed, T. H., and R. E. Clausen. 1916. Mendelian factor differences versus reaction system contrasts in heredity. Am. Nat., 51:31–101.

Gottschewski, G. 1934. Untersuchungen an Drosophila melanogaster über die Umstimmbarkeit des Phänotypus und Genotypus durch Temperatureinflüsse. Z. ind. Abst., 67:477–528.

Green, C. V. 1931. Size inheritance and growth in a mouse species cross (Mus musculus × Mus bactrianus). III. Inheritance of adult quantitative characters. IV. Growth. J. Exp. Zool., 59: 213–263.

1935. Quantitative characters in reciprocal hybrids. Am. Nat., 69:278–282.

1938. Homologous and analogous morphological mutations in rodents. Biol. Rev. Cambridge Philos. Soc., 13:293–306.

Gregor, J. W. 1938. Experimental taxonomy: II. Initial population differentiation in Plantago maritima L. of Britain. New Phytologist, 37:15–49.

GRINNELL, J. 1928. A distributional summation of the ornithology of Lower California. U. Calif. Publ. Zool., 32:1–300.

GROSS, F. 1932. Untersuchungen über die Polyploidie und die Variabilität bei Artemia salina. Naturwiss., 51:962–967.

GUDERNATSCH, J. F. 1912. Feeding experiments on tadpoles. I. The influence of specific organs given as food on growth and differentiation. A. Entwicklmech., 35:457–483.

GULICK, A. 1932. John Thomas Gulick, evolutionist and missionary. University Chicago Press, Chicago. 556 pp.

1932*a*. Biological peculiarities of oceanic islands. Q. Rev. Biol., 7:405–427.

GULICK, J. 1905. Evolution racial and habitudinal. Carnegie Inst. Wash., Publ. 25. 269 pp.

GUPPY, H. B. 1906. Observations of a naturalist in the Pacific between 1896 and 1899. Vol. II. Plant dispersal. Macmillan & Co., London. 627 pp.

GUYÉNOT, E. 1930. La variation et l'évolution. Paris.

HAGEDOORN, A. C. AND A. L. 1917. Rats and evolution. Am. Nat., 51:385–418.

HALDANE, J. B. S. 1927. The comparative genetics of colour in rodents and carnivora. Biol. Rev., 2:199–212.

1932. The causes of evolution. Harper & Bros., New York and London.

1932*a*. The time of action of genes, and its bearing on some evolutionary problems. Am. Nat., 66:5–24.

HÄMMERLING, J. 1938. Fortpflanzung und Sexualität. Fortschritte d. Zool., 3:363–386.

HARLAND, S. C. 1933. The genetics of cotton. Part IX. Further experiments on the inheritance of the crinkled dwarf mutant of G. barbadense L. in interspecific crosses and their bearing on the Fisher theory of dominance. J. Gen., 28:315–325.

1934. The genetical conception of the species. Tropical Agriculture, 11:51–53.

1935. The genetical conception of the species. Biol. Rev., 11:83–112.

HARMS, J. W. 1932. Die Realisation von Genen und die consecutive Adaption. 2. Birgus latro als Landkrebs und seine Beziehungen zu den Coenobiten. Z. wiss. Zool., 140:167–290.

1934. Wandlungen des Artgefüges. Heine, Tübingen. 212 pp.

HARRISON, J. W. H. 1916. Studies in the hybrid Bistoninae. J. Gen., 6:95–161.

1917. Studies in the hybrid Bistoninae. II. J. Gen., 6:269–313.

1919. Studies in the hybrid Bistoninae. III. J. Gen., 8:259–265.
1919*a*. Studies in the hybrid Bistoninae. IV. J. Gen., 9:1–38.
1920. Genetical studies in the moths of the geometrid genus Oporabia (Oporinia) with a special consideration of melanism in the Lepidoptera. J. Gen., 9:195–280.

HARRISON, J. W. H., AND L. DONCASTER. 1914. On hybrids between moths of the geometrid sub-family Bistoninae with an account of the behaviour of the chromosomes in gametogenesis in Lycia (Biston) hirtaria, Ithysia (Nyssia) zonaria and in their hybrids. J. Gen., 3:229–248.

HEINCKE, E. 1897–98. Naturgeschichte des Herings. Abh. d. Deutsch. Seefischereivereins. I, 138 pp. II, 223 pp.

HERIBERT-NILSSON, N. 1918. Experimentelle Studien über Variabilität, Spaltung, Artbildung und Evolution in der Gattung Salix. Lund Univr. Aarskr. N. F. 14: No. 28. 145 pp.

HERSH, A. H. 1934. Evolutionary relative growth in the titanotheres. Am. Nat., 68:537–561.

HERTWIG, P. 1936. Artbastarde bei Tieren. Handbuch der Vererb. II, B, 140 pp.

HERTWIG, R. 1927. Abstammungslehre und neuere Biologie. G. Fischer, Jena. 271 pp.

HESSE, R. 1937. Ecological animal geography (edition prepared by E. C. Allee and K. P. Schmidt). New York, Wiley Sons.

HOLTFRETER, J. 1938. Differenzierungspotenzen isolierter Teile der Urodelengastrula. A. Entwicklmech., 138:522–656.

HONING, J. A. 1923. Canna crosses. I. Mededeelingen Landbouwhoogschool Wageningen, 26:1–55.
1928. Canna crosses. II. *Ibid.*, 32:1–14.

HORTON, I. H. (1939). A comparison of the salivary gland chromosomes of Drosophila melanogaster and D. simulans. Gen., 24:234–243.

HUBBS, C. L. 1926. The structural consequences of modifications of the developmental rate in fishes, considered in reference to certain problems of evolution. Am. Nat., 60:57–81.
1938. Fishes from the caves of Yucatan. Carnegie Inst. Wash., Publ. 491:261–295.

HUNDERTMARK, A. 1938. Über das Luftfeuchtigkeitsunterscheidungsvermögen und die Lebensdauer der drei in Deutschland vorkommenden Rassen von Anopheles maculipennis (atroparvus, messeae, typicus) bei verschiedenen Luftfeuchtigkeitsgraden. Z. angew. Ent., 25:125–151.

HUXLEY, J. S. 1932. Problems of relative growth. Methuen and Co., Ltd., London, 276 pp.

1935. Chemical regulation and the hormone concept. Biol. Rev., 10:427–441.

1938. Clines, an auxiliary taxonomic principle. Nature, 142:219–221.

1938*a*. Natural history, taxonomy and general biology. Trans. South-Eastern Union of Scientific Societies, South-Eastern Naturalist and Antiquary, 43:1–21.

ILIJIN, N. A. 1934. Segregation in crosses between a wolf and a dog and material on the genetics of the dog. Trudy Dinam. Razvit., 8:105–166. (Russian)

JACKSON, D. J. 1933. Observations on the flight muscles of Sitona weevils. Ann. Appl. Biol., 20:731–770.

JENKINS, J. A. 1939. The cytogenetic relationships of four species of Crepis. U. Calif. Publ. Agr. Sc., 6:369–400.

JOHANNSEN, W. 1923. Some remarks about units in heredity. Hered., 4:133–141.

JORDAN, K. 1905. Der Gegensatz zwischen geographischer und nichtgeographischer Variation. Z. wiss. Zool., 83:151–210.

KAWAGUCHI, E. 1928. Zytologische Untersuchungen am Seidenspinner und seinen Verwandten. I. Gametogenese von Bombyx mori L. und B. mandarina M. und ihrer Bastarde. Z. Zellf. mikr. Anat., 7:519–549.

KAMMERER, P. 1904. Beitrag zur Erkenntnis der Verwandtschaftsverhältnisse von Salamandra atra und maculosa. A. Entwicklmech., 17:165–264.

1912. Experimente über Fortpflanzung, Farbe, Augen und Körperreduktion bei Proteus anguineus Laur. A. Entwicklmech., 33:349–461.

KEILIN, D. 1913. Diptères. Belgica Antarctica Jacobs. II. Expéd. Antarctique Française. Paris. 217 pp.

KERKIS, J. (1936). Chromosome conjugation in hybrids between Drosophila melanogaster and Drosophila simulans. Am. Nat., 70:81–86.

KINSEY, A. C. 1929. The gallwasp genus Cynips. Indiana U. Stud. 84–86, pp. 1–577.

1936. The origin of higher categories in Cynips. Indiana U. Publ., Science Series No. 4, 334 pp.

1937. An evolutionary analysis of insular and continental species. Proc. Nat. Acad. Sci. Wash., 23:5–11.

Klatt, B. 1928. Eine melanistische Mutation beim Schwammspinner. Zool. Anz., 78:257–260.

Klebs, G. 1907. Studien über Variationen. A. Entwicklmech., 24: 29–113.

Kleinschmidt, O. 1930. The formenkreis theory and the progress of the organic world. London. (Transl. from German 1927.)

Knapp, E., and I. Hoffmann. 1939. Geschlechtsumwandlung bei Sphaerocarpus durch Verlust eines Stückes des X-Chromosoms. Chromosoma, 1:130–146.

Kölliker, A. 1864. Über die Darwinsche Schöpfungstheorie. Z. wiss. Zool., 14:174–186.

Kosswig, C. 1935. Die Evolution von "Anpassungs"-Merkmalen bei Höhlentieren in genetischer Betrachtung. Zool. Anz., 112: 148–155.

1939. Die Geschlechtsbestimmung in Kreuzungen zwischen Xiphophorus und Platypoecilus. Rev. Fac. Sci. de l'Univ. d'Istanbul, 4:1–54.

Kosswig, C. and L. Kosswig. 1936. Ueber Augenrück-und-Missbildung bei Asellus aquaticus cavernicolus. Verh. Deutsch. Zool. Ges., 1936, 274–281.

Krumbiegel, I. 1932. Untersuchungen über physiologische Rassenbildung. Zool. Jahrb. (Syst.), 63:183–280.

Kühn, A. 1934. Genwirkung und Artveränderung. Der Biologe, 3:217–227.

Kühn, A. and M. von Engelhardt. 1937. Über eine melanistische Mutante von Ptychopoda seriata Schrk. Biol. Zentralbl., 57: 329–347.

Kühne, K. 1932. Die Vererbung der Variationen der menschlichen Wirbelsäule. Z. Morph. u. Anthrop., 30:1–221.

Lammerts, W. E. 1934. Derivative types obtained by backcrossing Nicotiana rustica-paniculata to N. paniculata. J. Gen., 29:355–366.

Lang, A. 1906. Über die Mendelschen Gesetze, Art- und Varietätenbildung, Mutation und Variation, insbesondere bei unseren Hain- und Gartenschnecken. Vortrag. 3. Tahresvers. Schweiz. Naturforsch. Ges., Luzern.

1911. Fortgesetzte Vererbungsstudien. Z. ind. Abst., 5:97–138.

Langlet, O. 1937. Study of the physiological variability of pine and its relation to the climate. U. S. Forest Service Translation No. 293, Washington. 89 pp. Original in Meddelanden från Statens Skogsförsöksanstalt, 29:421–470. 1936.

Larsén, O. 1931. Beitrag zur Kenntnis des Pterygopolymorphismus bei den Wasserhemipteren. Lunds Univ. Årsskr., N.F. 27: No. 8, 30 pp.

Le Gallien, O. 1935. Recherches expérimentales sur le dimorphisme évolutif et la biologie de Polystomum integerrimum Fröhl. Trav. de la Station Zool. de Wimereux, 12:1–176.

Lenz, F. 1926. Ein mendelnder Artbastard. A. Rass. Gesellschaftsbiol., 18:129–151.

1928. Ein weiterer mendelnder Artbastard Epicnaptera tremulifolia xilicifolia. Verh. V. Int. Kong. f. Vererb., Berlin, 2:984–986.

Levan, A. 1935. Zytologische Studien an Allium Schoenoprasum. Hered., 22:1–126.

Li, Ju-Chi. 1936–37. Studies of the chromosomes of Ascaris megalocephala trivalens. I. The occurrence and possible origin of nine chromosome forms. Peking Nat. Hist. Bull., v. 2, pt. 4.

Lloyd, R. E. 1912. The growth of groups in the animal kingdom. London. 185 pp.

Lotsy, E. 1916. Evolution by means of hybridization. The Hague.

Lundegaard, H. A. 1930. Klima und Boden in ihrer Wirkung auf das Pflanzenleben. Jena, Fischer.

Mangelsdorf, P. C. and R. G. Reeves. 1939. The origin of Indian corn and its relatives. Tex. Agr. Exp. Sta. Bull. 574:1–315.

Marey, E.-J. 1887. Recherches expérimentales sur la morphologie des muscles. C.R., Paris, 105:446–451.

Martini, E., A. Missiroli, and L. W. Hackett. 1931. Versuche zum Rassenproblem des Anopheles maculipennis. A. Schiff. Tropenhyg. 35:622–643.

Martini, E., und Teubner, E. (1933). Über das Verhalten von Stechmücken besonders von Anopheles maculipennis bei verschiedenen Temperaturen und Luftfeuchtigkeiten. Beih. Arch. Schiffs-Tropenhyg., 37:1–80.

Marx, L. 1935. Bedingungen für die Metamorphose des Axolotls. Ergeb. d. Biol., 11:244–334.

Mayr, E. (1924–1938). Birds collected during the Whitney South Sea Expedition. Parts I–XXXIX. Am. Mus. Novit. Nos. 115, 124, 149, 322, 337, 350, 356, 364, 365, 370, 419, 469, 486, 488, 489, 502, 504, 516, 520, 522, 531, 590, 609, 628, 651, 655, 666, 709, 714, 820, 828, 912, 915, 933, 939, 947, 977, 986, 1006.

Mehnert, E. 1897. Kainogenesis als Ausdruck differenter phylogenetischer Energien. G. Fischer, Jena.

MEISENHEIMER, J. 1928. Experimentelle Studien zur Soma- und Geschlechtsdifferenzierung. Zool. Jahrb. (Aug. Zod), 41:1–90.

MELCHERS, G. 1932. Untersuchungen über Kalk- und Urgebirgspflanzen, besonders über Hutchinsia alpina (L.) R. Br. und H. brevicaulis Hoppe. Österr. Bot. Zeit., 81:81–107.

1939. Genetik und Evolution. Deutsche Ges. f. Vererb. (Bericht), 229–259.

MERRIFIELD, F. 1912. Experimental embryology, factors in seasonal dimorphism. 1. Congr. Int. Entom., 1910, 1:433–448.

METZ, C. W. 1938. Observations on evolutionary changes in the chromosomes of Sciara (Diptera). Coop. Res., 501:275–294.

MILLER, A. H. 1931. Systematic revision and natural history of the American shrikes (Lanius). U. Calif. Publ. Zool., 38:11–242.

MILLER, M. A. 1938. Comparative ecological studies on the terrestrial isopod crustacea of the San Francisco Bay region. U. Calif. Publ. Zool., 43:113–142.

MISSIROLLI, A. L. W. HACKETT, AND E. MARTINI. 1933. Le razze di Anopheles maculipennis e la loro importanza nella distribuzione della malaria in alcune regioni d'Europa. Riv. Malar., 12:1–58.

MOHR, O. L. 1934. Heredity and disease. W. W. Norton, New York. 253 pp.

MORDVILKO, A. 1937. Artbildung und Evolution. Biol. Gen., 12: 271–298.

MÖWUS, F. 1933. Untersuchunge über die Sexualität und Entwicklung von Chlorophyceen. A. Protistenkde., 80:469–524.

MUENTZING, A. 1936. The evolutionary significance of autopolyploidy. Hered., 21:263–378.

1939. Chromosomenaberrationen bei Pflanzen und ihre genetische Wirkung. Deutsche Ges. f. Vererb., 323–350.

MÜLLER, F. 1864. Für Darwin. Leipzig.

MULLER, H. J. 1937. The biological effects of radiation with especial reference to mutation. Actualités Scient. et Indust. Reunion Int. de Phys.-Chim.-Biol., Congrès du Palais de la Découverte. Hermann et Cie., Paris.

1938. The remaking of chromosomes. Collecting Net, 13:181–195, 198.

MÜLLER, W. 1937. Die angeborenen Fehlbildungen der menschlichen Hand. Berlin.

MURPHY, R. C. 1938. The need for insular exploration as illustrated by birds. Sci., 88:533–539.

NAEF, A. 1917. Die individuelle Entwicklung organischer Formen als Urkunde ihrer Stammesgeschichte. G. Fischer, Jena.

NAVASCHIN, M. 1927. Ein Fall von Merogonie infolge Artkreuzung bei Kompositen. Ber. Deutsch. Bot. Ges., 45:115–126.

OEHLKERS, F. 1938. Bastardierungsversuche in der Gattung Streptocarpus Lindl. Z. Bot., 32:305–393.

OERTEL, R. 1924. Studien über Rudimentation, ausgeführt an den Flügelrudimenten der Gattung Carabus. Z. Morph. Oek., 1:38–120.

OPPENHEIMER, J. M. 1939. The capacity for differentiation of fish embryonic tissues implanted into amphibian embryos. J. Exp. Zool., 80:391–416.

OSGOOD, W. H. 1909. Revision of the mice of the American genus Peromyscus. U. S. Dept. Agr. N. Am. Fauna, 28, 285 pp.

PANTEL, J. 1917. A propos d'un Anisolabis ailé. Mem. R. Ac. Sc. Barcelona, 14:1–160.

PANTIN, C. F. A. 1932. Physiological adaptation. Linn. Soc. J.-Zool., 37:705–711.

PÄTAU, K. (1935). Chromosomenmorphologie bei Drosophila melanogaster und Drosophila simulans und ihre genetische Bedeutung. Naturwiss., 23:537–543.

PATTERSON, J. T. 1938. Aberrant forms in Drosophila and sex differentiation. Am. Nat. 72:193–206.

PAUL, H. 1937. Transplantation und Regeneration der Flügel zur Untersuchung ihrer Formbildung bei einem Schmetterling mit Geschlechtsdimorphismus, Orgyia antiqua L. A. Entwicklmech., 136:64–111.

PHILIPTSCHENKO, J. 1923. Variabilitätslehre und die Methoden der biologischen Variationsstatistik. Moscow, 1923. 233 pp.

PHILLIPS, I. C. (1921). A further report on species crosses in birds. Gen., 6:366–383.

POISSON, R. 1924. Contribution a l'étude des Hémiptères aquatiques. Bull. Biologique, 58:49–305.

PRZIBRAM, H. 1907. Differenzierung des Abdomens enthäuster Einsiedlerkrebse. A. Entwicklmech., 23:579–595.

RAPOPORT, J. A. 1939. Specific morphoses in Drosophila induced by chemical compounds. Bull. Biol. et Méd. exp. de l'URSS. 7:415–417.

RASMUSSON, J. 1933. A contribution to the theory of quantitative character inheritance. Hered., 18:245–261.

REDFIELD, A. C. 1936. The distribution of physiological and chemical peculiarities in the "natural" groups of organisms. Am. Nat., 70:110–122.

Reinig, W. F. 1938. Elimination und Selektion. Jena.

1939. Die genetisch-chorologischen Grundlagen der gerichteten geographischen Variabilität. Deutsche Ges. f. Vererb., 260–308.

Remane, F. 1927. Art und Rasse. Verh. Ges. Phys. Anthr., 2–33.

Renner, O. 1917. Versuche über die gametische Konstitution der Oenotheren. Z. ind. Abst., 18:121–294.

1929. Artbastarde bei Pflanzen. Handb. d. Vererb., v. 2. 161 pp.

Rensch, B. 1929. Das Princip geographischer Rassenkreise und das Problem der Artbildung. Berlin, Bornträger. 206 pp.

1932. Über den Unterschied zwischen geographischer und individueller Variabilität und die Abgrenzung von der ökologischen Variabilität. A. Naturg., 1:95–113.

1932a. Über die Bedeutung des Princips geographischer Rassenkreise. Geogr. Z., 38:157–166.

1933. Zoologische Systematik und Artbildungsproblem. Verh. deutsch. Zool. Ges., 19–83.

1934. Kurze Anweisung für zoologisch-systematische Studien. Leipzig Akad. Verl. 116 pp.

1936. Studien über Klimatische Parallelität der Merkmalsausprägung bei Vögeln und Säugetieren. A. Naturg., 5:317–363.

1937. Untersuchungen über Rassenbildung und Erblichkeit von Rassenmerkmalen bei sizilischen Landschnecken. Z. ind. Abst., 72:564–588.

Robb, R. C. (1929). On the nature of hereditary size limitation. II. Brit. J. Exp. Biol., 4:310–321.

Robson, G. C. 1928. The species problem. Oliver and Boyd.

Robson, G. C., and O. W. Richards. 1936. The variation of animals in nature. London.

Roux, W. 1883. Über die Bedeutung der Kernteilungsfiguren. Leipzig.

Rowntree, L. G., T. H. Clark, and A. M. Hanson. 1935. Biological effects of thymus extract. A. Int. Med., 56:1–29.

Rüschkamp, F. 1927. Der Flugapparat der Käfer. Zoologica, 75: 1–88.

Schepotieff, A. 1913. Die biochemischen Grundlagen der Evolution. Ergeb. u. Fortschritte d. Zool., Jena, 4:285–338.

Schiemann, E. 1932. Entstehung der Kulturpflanzen. Handb. Vererbg. 3. Berlin.

1939. Gedanken zur Genzentrentheorie Vavilovs. Naturw., 27: 377–401.

SCHINDEWOLF, O. H. 1936. Palaeontologie, Entwicklungslehre und Genetik. Berlin, Bornträger. 108 pp.

SCHNAKENBECK, W. 1931. Zum Rassenproblem bei den Fischen. Z. Morph. Oek., 21:409–566.

SCHRADER, F. 1926. Notes on the English and American races of the greenhouse white-fly (Trialeurodes vaporariorum). Ann. Appl. Biol., 13:189–196.

SCHREIBER, G. 1939. Ricerche sperimentali sulla neotenia degli urodeli. A. zool. ital., 27:181–215.

SCHULZE, P. 1922. Ueber nachlaufende Entwicklung (Hysterotelie) einzelner Organe bei Schmetterlingen. Naturgesch., 99A:109–114.

SEDGWICK, A. 1910. The influence of Darwin on the study of animal embryology, *in* "Darwin and modern science," ed. by A. Seward. Cambridge.

SEIDEL, F. 1936. Entwicklungsphysiologie des Insekten-Keims. Referat. Verh. deutsch. Zool. Ges. E. V. 291–336.

SEILER, J., AND C. B. HANIEL. 1921. Das verschiedene Verhalten der Chromosomen in Eireifung und Samenreifung von Lymantria monacha L. Z. ind. Abst., 27:81–103.

1938. Ergebnisse aus der Kreuzung einer diploid-parthenogenetischen Solenobia triquetrella mit Männchen einer bisexuellen Rasse. Rev. Suisse Zool., 45:405–412.

SEMENOV-TIANSHANSKY, A. P. 1910. Die taxonomischen Grenzen der Art und ihrer Unterabteilungen. R. Friedlander, Berlin. 34 pp.

SETCHELL, W. A. 1935. Pacific insular floras and Pacific paleogeography. Am. Nat., 69:289–309.

SEVERTZOFF, A. N. 1931. Morphologische Gesetzmässigkeiten der Evolution. Jena, Fischer. 369 pp.

SEXTON, E. W. AND A. R. CLARK. 1936. A summary of the work on the amphipod Gammarus chevreuxi Sexton carried out at the Plymouth Laboratory (1912–36). J. Marine Biol. Assoc. U. Kgdom., 21:357–414.

SHULL, A. F. 1928. Time of determination and time of differentiation of aphid wings. Am. Nat., 72:170–179.

1928a. Duration of light and the wings of the aphid Macrosiphum solanifolii. A. Entwicklmech., 113:210–239.

1937. The production of intermediate winged aphids with special reference to the problem of embryonic determination. Biol. Bull., 72:259–286.

SINNOTT, E. W., A. F. BLAKESLEE, AND H. E. WARMKE. 1939

The effect of colchicine-induced polyploidy on fruit shape in cucurbits. Gen., 24:84–85.

SPEMANN, H. C. 1938. Embryonic development and induction. Yale University Press, New Haven, 401 pp.

STANDFUSS, M. 1896. Handbuch der paläarctischen Gross-Schmetterlinge für Forscher und Sammler. G. Fischer, Jena. 392 pp.

STEINER, H. 1938. Der Archaeopteryx-Schwanz der Vogelembryonen. Viertelj. naturf. Ges. Zürich, 83:279–300.

STEINIGER, F. 1938. Die Genetik und Phylogenese der Wirbelsäulenvarietäten und der Schwanzreduktion. Z. f Mensch. Vererb. u. Konstit., 22:583–668.

STEINMANN, P. 1933. Vitale Färbungsstudien an Planarien. Rev. Suisse Zool., 40:529–558.

STERN, C. 1936. Interspecific sterility. Am. Nat., 70:123–142.

STOCKARD, C. R. 1931. The physical basis of personality. W. W. Norton, New York. 320 pp.

STRESEMANN, E. 1923–26. Mutationsstudien I–XXV. See J. f. Ornith., 74:377–385.

1936. The Formenkreis theory. Auk, 53:150–158.

1936a. Zur Frage der Artbildung in der Gattung Geospiza. Org. Club. Nederl. Vogelk., 9:13–21.

STURTEVANT, A. H. 1920. Genetic studies on Drosophila simulans. I. Gen., 5:488–500. II. *Ibid.*, 6:43–64.

1921. dto. III. *Ibid.*, 6:179–207.

STURTEVANT, A. H., AND TH. DOBZHANSKY. 1936. Inversions in the IIIrd chromosome of wild races of D. pseudo-obscura. Proc. Nat. Acad. Sci. Wash., 22:448–450.

STURTEVANT, A. H., AND C. C. TAN. 1937. The comparative genetics of D. pseudo-obscura and D. melanogaster. J. Gen., 34: 415–432.

SÜFFERT, F. 1924. Bestimmungsfactoren des Zeichnungsmusters beim Saisondimorphismus von Araschnia levana-prorsa. Biol. Centr., 44:173–188.

1929. Die Ausbildung des imaginalen Flügelschnittes in der Schmetterlingspuppe. Z. Morph. Oek., 14:338–359.

SUMNER, F. B. 1915. Genetic studies of several geographic races of California deer-mice. Am. Nat., 49:688–701.

1917. The role of isolation in the formation of a narrowly localised race of deer-mice (Peromyscus). Am. Nat., 51:173–185.

1918. Continuous and discontinuous variations and their inheritance in Peromyscus. *Ibid.*, 52:177–208.

1920. Geographic variation and mendelian inheritance. J. Exp. Zool., 30:369–402.

1923. Some facts relevant to a discussion of the origin and inheritance of specific characters. Am. Nat., 57:238–254.

1923a. Results of experiments in hybridizing subspecies of Peromyscus. J. Exp. Zool., 38:245–292.

1924. The stability of subspecific characters under changed conditions of environment. Am. Nat., 58:481–505.

1926. An analysis of geographic variation in mice of the Peromyscus polionotus group from Florida and Alabama. J. Mamm., 7:149–184.

1928. Observations on the inheritance of a multifactor color variation in white-footed mice (Peromyscus). Am. Nat., 62:193–206.

1929. The analysis of a concrete case of intergradation between two subspecies. Proc. Nat. Acad. Wash., 15:110–120, 481–493.

1930. Genetic and distributional studies of three subspecies of Peromyscus. J. Gen., 23:275–376.

1932. Genetic, distributional, and evolutionary studies of the subspecies of deer-mice (Peromyscus). Genetica, 9:1–106.

Sumner, F. B. and R. R. Huestis. 1925. Studies of coat color and foot pigmentation in subspecific hybrids of Peromyscus eremicus. Biol. Bull., 48:37–55.

Sumner, F. B. and H. S. Swarth. 1924. The supposed effects of the color tones of the background upon the coat color of mammals. J. Mamm., 5:81–113.

Tan, C. C. and J. C. Li. 1934. Inheritance of the elytral color patterns of the ladybird beetle Harmonia axyridis. Am. Nat., 68.

Thompson, D'Arcy W. 1917. On growth and form. Univ. Press, Cambridge. 793 pp.

Timofeeff-Ressovsky, N. W. (1930). Das Genovariieren in verschiedenen Richtungen bei Drosophila melanogaster unter dem Einfluss der Röntgenbestrahlung. Naturwiss. 18:434–437.

Timofeeff-Ressovsky, N. 1932. The genogeographical work with Epilachna chrysomelina. Proc. Int. Co. Gen., Ithaca, 2:230–232.

1934. Über die Vitalität einiger Genmutationen. Z. ind. Abst., 66.

Tower, W. 1918. The mechanism of evolution in Leptinotarsa. Carnegie Publ. Wash., 263. 384 pp.

Turesson, G. 1922. The genotypical response of the plant species to the habitat. Hered., 3:211–350.

1923. The scope and import of genecology. Hered., 4:171–176.

1925. The plant species in relation to habitat and climate.

Hered., 6:147–236.
1926. Die Bedeutung der Rassenökologie für die Systematik und Geographie der Pflanzen. Rep. spec. nov. reg. veg., 41:15–38.
1927. Contributions to the genecology of glacial relics. Hered., 9: pp. 81–101.
1929. Zur Natur und Begrenzung der Arteinheiten. Hered., 12: 323–334.
1930. The selective effect of climate upon the plant species. Hered., 14:99–152.
1931. Über verschiedene Chromosomenzahlen in Allium schoenoprasum L. Bot. Notiser, pp. 15–20.
1931a. The geographical distribution of the alpine ecotype of some Eurasiatic plants. Hered., 15:329–346.
1932. Die Genenzentrumtheorie und das Entwicklungscentrum der Pflanzenart. K. Fys. Sällsk. Lund För., 2:1–11.
1932a. Die Pflanzenart als Klimaindicator. *Ibid.*, 4th. p. 1–35.
1936. Rassenökologie und Pflanzengeographie. Bot. Notis. Lund, pp. 420–437.
1938. Chromosome stability in Linnean species. Ann. Agr. Coll. Sweden, 5:405–416.

Ungerer, E. 1936. Die Hypothese der Keimgangmutationen. Act. Biotheor., 2:23–58.

Uvarov, B. P. 1931. Insects and climate. Tr. Ent. Soc. London, 79:1–247.

Valadares, M. 1937. Declanchement d'une haute mutabilité chez une lignée pure de Drosophila melanogaster. Revista Agronomica, 25:363–383.

Vandel, A. 1930. La production d'intercastes chez la fourmie Pheidole pallidula sous l'action de parasites du genre Mermis. C. R., Paris, 180:770–773.
1936. L'Evolution de la parthénogenèse naturelle. C.R. XII Cong. Int. Zool., Lisbonne, 1935, 1:440–442.
1938. Chromosome number, polyploidy and sex in the animal kingdom. Proc. R. Soc. London, A. 107:519–541.
1938a. Contribution à la génétique des isopodes du genre Trichoniscus. Bull. Biol., 72:121–146.
1938b. Contribution à la génétique des isopodes du genre Trichoniscus. I. Bull. Biol. France Belgique, 72.

Vavilov, N. 1922. The law of homologous series in variation. J. Gen., 12:47–69.
1928. Geographische Genzentren unserer Kulturpflanzen. Z. ind. Abst., Suppl. 1:342–369.

WASMANN, E. 1903. Die Thorakalanhänge der Termitoxeniidae, ihr Bau, ihre imaginale Entwicklung und phylogenetische Bedeutung. Verh. deutsch. Zool. Ges., 113–120.
WEBER, H. 1931. Lebensweise und Umweltbeziehungen von Trialeurodes vaporariorum (Westwood) (Homoptera-Aleurodina). Erster Beitrag zu einer Monographie dieser Art. Z. Morph. Oek., 23:575–753.
WEISMANN, A. 1875. Studien zur Descendenztheorie. Leipzig, Engelmann.
1876–79. Beiträge zur Naturgeschichte der Daphniden. Z. wiss. Zool. Separat, Leipzig.
WELCH, D'A. A. 1938. Distribution and variation of Achatinella mustelina Mighels in the Waianae Mountains, Oahu. Bishop Mus. Bull., 152:1–164.
WETTSTEIN, R. VON. 1898. Grundzüge der geographisch-morphologischen Pflanzensystematik. Jena.
WHEELER, W. M. 1928. Mermis parasitism and intercastes among ants. J. Exp. Zool., 50:165–237.
WIGGLESWORTH, V. B. 1934. The physiology of ecdysis in Rhodnius prolixus. II. Factors controlling moulting and metamorphosis. Q. J. Micr. Sci., 77:191–222.
WILLIS, J. C. 1922. Age and area. Cambridge. 259 pp.
1923. The origin of species by large, rather than by gradual change and by Guppy's method of differentiation. Ann. Bot., 37:605–628.
WINGE, Ö. 1917. The chromosomes, their numbers and general importance. C.R. des Travaux du Lab. Carlsberg, 13:131–275.
1922. A peculiar mode of inheritance and its cytological explanation. J. Gen., 12:137–144.
1937. Goldschmidt's theory of sex determination in Lymantria. J. Gen., 34:81–89.
1938. The genetic aspect of the species problem. Proc. Linn. Soc., 150:231–238.
WINIWARTER, H. DE. 1937. Les chromosomes du genre Gryllotalpa. Cytol. Fuji Tub. vol. pp. 987–994.
WITSCHI, E. 1923. Ueber geographische Variation und Artbildung. Rev. Suisse Zool., 30:457–469.
WOLTERECK, R. 1919. Variation und Artbildung. Analytische und experimentelle Untersuchungen an pelagischen Daphniden und anderen Cladoceren. I. Teil. Morphologische, entwicklungsgeschichtliche und physiologische Variations-Analyse. A. Francke, Bern. 151 pp.

WRIGHT, S. 1931. Evolution in Mendelian populations. Gen., 16: 97–159.

ZARAPKIN, S. R. 1934. Zur Phaenoanalyse von geographischen Rassen und Arten. A. Naturg., 3:161–186.

INDEX

SILLIMAN VOLUMES IN PRINT

William Bateson, *Problems of Genetics*
Jacob Bronowski, *The Origins of Knowledge and Imagination*
S. Chandrasekhar, *Ellipsoidal Figures of Equilibrium*
Theodosius Dobzhansky, *Mankind Evolving*
René Dubos, *Man Adapting*
Anne McLaren, *Germ Cells and Soma: A New Look at an Old Problem*
John von Neumann, *The Computer and the Brain*
Lyman Spitzer, *Searching Between the Stars*
Karl K. Turekian, *Late Cenozoic Glacial Ages*